Robot Design Handbook

Robot Design Handbook

SRI International

Gerry B. Andeen Editor-in-Chief

McGraw-Hill Book Company

New York St. Louis San Francisco Auckland
Bogotá Hamburg London Madrid Mexico
Milan Montreal New Delhi Panama
Paris São Paulo Singapore
Sydney Tokyo Toronto

Library of Congress Cataloging-in-Publication Data

Robot design handbook/SRI International; Gerald B. Andeen [editor
    -in-chief].
        p.   cm.
    Bibliography: p.
    Includes index.
    ISBN 0-07-060777-X
    1. Robotics.   2. Manipulators (Mechanism)—Design and
construction.   I. Andeen, Gerry B.   II. SRI International.
TJ211.D47 1988
629.8′92—dc19

1234567890   DOC/DOC   89321098

ISBN 0-07-060777-X

The editors of this book were Betty Sun and Rita T. Margolies, the
designer was Naomi Auerbach, and the production supervisor was
Richard A. Ausburn. It was set in Baskerville by University
Graphics, Inc.

Printed and bound by R. R. Donnelley & Sons Company.

Contents

Contributors   vii

Preface   ix

Chapter   1. Introduction   1.1

Chapter   2. Manipulator Specifications   2.1

Chapter   3. Design Procedure   3.1

Chapter   4. Structural Design Methods   4.1

Chapter   5. Mechanical Design Methods   5.1

Chapter   6. Servo Design Methods   6.1

Chapter   7. Manipulator Simulation   7.1

Chapter   8. Digital Electronics and Electrical Design Methods   8.1

Chapter   9. Software Design Methods   9.1

Chapter 10. Actuators   10.1

Chapter 11. Transmissions and Joints   11.1

Chapter 12. Sensors   12.1

Chapter 13. Grippers   13.1

Chapter 14. Computations and Derivations (Robot Mathematics)   14.1

Chapter 15. Research Thrusts   15.1

References and Bibliography   R.1

Index   I.1

Contributors

Editor-in-Chief **Gerry B. Andeen**

M. B. Adams
S. Dubowsky
G. K. Durfey
J. S. Eckerle
B. J. Grossi
W. Park
L. E. Schwer

Preface

This book was originally titled *Robot Handbook*. It was privately published by SRI International in February 1983 for the Mitsubishi Electric Company (MELCO). The *Robot Design Handbook* is an updated and revised version of the original book.

MELCO had requested a handbook that would serve two purposes: to show the role and interaction of different disciplines in designing robotic manipulators and to critically survey the technologies involved, highlighting problem areas and pointing out research directions. As a result, the book follows a perilous course between trivial overview and esoteric rigor. Hopefully, the book will be useful to both the beginner and the experienced designer.

I would like to thank those who helped in preparation, especially V. K. Maslak, S. Sperry, and J. H. S. Kealoha, and those who reviewed and encouraged, especially S. Hagihara, F. J. Kamphoefner, P. M. Newgard, and N. R. Nielsen.

Gerry B. Andeen

Introduction

1.1 Purpose

The purpose of this book is to:

- Provide basic knowledge in a form useful for manipulator development
- Provide a common base of knowledge to promote mutual understanding among participants on a robot design team
- Highlight areas for development and future research

Robotics is a composite subject. The technologies involved and the skills required to design a manipulator are as broad ranging as with any subject. The skills required range from familiarity with the use of robots (manufacturing engineers) through the mechanical and electrical design and analysis skills (civil, mechanical, and electrical engineers) to computer interfacing (control engineers, programmers). Finally, those individuals concerned with production, finance, marketing, and even inventory control must know something about robotics to make their contribution. All members of a team have their own perspectives, but at the same time, they must understand everyone else's contributions.

In a sense, this book is a survey of robotic manipulators. It provides information that can be immediately useful. At the same time, it was our intention that it be a critical commentary on robotics. It discusses limitations and sometimes the research directions and is intended to stimulate thought about problems and their solutions.

1.2   Background

Robotics is closely tied to the continuing development of technology on a broad front in such diverse fields as materials, mechanics, controls, and computers. The topic would not exist were it not for the concepts of mass production and automation that are an integral part of western culture. Thus, although one can list some important events more or less in an historical line, as is done in Table 1.1, a real understanding requires an overview of the industrial revolution and its current extension into the information revolution.

The industrial revolution was essentially an amplification of our capabilities in three areas: power, organization, and automation. Power was the

TABLE 1.1   Events in the History of Robotics

To 3000 BC	Egyptians build water-powered clocks. Greeks and Chinese build water- and steam-powered toys.
1769	Efficient steam engine patented; beginnings of the industrial revolution (James Watt).
1770	Swiss make lifelike automata that can draw and play instruments under cam control (Pierre and Henri Jacquet-Droz).
1793	Cotton gin automation (Eli Whitney).
1801	Punch-card-controlled loom (Joseph Jacquard).
1830	Cam programmable lathe (Christopher Spencer).
1892	Motorized crane (Seward Babbitt).
1921	"Rossum's Universal Robots (R.U.R.)"—a play; the first robots in literature (Karel Capek).
1930	Analog computer invented (Vannevar Bush).
1944	Digital computer Mark I demonstrated (Howard Aiken).
1946	Programmable automation by magnetic process controller (George Devol).
1948	*Cybernetics*, a book bringing attention to the concept of feedback control (Norbert Wiener).
1951	Teleoperator remote manipulator equipment developed for radioactive situation (Raymond Goertz).
1952	Numerically controlled machine tool (MIT Servomechanism Laboratory).
1954	Programmable robot (George Devol).
1959	First commercially available robot (Planet Corp.).
1960	Unimation, Inc. founded.
1968	Intelligent robot using vision, Shakey, built at SRI (Charles Rosen).
	Walking teleoperator at G.E. (Ralph Mosher).
1970	Stanford Arm demonstrated (Victor Sheinman).
1976	Programming language AL developed (Ralph Bolles).
1978	PUMA (Programmable Universal Machine for Assembly) introduced by Unimation.
1981	Direct-drive manipulator (H. Asada and T. Kanade).
1984	Pogo stick and running robots (Marc Raibert).

most obvious spur to the industrial revolution. We no longer had to provide the power for a process from our own bodies or from animals, but rather could harness the rivers and the burning of coal. But the industrial revolution was more than power; it was the organization of methods of production in which specialization could be used to advantage. The assembly line is an example of specialization, which perhaps became dehumanizing beyond being a means of increasing production. The third ingredient was automation, making tools function automatically, i.e., without human control.

The amplification of our capabilities by organization and automation was most easily applied in mass production. There the investment in organization and automation could return a large volume of production. It did not make sense to organize trivial tasks, nor to build complicated automatic equipment for chores that were performed sporadically. However, this is changing as a result of the computer, and it is the essence of robotics.

Two trends are clear in manufacturing today:

- Diversification within mass production
- Extension of organization and automation to smaller production lot sizes

Mass production has been undergoing diversification for some time. At one time the hallmark of mass production was sameness, e.g., all cars were black. The consumer demand for diversity has been answered by making production lines increasingly flexible. In the automobile industry today, the consumer has so many different choices that it is possible that two identical automobiles might not be built in the same year. The extension of automation to smaller production runs has also been proceeding for some time. The stress has been on using equipment that is more flexible so that it can be used in different production situations.

The key to diversification and extension in manufacturing has been the computer, with its capability to organize information from sales through production, inventory, and shipping. Clearly, the role of the computer is to improve organization, while allowing for diversity of parts.

The role of the computer in automation, which is discussed in great detail in this book, has been called "softening the automation." Classical automation is "hard," in the sense that hardware is set up to perform the automated processes. Once set up, it can perform only a limited number of functions. Changing the automated processes requires changing the hardware. The current trend is toward more universal equipment, hardware that performs different functions under the control of a computer. The automation now resides in the computer software, and not in the hardware, hence the use of the term "soft."

Robotics in manufacturing is often properly traced back to numerical control machines, which followed the same trend described above. The earlier automatic machines operated on the basis of cams or other following devices. A change in the product required a change in the hardware, at least a change in the pattern being followed. Numerical control then substituted tape for the cams. More recently, as numerical control machines have come under direct control of computers, the machines have become much more general devices, capable of producing a variety of products. The actual product to be produced is determined by the computer software. Such a machine either allows diversification within high volume or encourages use of automation on shorter runs.

1.3   Emphasis

The subject is robotic manipulators, a small subcategory of robots. Robots include science-fiction humanoids, walking machines, numerical control machines, and simple pick-and-place equipment as well as manipulators. The term *robotic* has come to refer to the behavior of any device that can alter its operation in response to a command or a change in its environment.

Manipulators are devices that are attached to foundations (which may be mobile) and reach out to manipulate objects. In a sense, the manipulators are the arms of a robot. "Robotic," however, does not mean that the manipulators are attached to the body of a robot, but that the manipulators can do a variety of activities depending on commands or in response to the environment. Present-day robots are usually attached to fixed foundations.

State-of-the-art manipulators of the type useful in manufacturing are emphasized: open-chain (articulated) configuration, position controlled, and electric powered. Another emphasis is on the design approach and the technologies needed.

1.4   Philosophy of Manipulator Design

1.4.1   Design procedure

This book gives a procedure for the design of a manipulator, followed by a detailed discussion in several related areas. The procedure at first glance appears to indicate that manipulator design is an orderly process from specifications through a series of design decisions to a device. Such an organized view is only partially realistic.

Like the design of many other machines, the design of manipulators has evolved. Existing designs and conceived improvements serve as the basis for the next generation of design. Designs improve gradually.

The procedure presented in this book is an overlay to evolution of design. It is a way of evaluating proposed improvements. Furthermore, this book is intended to stimulate design alternatives. By following the procedure given here, a designer's ideas can be evaluated and compared to the alternatives. This is a useful test for the best ideas and a stimulant to improvement. Thus, although design must necessarily proceed on the basis of the past, having a design procedure is useful.

1.4.2 Trends and benchmarks

One of the trends in the evolution of robots already commented upon is the trend toward generality in the hardware and control of the function in the software. The evolution of tools first adopted power and then control. The steps might be stated this way: tools, machine tools, numerical control, computer numerical control, and general-purpose manipulators.

Another trend is the increasing concern with dynamics. In numerical control, position is the all-important consideration. Machines were built and powered so that dynamics were inconsequential. The early robots followed this trend. They were built heavy so that they would not vibrate, or at least so that the vibrations would not interfere with their motion. They were built stiff and rigid, with natural frequencies well above operating frequencies. Such an approach uses more material for construction, and much larger motors, than a flexible design would. The advantage of the rigid design is that it simplifies the control system. The trend toward more flexibility is based on increased computational capability and on advances in the computer and in control theory.

Although the development is evolutionary, the endpoint or benchmark of robotics has always been clear: to achieve the performance of biological systems. Trends toward generality are driven by the goal of being as general and all-purpose as the human arm. Trends toward flexibility are driven by the model of the arm as a flexible yet powerful manipulator. Trends toward increased sensing and even vision are driven by biological models that process touch, sight, sound, smell, and taste to achieve function. The comparison of machines to the biological processes was the heart of Wiener's famous *Cybernetics* (Wiener, 1961).The goal of robotics is to emulate the high performance of biological benchmarks.

Judged by the biological measure, especially the human being, robotics is still in its infancy. The rigid, sensor-poor point-to-point robots of today are just the beginning. Great improvements will eventually be possible at lower cost and with greater safety. The lighter, flexible manipulator with greater sensory feedback will be able to bump into objects without destroying itself or the objects. It will be able to accomplish tasks that require touch and force feedback. Such devices may eventually be able to learn coordination, and fail gracefully.

1.5   Organization

The 15 chapters of this book fall within six categories:

1. Introduction   Chapter 1
2. Overall design approach   Chapters 2 and 3
3. Analytical tools   Chapters 4 to 9
4. Available components   Chapters 10 to 13
5. Mathematics   Chapter 14
6. Research directions   Chapter 15

Manipulator Specifications

This chapter discusses the relationship between applications and specifications, and also looks at features that can be incorporated into a robotic manipulator to facilitate tasks.

Usually a manipulator designer has a use in mind. That use, whether it be remote manipulation in a hot or dangerous area, lifting heavy parts, or assembly, implies functional specifications. As much as we would like to have a universal manipulator that could serve in all applications, it is unlikely. Consider lifting capacity as an example: An arm that can manipulate automobile frames is unlikely to be the best for assembly of small electronic parts. Similar arguments can be made for reach, accuracy, and other characteristics. Furthermore, there are often tradeoffs between characteristics; for example, a manipulator may be made to move faster, but there will be some decrease in path-following accuracy. However, it is reasonable to expect a manipulator to be able to do a variety of similar tasks.

2.1 Capabilities and Definition of Priorities

The most obvious specifications determined by the use of a manipulator are the physical or functional specifications:

- Payload (maximum weight)

- Precision

    Absolute accuracy, ability to go to any specified point

    Repeatability, ability to repeat a location

    Resolution, the smallest position step

- Speed

    Velocity of the endpoint

    Cycle time, includes accelerations and decelerations

- Reach

- Stiffness

These specifications need not be a set of unrelated numbers, and, in fact, should not be. Speed and weight, speed and precision, or precision and reach may be interrelated. Certainly one should not expect heavy loads to be moved as rapidly as light ones. In a similar way, one should expect greater precision in some portions of the reach than in others.

The physical function is only the beginning of the specifications. How will the device be controlled? How will it be programmed? These questions must be addressed in operational specifications.

Then there are a set of usefulness specifications. These tell the user whether the device will be suitable in the larger context of operation in a manufacturing environment:

- Expected useful life

- Mean time between failure (MTBF)

- Repair time

- Space consumed

- Weight

- Safety

- Trainability (ease of operation)

- Cost

- Power

Finally there are environmental specifications, such as the temperature and the type of power available.

2.1.1 Requirements as a function of task

Table 2.1 shows a listing of specifications as a function of tasks in which robots have been or might be used. The table should be considered as

TABLE 2.1 Approximate Robot Characteristics

Robot function	Reach, m	Payload, kg	Speed, m/s	Accuracy, mm
Material handling (bin picking, positioning on belt, positioning in furnace, die casting, palleting)	3	100	1	±3
Machine loading (tools, parts)	1	10	1	±2
Welding				
Spot	2 to 3	50	2	±3
Seam	2	10	0.2	±1.5
Painting (spray)	2 to 3	5	2	±5
Grinding (rough)	2	100	0.5	±3
Machining	1	50	0.5	±0.01*
Assembly				
Insertion, light	0.3	1	1	±0.5†
Small	1	5	1	±1.0†
Large	2	10	0.5	±1.5†
Inspection				
Absolute metrology	Various	1	1	±0.01
Relative metrology	Various	5	1	±1.5
Humanoid	0.5	10	2	±0.5‡

*Not feasible without jigs.
†Requires compliance strategy for insertions.
‡Assumes visual feedback.

approximate information to be used as an initial guideline. (The humanoid function approximates the performance of a human in a task.)

2.1.2 Tradeoffs

Just as high speed and heavy-weight capacity are not expected to go together, several specifications trade off against one another. The designer must recognize that in several performance areas strengthening one specification can decrease performance in another. A few of these tradeoffs are:

- *Precision/life.* High-precision devices cannot tolerate wear. Hence, high precision and long life tend to be tradeoffs.

- *Strength/flexibility.* That a device is able to handle large forces generally implies that, at least when not handling large masses, the device is very rigid and has a high natural frequency.

- *High force (or torque)/speed.* High-speed devices are usually not capable of high forces. The function of variable transmissions is to make both possible, but we do not expect variable transmissions to be available on robots in the near future.

- *Ease of maintenance/sophistication.*   Simplicity and complexity are the opposites that describe this situation.

The designer's job is to reconcile these tradeoffs and develop a design compromise that suits the application.

2.2   Position, Force, and Stiffness

2.2.1   End-effector positioning characteristics

Resolution, repeatability, and accuracy are the three most important characteristics in evaluating how well an arm can position its end effector.

- Resolution is determined by the magnitude of arm motion that can be commanded reliably. The smaller the motions, the higher the resolution is said to be. High resolution is most important for making sensor-controlled motions when fitting parts together. The resolution cannot be smaller than the smallest arm motion that can be sensed. Roughly, the fewer the bits of data produced by a position feedback transducer at a joint, the lower the resolution of the arm. Sticking friction in the actuator and the transmission can also reduce resolution.

- Repeatability is determined by the arm's ability to return to a position. It is important in situations where the arm motions are to be trained, rather than programmed. Friction and backlash in the linkage between the position transducer and a joint reduce repeatability.

- Accuracy is determined by the arm's ability to go to a position specified by a set of coordinates in a reference frame fixed in the work space (usually a cartesian frame). It is most important when the arm motions are to be programmed, rather than trained or computed from sensor signals. Nonlinearity in a position feedback transducer reduces accuracy.

The customer is interested in an accuracy specification because it represents a worst-case positioning error in moving to an arbitrary position. However, since the repeatability (and resolution) of conventional arms is usually better than their accuracy, manufacturers usually quote the repeatability. Both repeatability and accuracy apply to static positioning when the arm is at rest. They do not describe dynamic position errors, which occur when the arm is in motion.

Accuracy is related to the way in which the control system represents numeric values. There are always two representations: external and internal. The external representation, which is the one that the user reads and writes, is usually in decimal notation with a certain number of significant digits. The internal representation, which is used by the control system, is usually a binary integer or floating-point value. It should be possible to convert from one representation to the other and back without introducing any errors. For example, in Unimation's VAL controller, the external representation reads to 0.01 mm and the internal representation is apparently a 16-bit scaled integer in which the least significant bit represents 0.1 mm. The accuracy of the arm cannot be any better than the representation formats allow.

It is also important to know the task's requirements for resolution, repeatability, and accuracy, because these three characteristics can vary greatly with the design of the arm. One arm design might have very high resolution but very low repeatability and accuracy. Another design might have very high repeatability but very low accuracy and resolution. A third could have high resolution and high repeatability but low accuracy. Yet another could be highly accurate and repeatable but have very low resolution.

To illustrate the latter case, consider an unusual arm design with linear joints that have mechanical detents spaced 1 mm apart, somewhat like the detents in a head-positioning mechanism of a disk drive. Such an arm would have a resolution of 1.0 mm, but because the detent mechanism could easily be built to have a backlash of 0.03 mm or less, the arm would have a very high repeatability. Similarly, its accuracy would depend only upon errors in placing the detents, which with care could also be less than 0.03 mm. So in that arm, the resolution would be 33 times worse than the accuracy or the repeatability. The worst-case error in moving to an arbitrary position would be about one-half the resolution (the detent spacing), or 0.5 mm. This is 15 times worse than the accuracy rating of that arm.

This arm design is not just an unrealistic example to illustrate a point. It would be very cost-effective for such tasks as inserting electric components into circuit boards in which the holes are located on a grid of points, with the the grid spacing equal to the detent spacing. The key idea is that in such a task, the arm does not have to be able to move to an *arbitrary* position, but only to a small number of specific positions over the holes. To further illustrate how minor details of how a task must be performed can determine which performance characteristics of a task are important, we will consider several different variations on a simple peg-in-a-hole insertion task. We will assume that the clearance between hole and pin is 0.1 mm, that the arm is very rigid, that no sensors are used, and that the position of the hole is not restricted in any way.

Case 1.   The manipulator can be trained to find the position of the hole. In this case, accuracy does not matter, but repeatability and resolution do. Accuracy does not matter because the controller can simply record the joint positions during training and play them back during production. The arm's repeatability, however, must be 0.1 mm or less for successful insertion. Because the hole might be anywhere, the resolution must also be 0.1 mm or less. There is one further constraint: The resolution and repeatability must also *total* less than 0.1 mm, because in the worst case they can act in the same direction.

Case 2.   The manipulator can be trained to find the position of the hole, but the tooling is designed so that the position of the object with the hole in it can be adjusted before starting the production run. In this case, neither the resolution nor the accuracy is important—only the repeatability. When setting up the equipment, the operator can first move the pin as close to the hole as the arm's resolution allows and then can adjust the object's position to bring the hole the rest of the way to the pin. All that is required then is that the arm's repeatability be 0.1 mm or less.

Case 3.   The manipulator cannot be trained to find the hole position, and its cartesian coordinates must instead be specified in the task program. They might appear as constants, for example, or they could be measured at run time by a vision system. In this case, all three characteristics are important. The accuracy must be 0.1 mm or less, and, for the same reasons as in case 1, the resolution and repeatability must total that amount or less.

Case 4.   The hole is moving. In this case, the dynamic characteristics of the arm would be crucial. Most manufacturers do not provide the information needed to determine if their arm can perform such a tracking task.

2.2.2 Compliance characteristics

No arm is completely rigid. A force or torque on the end effector will always produce some deflection or rotation. But often a force can result in some rotation and a torque can produce some linear motion. To completely describe arm compliance requires at least a 6×6 matrix of numbers. Each number is the ratio of a specific component of either the linear or angular deflection of the end effector to a specific component of either the force or torque acting on it. However, such a matrix describes only how the arm behaves in response to small and static (non-time-varying) forces and torques.

The total behavior is affected by:

- Amplitude and frequency of joint forces/torques
- Use of brakes on some joints to conserve energy
- Position and motion of the arm

- Load

- Externally applied forces and torques

Robot manufacturers almost never specify how rigid the arms of their robots are. It is very difficult to measure and to characterize arm compliance, other than high rigidity, in a way that can be used to determine whether the arm can perform a specific task. The manufacturer usually tries to make the arm as rigid as possible, because the more rigid it is, the more predictably it will behave under unpredictable loads.

It is important to realize that a complete treatment of compliance should include consideration of the compliance properties of the following pieces of equipment:

- The end effector

- The robot arm

- The structure that supports the arm above the factory floor

- The structure of the floor between the arm support and the tooling support

- The structure that supports the tooling

- The tooling

- The workpiece held in the tooling

These objects form a loop in space when the end effector touches the workpiece. Any of the elements around the loop can display undesirable compliance properties, especially dynamic compliance properties, that can make an assembly operation fail. A classic example is a case in which two manipulators could not be operated side by side, even on different tasks: the compliance of the floor between them coupled the motion of one robot to the other sufficiently to make them go into violent oscillation.

High rigidity is a very simple type of arm configuration, however. It is adequate for rudimentary assembly tasks, and it is useful in tasks such as grinding, in which the arm must resist large and unpredictable forces. More difficult tasks will require different types of compliance. There are two methods of achieving this, called "passive" and "active" accommodation.

2.2.2.1 Passive. *Passive accommodation* is simpler to implement but is not able to produce as wide a variety of compliance characteristics as active accommodation. It entails placing a compliant structure between the wrist of the arm and the end effector, or between the tooling that holds a workpiece and a rigid support.

A famous example of passive accommodation is the remote center compliance (RCC) tool, invented at the Charles S. Draper Laboratories (Whitney, 1979). The RCC tool is an arrangement of springs that makes it possible for an arm to insert a pin into a hole even though the clearance between the pin and hole is such that the accuracy of the arm would not normally allow the placement of the pin.

Draper Laboratories experimented with active compliance strategies that made use of force and torque sensing to insert a round peg into a round hole in a workpiece. After they had developed an effective compliance matrix for this task, they noticed that they could obtain the same compliance characteristics by mounting a certain arrangement of springs between the manipulator's wrist and the gripper that held the pin. They installed the springs, which they found performed as well as complex sensor-controlled computer algorithms. In fact, later versions of the device worked much faster than any software algorithm.

The operating principle depends upon the notion of a *center of compliance*. If a force is applied to a point on an object and if the force acts in the direction of the center of compliance, then it will displace the object without rotating it. A torque applied to the object around an axis passing through the center of compliance will rotate it without displacing it. Any other force or torque will both displace and rotate the object.

The RCC tool may be installed between the gripper and wrist of a manipulator. It then creates a compliance center a certain distance along the hand, usually beyond the fingertips. If the hand then grasps a pin so that the free end of the pin does not quite reach the compliance center, the pin can be inserted easily into a hole with small clearances.

The operation of the RCC tool is quite simple. Assume that a pin is to be pushed down into a vertical hole (Figure 2.1). If the top of the pin is tilted to the right, the right side of the pin will touch the side of the hole first. The point of contact will be above the compliance center, so the pin will pivot *counterclockwise* about it. This will bring the pin into alignment with the hole.

Without RCC, the natural center of compliance is usually somewhere in the wrist (Figure 2.2). In the above example, the pin would tend to rotate *clockwise* and become increasingly misaligned until it jammed.

RCC tools are currently available from at least two manufacturers (Lord Corporation and Barry Wright Corp.). They promise to be an important accessory for both manual and hard-automation assembly, as well as for robotics.

2.2.2.2 Active. *Active accommodation* involves sensing the force and torque being exerted on the end effector and moving it accordingly. It can produce any kind of compliance that passive accommodation methods can, as well as many others.

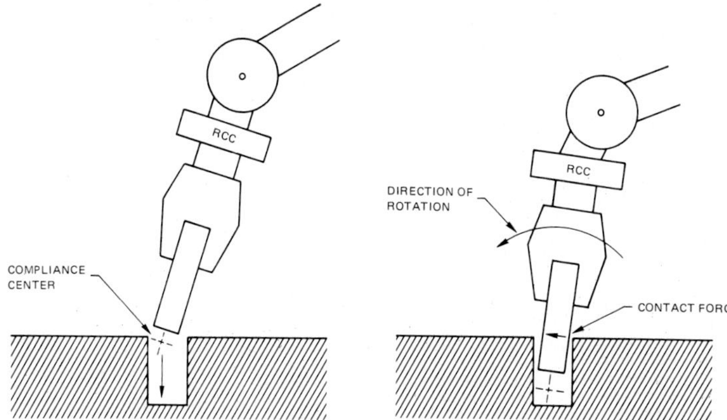

Figure 2.1 Insertion with RCC.

2.2.2.2.1 By joint effort regulation. Controlling the efforts exerted by the individual joints was one of the first methods of controlling the force and torque exerted by the gripper. The method is appealing because it requires no special wrist-mounted force/torque sensor and because small amounts of compliance in the arm links do not matter at slow speeds. However, there are significant practical difficulties in applying this technique.

For example, static and dynamic friction in the power transmission between the actuator and link can introduce significant errors in the net force and torque at the gripper. Asada, while at Carnegie-Mellon University, proposed the use of *direct-drive joints* to overcome this problem (Asada et al., 1982). In his design, high-torque dc servomotors are coupled directly to their joints without intervening gears. Joint torque is then very closely proportional to motor current, because the only friction is the small amount in the main joint bearing.

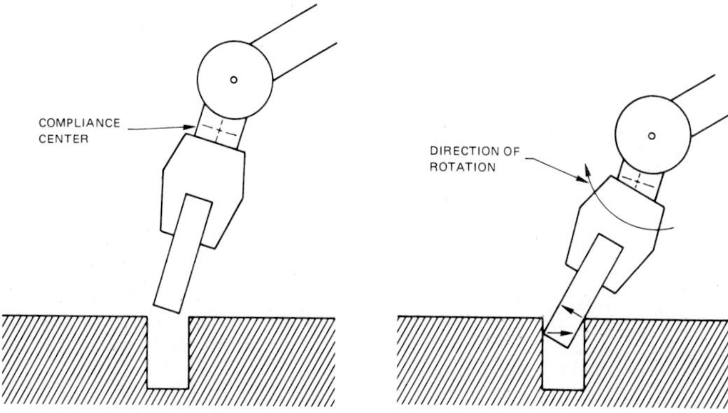

Figure 2.2 Jamming without RCC.

Another problem is that controlling the force and torque may be very difficult in certain postures, depending upon the kinematic design of the arm. These are postures in which none of the joints are in a good position to exert a force (or a torque) in the direction required.

As an example of such an awkward posture, consider a PUMA-style manipulator with its arm stretched out vertically almost to its full extent. Now suppose that it is supposed to push gently upward on something, such as the ceiling. In this position, the arm forms a classic toggle linkage. This means that a very small change in the shoulder or elbow torque will result in a very large change in the vertical force exerted by the hand. Thus, even if the shoulder and elbow joint servos are very high quality and provide very fine torque resolution, the effective resolution in the force control at the hand can be very coarse.

Once the kinematic design for an arm has been decided upon, it is possible to evaluate its force/torque control characteristics analytically for any given arm posture. One might choose to investigate a set of arm postures that are representative of postures that the arm will assume when performing a particular task. Alternatively, one might use physical intuition to predict which postures are likely to result in poor force/torque control and evaluate them. For instance, postures in which two or more joint axes become approximately parallel, or in which the arm approaches a boundary or its working volume, are likely to degrade force control.

To evaluate a particular arm posture for its force/torque control characteristics, we must determine the relationship between individual joint efforts and the six cartesian components of the net force and torque that they produce at the end effector. We then look for the following cases that can lead to problems:

- A cartesian component that is extremely *sensitive to some* of the joint efforts. This can result, for example, from a toggle situation as described above. It makes precise control over that component very difficult.

- A cartesian component that is very *insensitive* to *all* of the joint efforts. This means that the joints will have to be driven very strongly if the end effector must exert even a small force or torque in that direction. The arm is weak in that direction.

- Coupled cartesian components which are related to the individual joint efforts in approximately the same way. This makes it difficult for the arm to exert certain combinations of net force and torque.

2.2.2.2.2 Force control by compression of a spring. It is difficult to control the force that a stiff, position-servoed manipulator exerts on an object. If, for example, the object and the end effector are both made of very rigid materials, such as metal, and the joint servos are very stiff, the contact force will rise almost instantly to a very large value as soon as the end effector touches the object. If the arm is then retracted in an attempt to reduce the force, the least motion that the arm is able to make may pull the end effector completely away from the object. The force will then drop abruptly to zero. If there is a servo loop that is attempting to maintain a certain force on the object, it will simply oscillate, driving the end effector repeatedly into the object. This will result in severe vibration, if not damage to the arm and object.

In practice, there is usually a small amount of compliance in the arm that makes possible some crude control over the force that the arm exerts. This is true in practice even if the manipulator has very stiff hydraulic servos to drive its joints. The amount of this inherent compliance is often sufficient to prevent the kinds of oscillations described above. Several factors contribute to this compliance—flexibility in the arm links, backlash in power transmissions, flexibility in the support on which the arm is mounted, and some small amount of compliance in the servos themselves.

The quality of force control available from a stiff, position-controlled manipulator can be improved by artificially increasing the compliance as it would be measured at the end effector. A simple way to accomplish this is to install a compliant member between the wrist and the end effector. This can be a metallic spring or a block of elastomer, for example. Then, when the end effector makes contact with an object, any further motion of the wrist toward the object will result in a smoothly increasing force on the object as the compliant member is compressed. This gives the joint servos an opportunity to stop moving when the force reaches the desired value.

If the distance that the wrist can move before the member is fully compressed is relatively large compared to the resolution with which the wrist position can be controlled, then the quality of force control will be acceptable for simple force-control tasks. For example, if the compliant member compresses a distance of 5 mm when the force on it is 35 lb, and if the manipulator control system can reliably move the wrist in steps of 0.4 mm, then the effective resolution with which the force can be controlled is (35 lb/5 mm) (0.4 mm) = 2.8 lb. This is too coarse for delicate manipulation, but it might be adequate for some gross fitting and insertion operations.

In general, the compliant member will have some compliance in any direction, so it can be used to exert controlled forces in any direction, as well as torques. However, the situation rapidly becomes complex, because the force and torque exerted on the object depend in a complex way on the member's compliance properties and on its position relative to the object. As an extreme example, it might be possible in some cases for the end effector to buckle at the compliant member, in somewhat the same way as a column fails under excessive load. This is definitely to be avoided.

The above discussion also shows why it may be useless to attach a high-quality force sensor to a typical position-servoed manipulator such as a Unimation PUMA. Such sensors are usually quite stiff. If there is no other compliance in the system, the force resolution may be quite poor. In effect, the force sensor would be acting like the compliant member in the above example. However, because it is so stiff, the smallest motion that the hand can make will result in a change of many pounds in the force that the hand exerts; the sensor's ability to detect force changes of a small fraction of an ounce would therefore be wasted.

One expedient that can be applied in such a situation is to disable integration of position errors in the joint servos. This is a standard option in the VAL control system, for example. Normally, joint position errors are integrated over time to ensure accurate positioning in the presence of unpredictable but slowly varying disturbing torques such as those that arise from gravity loads. When this integral feedback path is disabled, each joint displays a certain amount of springiness. That is, each joint now shows a steady-state position error proportional to the load on it. The net result is additional compliance at the wrist that allows use of more of the resolution available in a good-quality force sensor. The situation is actually not so simple in a manipulator such as the PUMA, because when a joint reaches its setpoint, the VAL control system engages a brake in the joint so that it can turn off the joint's drive motor and allow it to cool. This makes the joint very rigid instead of springy and can greatly interfere with force servoing.

A better solution to the problem of how to obtain good force control is to simply design the joint servos to be able to exert a specified torque instead of to hold a specified position. One design that has been tried successfully is to place a torque sensor (such as a strain gauge) at the joint and use its signal as feedback to regulate motor drive current for that joint (Wu and Paul, 1980).

2.3   Desired Features

Despite the enormous impact that robots have already had on factory productivity, most robots in daily use are severely "handicapped." They are like a person who is blind, deaf, dumb, numb; who has only one arm, two fingers, and no legs; and who does not think. We would be surprised if such a person could perform assembly tasks. Such a robot, however, can repeat a motion more tirelessly, powerfully, and accurately than a person can. As robots' handicaps are removed by giving them the equivalent of eyes, ears, touch, and increased intelligence, they will be able to perform more and more complex and useful tasks. This section describes a number of specific features that may be needed in an assembly robot in order to perform a particular task or to perform it more efficiently.

2.3.1 Sensing

Sensing, together with the decision-making power of the computer, makes a robot quantitatively different from a numerically controlled machine tool or a hard automation machine. It allows the robot to accomplish far more complex and difficult tasks than ever before by enabling it to react "intelligently" to changing conditions. For example, with sensing a manipulator may be able to find parts in inexpensive bins or trays thus eliminating the need for feeders, fixtures, and jigs. The robot could also inspect each part before installing it and verify that it has been properly installed.

The two most important sensors for a robot are *vision* and a *sense of touch*. A number of others, such as proximity sensing, can also be useful. Such senses can perform the following important functions in assembly tasks:

- *Detecting objects.* In some tasks it will be necessary for the robot to wait for the arrival of workpieces. If the robot can see the area where they arrive, there is no need to install a special sensor.

- *Identifying objects.* An assembly robot may have to identify a workpiece:

  To make sure that it is the correct workpiece.

  Because it may be more economical to mix different workpieces in a part feeder, bin, or conveyor than to have a separate one for each kind of workpiece

- *Measuring an object's position and orientation.* In many tasks it will be uneconomical to provide special tooling to orient a workpiece and present it to the robot in a particular position and orientation. If the robot can sense the workpiece's location and orientation by itself, it can grasp that workpiece successfully.

- *Inspection.* For reliable automatic assembly, it is desirable that as many inspections as possible be performed on the components before they are installed, to reduce scrap and the need for rework. A robot equipped with appropriate sensors can inspect for part integrity, finish quality, internal voids in castings, malformed screw threads, and many other important defects.

- *Motion control.* Sensors will be very useful for detecting obstacles in the path of a moving arm, for implementing active compliance for fitting operations, for following a surface or edge of a workpiece, for tracking moving workpieces, and for correcting positioning errors that arise from inherent inaccuracies in the joint servos.

The ability to see and feel can be made an integral part of the robot, it can be offered as an option, or it can be packaged as a separate modular accessory. Making sensors an integral part of the robot would increase the minimum cost of the robot and could price it out of a certain segment of the market. Whatever approach is taken, it is certain that extremely sophisticated visual and tactile sensing systems will become available in the next few years.

2.3.1.1 Visual.   A robot can see by means of a video camera that is connected to a computer. However, the lighting, the background, and the way the objects are presented to the camera must be controlled; otherwise it is too difficult for today's image-processing techniques and today's computers to interpret the pictures.

Visual sensing is very useful because it does not require contact with the workpiece, it can take in information from a large volume of space, and it can provide a large amount of information at high rates. It can be used to locate, identify, measure, and inspect objects. Its accuracy is limited only by the resolution of the image-forming device in the camera and the magnification of the lens.

There are many ways of obtaining a visual image other than with a TV camera. There are cameras that are sensitive to infrared and x-rays, cameras that take in a whole picture at once, and cameras that see only a narrow slice of a scene. One can even obtain images from an array of photocells in a tabletop.

Solid-state video cameras use an integrated circuit behind a glass window in the case as their image-forming device. They have many advantages over older video cameras that use a vacuum tube image sensor such as a Vidicon. The solid-state video cameras are smaller, more rugged, have less image distortion, and require lower operating voltages and simpler interface circuitry. They are available as both linear and planar arrays, and most are sensitive to visible light. Some, however, are sensitive to infrared light and some to x-rays. The cameras are available in monochrome or color, the signal-to-noise ratio of the image is from 20 to 40 dB, and the resolution approaches that of conventional cameras. Many cameras are made specifically for image processing and are easily used.

Complete computer-controlled vision systems, which include the camera, a computer, and complex image-processing software are available. These systems can be trained to recognize and locate a number of different workpieces. Some can even be programmed to inspect a workpiece or to measure its dimensions precisely. Most vision systems today operate on *binary images*; the image is either black or white at each point, without any grays. Such images are obtained by "thresholding" the gray-scale or continuous-tone image produced by the camera at a certain brightness level. Choosing the correct level is a difficult problem in many tasks, so there is increasing interest in more sophisticated vision systems that can use the gray-scale image, which contains more of the kinds of information that will be important for inspection.

2.3.1.2 Tactile.   Tactile sensing includes touch sensing of various levels of complexity and force/torque (FT) sensing.

If touch sensing is to be used effectively, the robot should have provision for a number of binary input signals from touch sensors. The most useful location for the sensors is in the end effector, but they can also be placed in the tooling to suit particular task requirements. The robot control system should be able to accept new touch sensors if they are needed for a task.

The task programmer will want to be able to turn specific touch sensors on and off, to indicate whether they are expected to be activated or not during any particular activity, and to use the touch signals in conditional expressions to control what the robot does. It will also be useful to the task programmer if the robot control system does not have to be told how to use the touch sensors for certain activities that will occur frequently in assembly tasks. One such activity might be locating an object precisely by touching it from different directions; another might be to activate appropriate touch sensors during an arm motion to detect collisions (an example of what IBM calls a "guarded move").

2.3.1.2.1 Contact sensing.   One of the simplest forms of tactile sensing is detecting when the gripper or tool touches a workpiece. Many different transducers, including microswitches and strain gauges, can be used. A method that requires no transducer uses the contact between the end effector and the workpiece to complete an electric circuit through the worktable and arm. For that method the workpiece must be electrically conductive.

2.3.1.2.2 Tactile arrays.   A touch sensor can also be built that measures the pressure exerted on it and even the shear force on it. Such tactile sensors can also be arranged in a grid pattern to provide a *tactile array* (TA). A TA can provide a kind of image showing the distribution of pressure or shear forces over the whole surface of the array at once. A small TA can be placed on a manipulator's fingertip to give it a sense of touch more like that of a human assembler so that it will be able to handle small objects like wires and screws or inspect surfaces for cracks. Larger TAs can be built into a tabletop so that the robot will know where objects are located. Image processing techniques may be needed to use TAs effectively.

2.3.1.2.3 Force and torque sensing.   A manipulator can be equipped with devices that measure forces and torques applied to its end effector. These force/torque (FT) sensors usually use strain gauges as transducers and are mounted between the wrist and the end effector. They can also be placed in fixed positions on the worktable or built into tools that the manipulator can pick up. Salisbury (1985) has used small FT sensors in fingertips of a gripper.

An FT sensor usually measures the force components along three orthogonal axes and the torque components around those same axes, for a total of six simultaneous measurements. Simpler models sense only some of these force and torque components.

The FT sensor is especially useful in determining global properties, such as weight, of objects being held and in sensing forces and torques applied to objects being held. Use of this force-on-the-object information can assist in orienting the object to a surface, inserting, and following edges, generally referred to as "constrained motion problems"; the motion of the object is constrained in some way or another by the environment. Other examples of constrained motion include turning a doorknob, opening a door, and turning a crank.

Force sensing is more complicated. The robot control hardware should have provision for accepting the signals from an FT sensor (usually voltages on the order of 10 V; amplified from millivolt strain gauge signals), converting them into force and torque values, and then transforming the coordinate frame in which they were measured to a different frame.

Coordinate transformation ability is very important because the FT sensor will usually not be located in the tool or gripper or object that the arm is moving. An example is inserting a pin into a hole. The FT sensor is usually mounted at the wrist of the arm, but the force and torque must be measured about a point that is located a little past the free end of the pin. It is impossible to put the FT sensor there, so its readings must instead be transformed mathematically. The transformation is simply a matrix multiplication, which is not difficult for the control computer. With reference to Figure 2.3, the transformation matrix to go from the sensor to the tip is

$$
\begin{bmatrix}
1 & 0 & 0 & 0 & 0 & 0 \\
0 & 1 & 0 & 0 & 0 & 0 \\
0 & 0 & 1 & 0 & 0 & 0 \\
0 & +c & -b & 1 & 0 & 0 \\
-c & 0 & +a & 0 & 1 & 0 \\
+b & -a & 0 & 0 & 0 & 1
\end{bmatrix}
\begin{bmatrix}
F_x \\ F_y \\ F_z \\ T_x \\ T_y \\ T_z
\end{bmatrix}_{\text{at } A}
=
\begin{bmatrix}
F_x \\ F_y \\ F_z \\ T_x \\ T_y \\ T_z
\end{bmatrix}_{\text{at } B}
$$

2.3.1.2.4 Proximity sensors. The two most useful characteristics of proximity sensors are that they are small enough and inexpensive enough so that many of them can be used and that they do not have to make physical

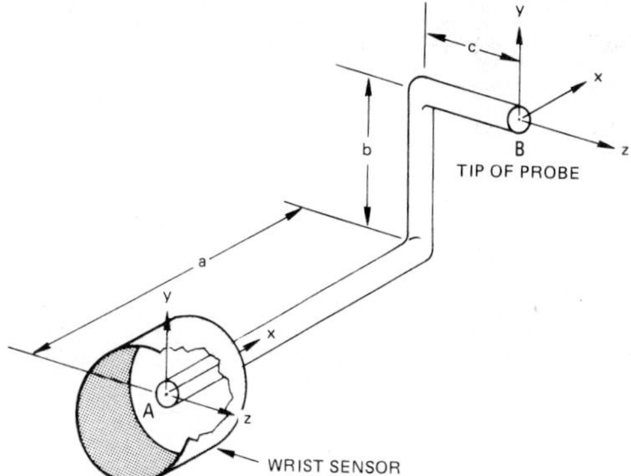

Figure 2.3 Wrist sensor to tool tip orientation.

contact with an object to sense it. Thus, proximity detectors will be very useful in assembly tasks in which visual and tactile sensors are uneconomical or impractical.

The main disadvantage of most available proximity sensors is poor spatial resolution; they cannot locate objects as accurately as visual or tactile sensors. They also have to be brought physically close to the object, and this may have to be done slowly and carefully. Finally, they usually do not sense all objects equally well. For example, acoustic sensors may not "see" a soft object or an object that reflects the emitted sound away from the sensor. An optical proximity detector may not be able to see a black object or a transparent one.

Proximity sensors placed at various locations on the structure of the arm itself would be more useful than tactile sensors for detecting collisions. By the time a tactile sensor is activated, the collision has already occurred. A proximity sensor gives the control system more time to stop the arm without damaging levels of deceleration.

2.3.1.3 Other sensors. A wide variety of other sensors will be needed for particular tasks. Some of these will be quite specialized and useful only in one particular step in assembling a particular product. Others will be more generally useful—for handling certain types of materials, workpieces of a certain shape, or very small workpieces. Some more-or-less specialized sensors that will be useful in some assembly tasks include:

- Optical character readers
- Bar code readers
- Occlusion detectors
- Eddy current detectors
- Thermocouples

- Photocells, perhaps with color filters

- Hall-effect sensors

- Infrared reflectance detectors

- Contact microphones

- Thermal conductivity detectors

- Instrumented tools, such as torque-sensing wrenches

- Probes for voltmeters, frequency generators, and other digital and electrical test equipment

In general, the noncontact sensors will be useful when the object to be sensed is very delicate and could be broken by the force that a contact sensor would exert. They are also useful when a small force might knock over the object, cause it to roll away, or make it jump out of a position in which it was being held by a compressed spring.

2.3.2   Communication

A robot manipulator in a factory must be able to communicate with people, computers, conventional machinery, sensory subsystems, and databases.

2.3.2.1   With people (operator interface).   A robot will have to communicate with people. Even a highly automated factor will have human supervisors and repairers. In more conventional factories, the robots will interact with supervisors and human coworkers more frequently.

There are two basic methods of communication between a robot and a person: symbolic and kinesthetic. *Symbolic* communication uses devices such as keyboards, printers, push buttons, annunciator boards, graphic displays, and voice input/output equipment. *Kinesthetic* communication uses devices such as joysticks, trackballs, mice,* and teleoperator master arms.

People communicate three types of information to robots: algorithms, data, and real-time control signals. Algorithms define the procedure to be followed in performing a task; data are items of numeric, logical, or textual information; and real-time control signals are used to move the manipulators (or operate other equipment) under remote control for setup or repair.

In a highly automated factory, the algorithm and data for a task will be generated automatically by large computers and then downloaded into

---

* A mouse is a handheld x,y position-indicator device that can control a cursor or other computer function.

each robot. People will interact with the robot only when a problem develops. In a partially automated factory, people will do some or all of the programming for each robot, so more communication will be required.

2.3.2.1.1 Manual control of arm motion. The operator should be able to intervene in robot control, that is, to take over from a computer. This ability can be especially important in training and debugging procedures. It may also be important in case of errors or power failure to reestablish proper positions to resume functions.

The choices in operator control are whether the manipulator should be active or passive and whether the control exerted by the operator should be remote or hands-on. Passive implies that the operator can move the robot to whatever position is desired. Passivity is difficult to design because of gravity loads that must be compensated, use of brakes, and high gear ratios. The advantage is that the passive device is safer than the active one: It can be moved without power.

The remote- versus hands-on-control choice is a tradeoff of safety versus convenience. Operators prefer to be very close to their work, within striking range of a robot. They would like to guide the robot into position by a controller attached to the manipulator. Various safety features such as a deadstick power switch and double power switches should be used if the hands-on method is chosen.

2.3.2.1.2 Training. *Training* is a simple method of programming a robot with actions, instead of computer commands. It is most suitable for simple tasks with no alternate or repeated actions. It can also be used to fill in the position information needed in a computer program.

In training, a person uses the robot arm as a measuring machine. The arm is moved into a position, and a computer or other device records the position. Some methods for moving the arm include remote control, making the arm limp and moving it manually, and steering the "live" arm with special control handles (on the arm). Remote controls include button boxes, joy sticks, voice input devices, and teleoperator masters. The robot arm may also be simulated in a computer and moved by means of commands to the simulation program.

2.3.2.2 With other computers. In a highly automated factory, a robot will receive its programming from a supervisory computer and will submit periodic reports to that computer for productivity evaluation, trend analysis, and failure prediction. Robots will probably have to communicate with each other in many tasks, too. This is especially true if the task requires two or more robots to work on the same product or to share the same area or tools. Finally, robots will have to be able to request and receive information from subsidiary computers in equipment such as vision systems, tactile feedback devices, and smart tools.

2.3.2.3 With databases. Robots in a highly automated factory may have to query and update large databases. Some information that might be stored in such a database includes geometric shape descriptions of the components in an assembly, a model of the workplace, and a library of discretionary manufacturing procedures.

2.3.3 Joint coordination issues

Many different types of motion have been devised. Some of the important characteristics of these different motions include the degree of automatic control over the motions, the method used to coordinate the motions of the individual joints, and the terms in which the motion is described.

2.3.3.1 Degree of automatic control. Three degrees of automatic control are important: fully automatic, manually controlled, and computer-augmented manual control.

Fully automatic motions are used in normal manufacturing operations. The control computer or computers are in full control of the manipulator. Manual, or remote, control is used for training, calibration, and recovery from failures.

Computer-augmented manual control has not been used often in industrial robotics. In this mode, the computer helps a person move the manipulator more safely, more accurately, or more easily than could be done by controlling it directly.

For safety, for example, the computer might prevent the operator from driving the arm into the worktable. For accuracy, it could constrain the end effector's position so that the operator could only move it back and forth along a precise preprogrammed path—such as to go through a narrow opening. To make a job easier, the operator could use a teleoperator master arm whose shape is convenient in order to control a manipulator with a different shape.

The human and the computer could also control the manipulator alternately. The computer could perform the routine portions of a task, such as exchanging end effectors, so that the human could concentrate on the more difficult portions.

2.3.3.2 Method used to coordinate joint motions. Three different methods are used to coordinate joint motion in robots today: uncoordinated joint motion, coordinated motion in joint space, and coordinated motion in cartesian space. Of these, the last is probably the most useful for assembly tasks. Unimation's VAL controller allows the task programmer to select either of the last two types of motion.

2.3.3.2.1 Uncoordinated joint motion. In this method, each joint of the manipulator moves independently, without coordinating its motion with that of the other joints. In general, each joint takes a different amount of time to reach its commanded position, and such control systems usually wait until all joints have stopped moving before going on to the next point. This is also called *point-to-point* motion and is typical of the simplest manipulator control systems.

When joint motions are not coordinated, the precise path followed by the end effector is difficult to predict, and it can vary with the weight of the load and the speed at which the motion is made. Thus, there is a danger of collision.

2.3.3.2.2 Coordinated motion in joint space.

If the individual joints move so that all of them reach their commanded positions simultaneously, we say their motion is *coordinated* in joint space. To reduce peak acceleration requirements and to reduce the sharp jolt experienced by the load being carried, most manipulator control systems increase the speed of each joint smoothly at the start and decrease it smoothly to a stop at the end of each motion.* In this method, the precise path followed by the end effector is difficult to predict, but at least it is repeatable. Thus, it is to be preferred over uncoordinated joint motion if the manipulator must pass close to an obstacle.

Coordinated motion is usually synchronized by a clock signal, as in Unimation's arc-welding control systems. A variation of this method uses an external signal to synchronize the motion. For example, Unimation's 2000-series manipulators can use the position of a car body passing on a conveyor as a synchronizing signal for spot welding. (Unimation calls this "point-to-point, controlled" motion.) Cincinnati Milacron has a similar method of control for their T3 manipulators.

2.3.3.2.3 Coordinated motion in cartesian space.

This is the usual method of control used in modern computer-controlled manipulators such as Unimation's PUMA, Unimation 2000-series manipulators with the VAL control system, and Milacron's T3. It is sometimes called *continuous-path* (CP) control, because the end effector moves in a straight line at a controlled speed between programmed positions.

This method of control depends upon the ability of the computer to determine the joint positions that will place the end effector at a specified position and orientation in the work space. This "arm solution" is a relatively easy computation for most manipulators, taking less than 25 ms in an LSI-11 microcomputer.

The CP method is necessary for tasks such as arc welding, spray paint-

---

* The speed at which the end effector moves can be controlled somewhat by specifying the duration of the motion and by placing limits on the velocities at which each joint can move.

ing, flame cutting, and deposition of glue or sealant. In these tasks the end effector must follow a curved path smoothly. Some tasks, like arc welding, require more precision than others, such as spray painting. Though this method is sometimes useful in assembly, it is not crucial. But, because the computations are so easy, it probably will serve as the normal method of operating a manipulator for assembly tasks.

The arm solution makes possible a variety of useful types of motion that robot manufacturers have not yet even used, such as the ability to move the hand along a circular path, to track visually an object moving along an unpredictable path, or to make one manipulator follow the motion of a second when they are both holding the same object.

Path following requires that the servos be able to track a moving position setpoint accurately. This is difficult: A manipulator will generally not be able to move along a path as precisely as it is able to go to a stationary position.

2.3.3.3   Terms in which motion is described or controlled.

There are also several different ways to describe manipulator motions. The major characteristics of motion descriptions are the coordinate system in which they are expressed, whether they are absolute or relative, and the complexity of the motion description.

2.3.3.3.1   Coordinate system.

The two most important coordinate systems are joint angles and cartesian (X, Y, Z) coordinates. A few robots support cylindrical coordinates, but these are probably not as useful as cartesian, because they are manipulator-centered rather than work-space–centered.

The form of a joint angle description is just a list of the joint angles. In all manipulator designs, any such list corresponds uniquely to one position and orientation of the end effector. This is typical of simple control systems.

The form of a cartesian description is always six numbers. Three numbers (usually called "X," "Y," and "Z") specify the position of a point in the end effector with respect to a set of orthogonal coordinate axes. The other three specify the orientation of the end effector about this point. There is no general agreement among robot manufacturers on how to describe the orientation. (This is not surprising—the machine tool industry has wrestled with the same problem for 20 years in connection with five-axis numerical control tools.) The VAL controller calls the angles "approach," "orientation," and "twist" ("A," "O," and "T"). Other control systems use different sets of angles. Sometimes these are called "Euler angles," represented by the Greek letters alpha, beta, and gamma. However, different sets of angles have been called Euler angles by different people.

If the reference frame in which the cartesian coordinates are measured is fixed in position in the work space, it is called an *inertial* reference frame. A particular inertial frame is usually designated as the *default* reference frame. It is often called the "base" frame, which moves and turns with the object.

It is useful in assembly tasks to be able to specify positions and motions with respect to a tool, fixture, or workpiece. For this purpose, it should be possible to define a new reference frame fixed in an object. These auxiliary reference frames should not have to be aligned with the base frame.

A frame fixed in the end effector and rotating with it is also very useful in assembly tasks—for describing reaching motions, for example. Unimation calls this the "tool frame." Another useful moving frame moves with a conveyor belt.

2.3.3.3.2 Absolute and relative motions. Absolute motions are described in terms of cartesian coordinates with respect to a reference frame that is fixed in space. That frame is usually the base frame of the robot. Thus, the robot always puts its hand in the same place in the work space every time it makes that motion.

Relative motions are supported only by the more advanced robot control systems. Such motions are described with respect to a reference frame whose position and orientation can change. For example, suppose that the robot must install 100 bayonet-socket lamps on a three-dimensional framework and that each lamp socket is at a different position and orientation. Let us say that each installation requires the same sequence of 10 movements relative to the socket. If we use only absolute motions, then the program will require $100 \times 10 = 1000$ different absolute motions. It would take several days to train that many motions, which would take a significant space in the memory of the computer that controls the robot. It would be much better to train 100 socket locations as absolute positions and orientations, and to train the 10 installation motions as relative motions with respect to a socket position. The program would then contain only 110 positions and take about one-tenth as long to train.

The VAL controller supports both absolute and relative motions because the frame of reference for any motion can be specified. For example, a vision system can be used to measure the position and orientation of an object, and then a frame can be defined there. Any future arm motions that have to do with that object can then be made with respect to that frame. VAL allows a programmer to use as many different reference frames as needed for a task, subject to storage limitations.

2.3.3.3.3 Complexity. The least complex kind of motion is point-to-point motion, in which the end effector moves from rest to a specified position and then stops. Any robot can do it.

In the next most complex kind of motion, the end effector moves through a sequence of specified postions, stopping only at the last one. Unimation calls this a "continuous-path" motion and the intermediate points "via" (by way of) points (as in "move to A via B, C, and D").

An even more complex kind of motion is trajectory following, in which the programmer specifies the complete path that the end effector should follow. (When using via points, the robot control system chooses a path itself, which may not be appropriate for the task.) Spline curves or polynomials can be used to describe the trajectories.

No commercial robots support trajectory following. The only exceptions might be certain automative spray-painting arms that are programmed from computer-aided design (CAD) information about car body shape rather than being trained. Such motion would also be useful in arc-welding and glue/sealant deposition applications.

Probably the most complex kind of motion that has so far been demonstrated is partially constrained motion, which has not yet been used by any commercial robots. In this sort of motion, the programmer specifies limits on the position or velocity of the end effector along certain directions. These constraint relationships may be functions of time, the instantaneous position of the hand, or other quantities. The constrained directions of motion may be fixed, or they may change, or they may be directions in joint space. An example of a constrained motion is crank turning. One way to turn a crank is to drive the hand circumferentially and allow it to comply more or less freely in the radial direction.

2.3.3.3.4 Special-purpose motions. In addition to the general-purpose motions above, robot manufacturers sometimes allow the robot programmer to call for specialized types of motions that are useful in specific kinds of tasks. Robots for arc welding, for example, can usually weave from side to side automatically as they move from point to point. For inserting one part into another, Japanese researchers have found that it is useful to shake the part; this might be a useful type of motion to provide in an assembly robot. Some other useful automatic motions would require an appropriate tactile sense. These include following an edge, putting a plate with a hole in it onto a spindle, inserting a pin into a hole, and screwing one part into another.

2.3.4 Parallelism

Parallelism simply means that two or more processes take place simultaneously. This capability will be very important in manufacturing in order to maximize productivity. Parallelism of a trivial sort exists in any robot that can move more than one of its joints simultaneously. Two kinds of parallelism that will be important are the ability to describe sets of parallel actions in the programming language and the ability to perform different parts of a complex computation simultaneously.

Some examples of parallel actions that will be important in manufacturing include carrying an object with two or more arms, simultaneous manufacturing operations by multiple arms and sensors in a shared workspace, and grasping a moving object using visual tracking information. Parallelism may be absolutely necessary in order to accomplish certain tasks, but it will often be used simply to reduce the time required.

Certain kinds of repetitive computations that are necessary for sensor-controlled motions can be subdivided and performed simultaneously by different parts of a specially designed computer. Arm motion control and image analysis are examples of two bottleneck computations that could benefit greatly from this technique. The trend is toward specialized integrated circuit and very-large-scale integrated circuit chips to perform these computations.

2.3.4.1 Interrupts and condition monitors. It is often convenient to be able to say, in a robot programming language, what should be done when a certain event occurs during a procedure. Interrupts and condition monitors are two control mechanisms that can be used for this purpose.

An *interrupt* is a transfer of control on receipt of an electrical signal. The microcode of the computer detects the signal and causes the control transfer. Interrupts are most appropriate when an extremely fast response to the event is required—on the order of tens of microseconds—and where the event is indicated by a logical signal (e.g., a switch closure). Interrupts are implemented in the hardware of the computer, so they do not slow down execution of the robot's control program even if the computer is watching for many different interrupts simultaneously.

A typical event that might be used to trigger an interrupt is the activation of a touch sensor by contact with a solid object. The interrupt might transfer control to a routine that stops all arm motion instantly. This type of behavior is useful for finding objects by feeling for them, because it is necessary to stop the arm quickly so that it does not knock the object over or break it.

A *condition monitor,* on the other hand, is implemented in software. Thus it can detect a more general kind of event than an interrupt, but it cannot react as quickly. When the monitor detects that the event has occurred, it causes some action to occur in response. It might perform the action itself, it might call another routine to perform the action, or it might simply place some information in a global data area for use later by the normal robot control software. The condition monitor program may perform considerable computation; it is run automatically by the operating system of the robot control computer at regular intervals—perhaps 10 times per second. The use of condition monitors can therefore interfere with normal arm control (although an auxiliary computer could be provided just for monitors).

The programmer who writes the system software for controlling the robot will usually provide a small number of specific interrupts, which the task programmer can enable or disable as desired. In contrast, the robot programmer usually wants to be able to write condition monitor programs to suit the requirements of each task. So the job of the system programmer is to provide mechanisms in the programming language to define monitors and turn them on and off during execution. The system programmer also must design the real-time control system to run the monitors at appropriate intervals.

For example, consider a task that requires inserting a long, thin rod into a tube. There is friction, so the rod may buckle, break, or jam. The task programmer might decide to use a condition monitor to stop the insertion if any of these problems develops. To detect buckling, condition monitor code might be written to check for torques acting about the point at which the fingers grip the rod. To check for breakage, the task programmer might note whether the reaction force along the tube axis has been very small for more than a second or so. To detect jamming, excessive reaction force would be looked for.

2.3.4.2 COBEGIN-COEND constructs. Another important form of parallelism is the ability to specify that two or more procedures should be performed in parallel. COBEGIN and COEND statements are one way of grouping the statements of a program that are to be executed simultaneously. In operation, the real-time control system shares the processor between the different parallel instruction streams—on a time-sharing basis, for example.

2.3.4.3 Multihand coordination. Many assembly tasks can be performed better by more than one manipulator. There are, however, different degrees of difficulty. The most difficult tasks require accurate control of the position of several moving hands at every instant. The easiest tasks are the ones in which just one hand has to move at any time. Tasks in which the hand motions are simultaneous, but are simply overlapped in time without precise synchronization, are of intermediate difficulty.

Some examples of difficult multihand activities are:

- It may be easier to orient a long object accurately by grasping it at both ends with two hands rather than by grasping it in the middle with one hand.

- If an object is held in one hand and the only grasp points on it are far from its center of gravity, then gravity and inertial forces can exert excessive torques on the hand. If, instead, the object can be grasped by two hands on opposite sides of its center of gravity, the torques on the hands will be much smaller.

- It is very difficult to handle limp materials that must be kept taut. Situations in which the material must be kept slack at all times are somewhat easier, because relatively large positioning errors can then be tolerated.

- The object to be handled may simply be too heavy for one arm to lift.

The servos used in commercial manipulators today are not very well suited to handling objects with two or more hands simultaneously. This is because they are usually quite stiff and their dynamic characteristics are not well controlled. Thus, if two manipulators A and B holding the same rigid object do not move in exact synchronization with one another, they will exert large forces on the object and may damage it. If A, for example, should get slightly out of position, it will displace B by the same amount, since they are rigidly connected through the object that they are holding. If B's servodrives are stiff, it will resist this displacement strongly and thus exert a large force in the opposite direction on the object. The relative positions of the hands holding the object are quite likely to vary in this way, because dynamic characteristics vary considerably from one manipulator to another. They even vary from one posture to another in the same manipulator. So, in practice, it is very difficult to make one manipulator follow another's motion accurately unless they move very slowly.

Some tasks that require motion of only one hand at a time are:

- One hand giving an object to another. Either the giver can remain stationary until the receiver grasps the object, or the receiver can hold still while the giver puts the object into its hand.

- One manipulator holding an object while other manipulators work on it. (It might grasp the object, hold it down, or act as a backstop.) The stationary manipulator plays the role of a programmable jig or vise.

The simplest cases above, in which only one arm moves at a time, require only conventional *serial programming* facilities. The cases of intermediate and severe difficulty require *parallel programming* facilities in the robot programming language to describe the simultaneous activities, as well as in the run-time control software to carry them out.

Most commercial robot control systems, such as AutoMatix's RAIL and Unimation's VAL, support only serial programming. Of the few systems that support parallelism, most, like Stanford's AL, are experimental. IBM's AML is probably the only commercial system that does so. Even so, both AL and AML are adequate only for the multiple-manipulator control situations of intermediate difficulty described above. No system truly supports coordinated motion of multiple arms in any general way, such as for moving along curved paths or for tracking a moving object.

In AL, for instance, COBEGIN and COEND statements are used to describe sections of code (called *processes*) that are to be carried out in parallel. One AL program can have many different processes running simultaneously, and any process can operate any of the manipulators, but only one at a time. In contrast, IBM's robot control system allows up to four different AML programs to run simultaneously. Each AML program controls a different arm.

Both AL and AML support only simple synchronization primitives (shared variables and signal/wait operations). The AL system is more flexible, however, because the programmer decides the number of processes, and the AL run-time system can create and discontinue processes dynamically under program control.

Both the AL and AML systems run all parallel processes of a robot-control program in the same computer. This makes it easier to coordinate activities in each path. It would also be possible to run different processes in different computers, but then interprocess communication would require transmission and reception of messages over some relatively slow channel. The system would then have to deal with unpredictable transmission delays and occasional garbled messages. Coordination of multiple arm motions would therefore become much more difficult.

Another important form of parallelism is cooperation between different robots on a task. If they are both controlled by the same program and run-time system, then the COBEGIN-COEND method described above would serve. If each of the robots has its own control computer, task program, and run-time system, however, coordinating their actions is more difficult.

One approach would be to make it the task programmers' responsibility to make the various robots exchange coordinating messages among themselves over some communication channel. However, this would not allow quite enough coordination for certain tasks such as carrying an object with two hands. It would be possible to provide a programming mechanism for placing one robot under the control of the other robot. If the robots have FT sensors, for example, one arm might be made to follow the motion of the other while complying with any forces acting along the line between its hand and the hand of the other arm.

2.3.4.4 Pitfalls. Although parallelism is extremely useful in a robot programming system, it is also dangerous. It is much more difficult to verify that a program with parallel activities has been written correctly because it is impossible to predict the order in which events occur. Thus a new program may work correctly duri testing but fail later during a production run. Two arms might happen to move through the same space at the same time and collide, or they might both try to use the same screwdriver at once.

Methods have been developed to prevent, or at least detect, many of the problems that parallelism introduces. This is definitely still a subject for research, but the advantages of parallelism for efficient manufacturing may make it worth the risks.

2.3.4.5   Collision avoidance.   Collision avoidance is an important safety feature in a robot programming system. There are two important kinds: that which takes place when planning how a robot will perform a task and that which takes place while the robot is performing the task. No commercial robots now support collision avoidance, but it will be increasingly important as they are used for more complex assembly tasks.

2.3.4.5.1   Planning-time collision avoidance.   *Planning-time collision avoidance* is the process of deciding how to move an arm so that it avoids hitting something when it makes a motion. To do this automatically requires extremely sophisticated geometric modeling software of the type that is just now beginning to become available.

One can also manually check for the possibility of a collision by using computer graphics equipment to display an image of the robot moving in the work space. This is not a good method, because the images are very difficult for humans to interpret, and it is very easy to fail to notice an impending collision.

A better method is to use the computer to check for collisions automatically. Unfortunately, this is a time-consuming operation with existing software and hardware (from 1 to 10 min of VAX-11/780 time to check one arm position, for example). Special-purpose VLSI circuits could reduce this time by orders of magnitude.

2.3.4.5.2   Run-time collision avoidance.   *Run-time collision-avoidance* methods use sensors to look for obstacles in the path of the robot. Several kinds of sensors would be appropriate, such as ultrasonic range sensors or infrared reflectance proximity detectors. Little work has been done on this subject, however, and there are as yet no completely reliable methods. Even if reliable obstacle detectors can be developed, there is still the problem of stopping the arm rapidly enough to prevent a collision. Even a PUMA-500 or -600 can be damaged by an attempt to stop it suddenly when it is moving at a meter per second.

Nevertheless, such an ability would be very useful in a robot—especially in one that is equipped with sensors—because sensor-controlled motions are, of course, impossible to predict accurately at planning time (otherwise there would be no need for the sensing). This makes it more necessary to check for obstacles when performing the motion at run time. And then, there is the human factor: If people work near the robot, they can cause a collision at any time.

2.3.5 Safety

Safety is an important issue in robotics, one which will become increasingly important as factories install more robots on conventional manual production lines. One person has already been killed in Japan, amid lurid headlines around the world. In the United States, the manufacturer of a product can be sued for millions of dollars if the product injures someone—even if the person was misusing the product! This section discusses several safety features that could be built into a robot.

Robots will have to work in close cooperation with people in some factories. They will all have to be repaired by people, at least for the near future. Visitors to a factory can accidentally wander into the danger zone around a robot, and the robots themselves can throw objects for some distance outside that zone.

Sensors can detect the presence of a person near a robot and shut it down. A safety sensor can be as simple as an interlock switch on a gate that keeps people out of the robot's work area. Any such sensors, however, should be designed for fail-safe operation—that is, no matter what component fails in the system, the system will not allow the robot to operate when a person is present. Redundant systems in a robot increase its cost, but also its reliability and, indirectly, its safety. It is easier to make a safety system reliable than to make it undefeatable. For example, the person who was killed by the robot intentionally climbed over the fence that surrounded it.

A *time-out* is a general method for enhancing safety by detecting failures in equipment. The idea is that each piece of equipment turns itself off if it does not receive some sort of "keep-alive" signal from another piece of equipment at regular intervals, such as every 0.1 s. Thus, when any component fails, the whole system rapidly shuts down. This distributes the responsibility for system shutdown over the whole system rather than concentrating it in a single piece of equipment.

A complementary strategy is to dedicate one simple, highly reliable piece of equipment to monitoring the other pieces of equipment for correct operation. Such a device might, for example, monitor the magnitudes of the errors in all the servos in a work area. Too large an error would probably indicate a failure in the corresponding actuator, sensor, or control system. Excessive accelerations are another good indication of trouble. Each microprocessor in the work area could also be programmed to maintain an indication of its present status in a form accessible to the safety-monitor computer. If this status indication is designed so that it will change continually and predictably during normal operation, the monitor computer will be able to detect many kinds of problems in that computer. Error-correcting codes can be used to detect and correct multibit errors in data or programs.

Start-up procedures are particularly dangerous. A person usually must be present to ensure that the work area is in a state in which it is safe to begin a production run. For example, that person should check for debris, wear, and breakage, and may have to guide the robot arm to a safe "home" position or change the end effector. The robot control system should include features to simplify the setup procedures and to eliminate as far as possible the chance for errors.

Some purely physical features are important in a robot arm, too. Well-marked equipment mounting points on the robot are very convenient. Such points reduce the chance of the user drilling a screw hole into the arm's housing and damaging internal equipment. A robot device should not have dangling wires or hoses—they should all be routed through the joints to connectors on the wrist plate. It should be possible to change the end effector rapidly and automatically. This may necessitate some sort of auxiliary device to hold the end effectors that are not being used and to present them to the robot as needed.

Automatic calibration procedures are one of the most important features in a robot. To calibrate some robots now on the market, manufacturer's representatives must come to the user's plant with special equipment. They may spend an entire day mounting special jigs on the arm, running special calibration programs, and creating PROMs or floppy disks with the needed data. This disrupts production.

Mass storage should be available in the robot control system, but it need not be physically located near the robot. It could be at a remote computer that is connected through an appropriate communication link. Mass storage will be necessary to store data, programs, online documentation, calibration procedures, calibration data, text for dialogues with the operator, diagnostic information and programs, production statistics, and probably many other kinds of information.

2.4 Integration into Flexible Manufacturing System (FMS)

2.4.1 Role in factory

The manipulator will be at the production level in the factory. In one scenario, it will be a part of a production line working in series or parallel with other computer-controlled machine tools. The manipulator may do assembly, feed parts, do inspection, or other tasks. Its function is to increase the flexibility of the production line to produce several items merely with software changes.

In another scenario, a lower-volume-production situation, the manipulator will be a part of a cell of machines that can be programmed to do many tasks. The robot is likely to perform the functions of an operator, loading and unloading machines, packing and unpacking, possibly changing tools, and doing other operatorlike chores.

2.4.2 Levels of control

The manipulator will probably have several microcomputers associated with it as part of the sensing systems (such as vision or touch) and as part of the control systems to improve coordination and speed. Certainly there will be a small computer associated with the robot function, and perhaps also for controlling other nearby machines in a cell.

The manipulator computer receives its direction from a line or floor computer that coordinates production as its prime function. This computer, in turn, gets instructions from other computers charged with scheduling, orders, inventory, warehousing, shipping, and billing. The levels of control for a manipulator are shown in Figure 2.4.

A longer-term goal is that the computer system would include computer-aided design (CAD) and go directly to manufacturing.

2.4.3 Manufacturing automation protocol (MAP)

A concern in automation has been the coordination between all the equipment involved, particularly when a variety of control computers and systems are being used on different robots and machine tools. Interchangeability of equipment from different manufacturers is also desired. The solution is to establish standards, but the standards must not be so restrictive as to exclude good existing equipment or mitigate the development of superior equipment.

MAP is an example of an industrial standard designed to fill this need. MAP defines the kinds of interactions by level and specifies appropriate communication protocols at each level.

2.4.4 Failure and replacement

Robots will fail. The designer's objectives are to minimize the failures and to minimize the effect of failures.

Failures can be minimized by making the equipment more rugged, increasing the reliability, and increasing the mean time between failures. Improved maintenance on a regular preventative basis also minimizes failure, as is done in the aircraft business. Self-diagnosis is an aid to pre-

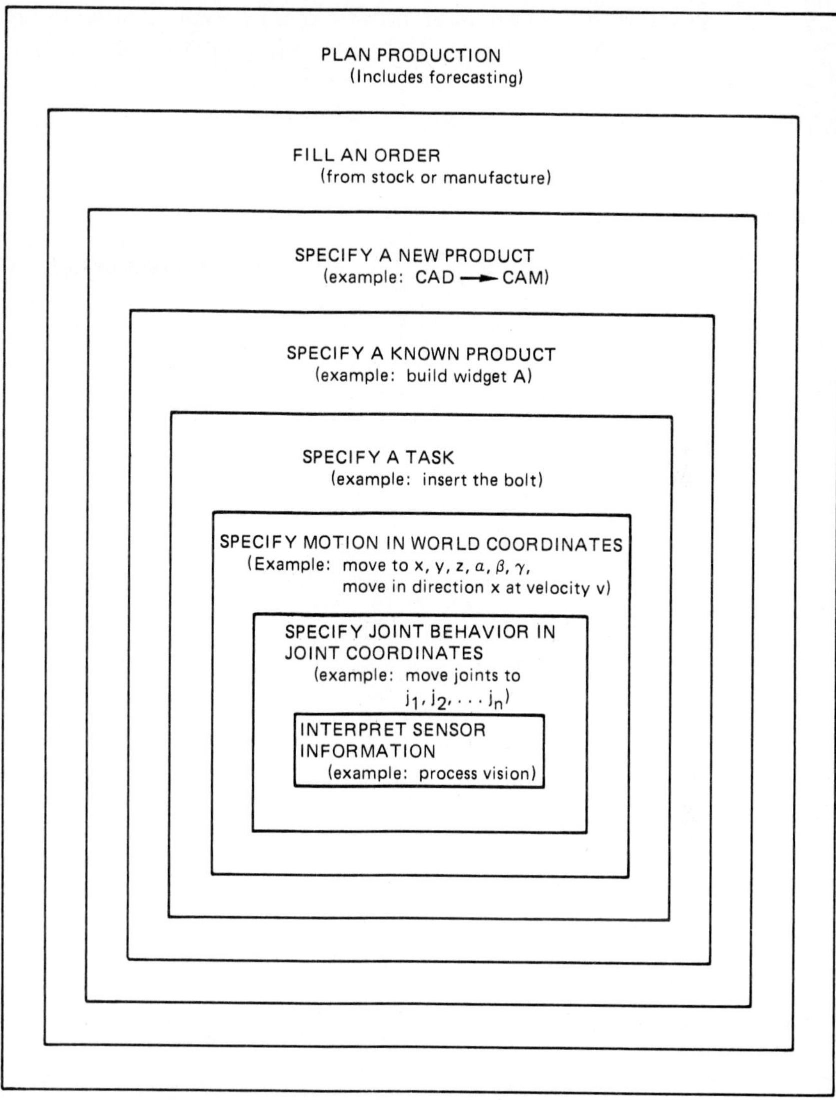

Figure 2.4   Levels of control.

ventive maintenance. The robot would have special circuits to monitor wear and to check on critical elements periodically, much like the warning lights on automobiles that signal overheating or low oil level.

Minimizing the effect of failure has two components: to "fail gracefully," or "fail soft," and to minimize the repair time. Graceful failure is a term that derives from the behavior of biological systems. Many biological failures are not catastrophic. They do not lead to immediate major interruption of the system as a whole and a sacrifice in quality of the product. The concept of graceful, or soft, failure, as important as it is, is only in its infancy. The methods have not been generalized. Yet the idea of a graceful failure, rather than the function/nonfunction dichotomy typical of machine failures, should be kept in mind by designers, who will surely find ways to accomplish a goal if they seek it.

Minimizing repair time has been addressed by having a replacement available and by speeding up the diagnosis of the failure. It is often useful to make components modular, which means that repairs can be made by replacing the modules. Only the modules, rather than the entire machine, need be taken offline. Self-diagnosis facilitates detecting failed components rapidly, which also minimizes repair time.

Design Procedure

Once the task of the manipulator has been defined and translated into performance requirements, design can begin. The *design process* is the bringing together of the knowledge of available techniques, insight, and the skills of analysis to arrive at equipment to satisfy the requirements. A variety of people will usually be a part of the design team; the design will be a result of their background, skills, and interaction. This chapter is intended to show the kinds of talents required and a structure or procedure for their interaction. Chapters 4 through 13 describe subtasks and available technology in more detail.

3.1 Overview

Early manipulators were designed by a single individual or a small design team (fewer than five engineers). Early teleoperator arms, the Stanford Arm (Paul, 1981; Scheinman, 1969), the Microbot MinoMover-5, and the Bridgeport Tiger 5 are examples. The design procedures were not rigorous or well organized; designers relied on intuition, experience in other fields, and trial and error. As experience with manipulators has increased, design choices have become better defined; in many cases a superior choice has become a standard. Emphasis is increasingly being placed on sophisticated improvements requiring larger, multidisciplinary teams and more structured procedures. A characteristic of the sophisticated design process is increased use of engineering prediction through modeling before hardware experimentation. The approach to robot design is becoming more like that used in the design of aircraft, automobiles, computers, and other complex electromechanical systems. A review of a formal procedure can be useful even when a simple robotic actuator is the goal.

Figure 3.1 shows a flowchart for robot design. The procedure need not be followed exactly, but any organized effort is likely to have similar features. In particular, the procedure will include considerable iteration to ensure that requirements have been met and that separate aspects of the design fit together. The procedure will be laid out to make the decisions with the greatest impact first. Design decisions having less far-reaching consequences, i.e., a detailed design, are not made until later in the process. This procedure is intended to minimize the need for major redesign except during the early stages when such redesign is less costly.

The design procedure begins with task and performance specifications, already covered in Chapter 2. The other tasks shown in the chart are discussed in this chapter.

3.2   Kinematic Design

Figure 3.2 shows examples of robots with different kinematic designs, different selections of kinds of joints and lengths of links that comprise the manipulator. Both a sketch of the manipulator and its kinematic representation are given. The choice of kinematic design is probably the most important choice in the design procedure, and the set of choices is very large. Yet there are only a few guiding principles.

3.2.1   Degrees of freedom

The position and orientation of a rigid part can be described uniquely by six parameters. Three parameters describe the position with respect to a reference, i.e., X, Y, and Z positions in cartesian coordinates. Three parameters describe the orientation, i.e., pitch, yaw, and roll. These six parameters (a vector) are the *degrees of freedom of the part*.

Robot manipulators usually consist of several joints interconnected by links. In the human arm, the shoulder, elbow, and wrist are joints connected by the upper-arm and forearm links. The degree of freedom of a joint is the number of parameters required to characterize the joint position. A hinge or rotary joint can be described by a single parameter, the angle of the hinge. The human shoulder joint is a ball joint with three degrees of freedom; the upper arm can be raised forward or sideways away from the body and can be twisted.

The number of degrees of freedom of a mechanism can be as many as the sum of the degrees of freedom of the joints, or may be reduced by links that connect in loops. For example, the degrees of freedom of a truss

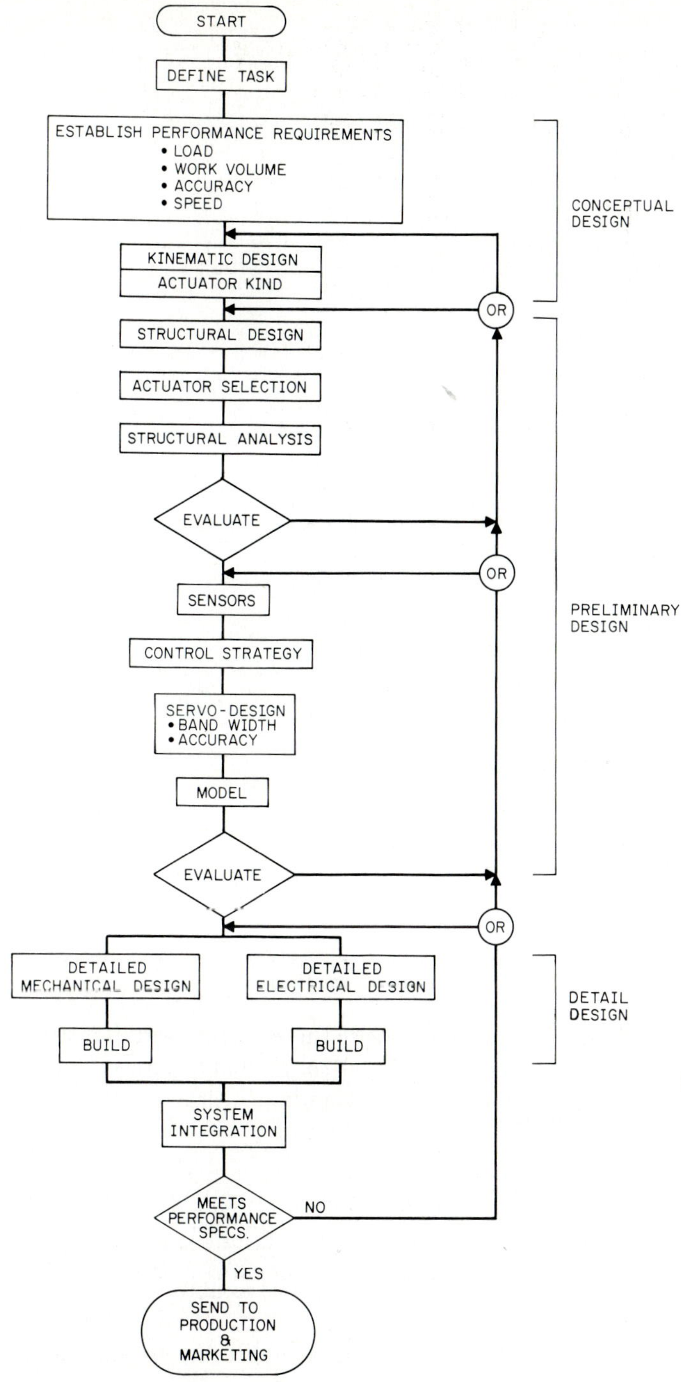

Figure 3.1 Flowchart for robot design.

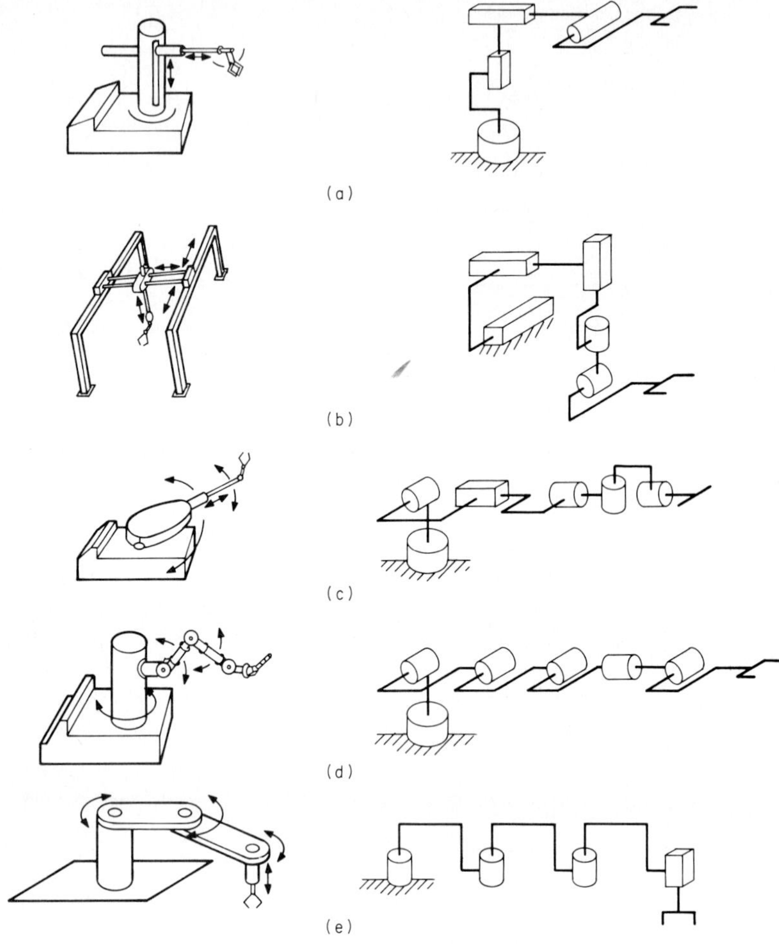

Figure 3.2 Robots showing examples of different kinematic designs.

formed by three hinged joints is zero because the links are held in relative position to one another. An actuator and control mechanism must be provided for each degree of freedom of the manipulator. Although the number of degrees of freedom of a mechanism can be arbitrarily large, each part can always be described by six parameters. If one desires to have six degrees of freedom for a part, it follows that the manipulator holding the part must have at least six degrees of freedom. A number larger than six may be used for reasons other than providing endpoint degrees of freedom. One reason may be the desire to configure the manipulator to reach around objects in its environment as well as position and orient the part. However, because additional degrees of freedom add to the machine complexity, the usual general-purpose manipulator has only six degrees of

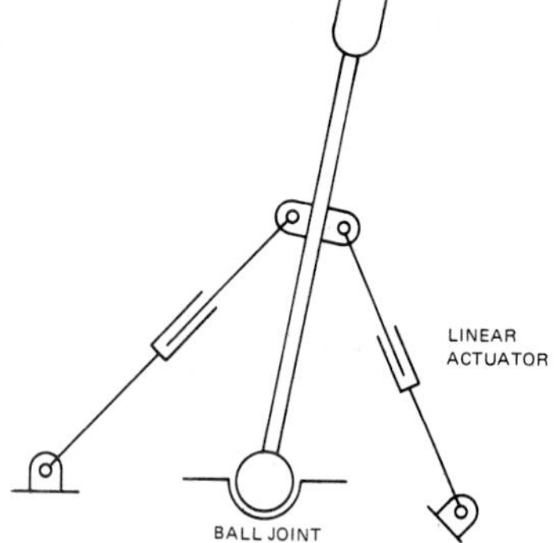

Figure 3.3 Two parallel linear actuators control a ball joint.

freedom. For specialized tasks where arbitrary position and orientation are not required, fewer than six degrees of freedom can be used. The advantage of fewer degrees of freedom again is decreased cost and complexity.

The conclusion that general-purpose manipulators should have six degrees of freedom must be treated with caution. The envelope of possible position orientations of the end effector may be complicated; particular position orientations well within the reach of the manipulator may not be possible. Furthermore there may be "degenerate" positions—orientations that can be reached by more than one combination of joint positions (Paul, 1981) that may cause problems of control. Finally, these comments do not deal with real factors such as accuracy of measurement and deflection of the links.

Much of the mathematics of robotics has to do with the relationship between manipulator degrees of freedom and the degrees of freedom of the part (or end effector) as it is described in relation to a fixed reference. A discussion of coordinate transformations is provided in Chapter 14.

3.2.2 Joint types, linkages

The robot designer should always be alert to the possibility of using unconventional joints and linkages in manipulator designs. Browsing through the pages of a compendium of mechanisms (Jones, 1930) gives a perspective on the various choices and may trigger a valuable design idea.

Of the many types of joints available, only a few have found use on manipulators. Figure 3.3, for example, shows a ball-and-socket joint with two degrees of freedom controlled by two linear actuators. Figure 11.5 shows an involute joint in which the center of rotation changes with position. By far the most common robotic joints are the simple hinge, or revolute, joint and the linear sliding, or prismatic, joint. The Stanford Arm

(Scheinman, 1969), shown in Figures 3.4 and 3.5, for example, is made up of five revolute joints and one prismatic joint (joint 3) chained together by links. Each of these joints has a single degree of freedom, and in this configuration, the number of joints corresponds to the degrees of freedom of the manipulator and the number of actuators.

The joints can be connected in any number of ways. As has been noted, the truss is three hinge joints connected in a loop by three links. The truss has no degrees of freedom. The parallelogram mechanism, Figure 3.6a, has four hinges, four links, and consequently one degree of freedom. A five-bar linkage is shown in Figure 3.7. This linkage has two degrees of freedom and therefore requires two actuators. Specification of any two of the joints specifies the structure; the actuators could be placed at any two of the five links.

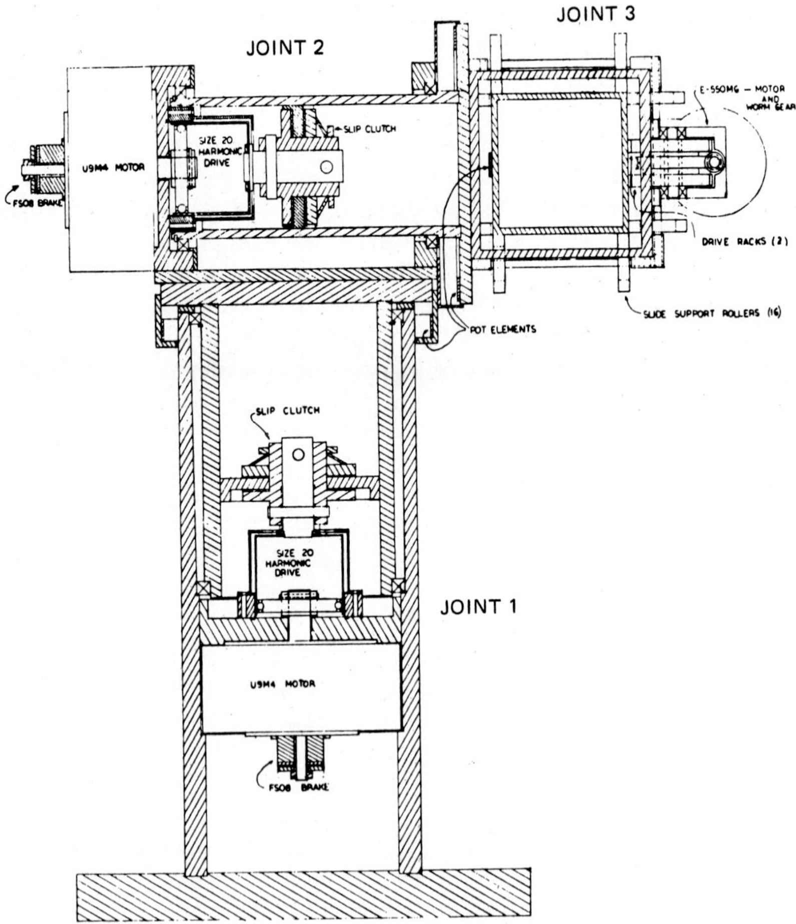

Figure 3.4   First three joints of the Stanford Arm, section view.

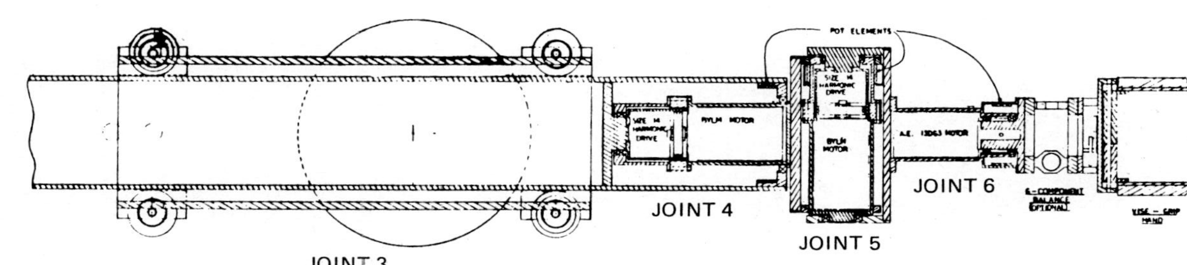

Figure 3.5 Last three joints of the Stanford Arm, section view.

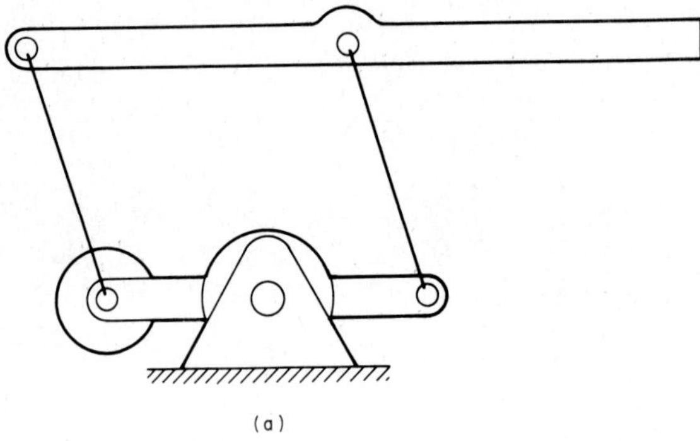

(a)

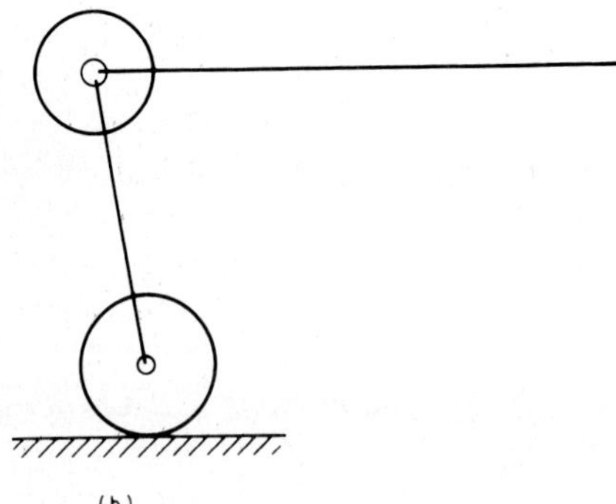

(b)

Figure 3.6 (a) Parallelogram linkage. (b) Kinematically equivalent chain design.

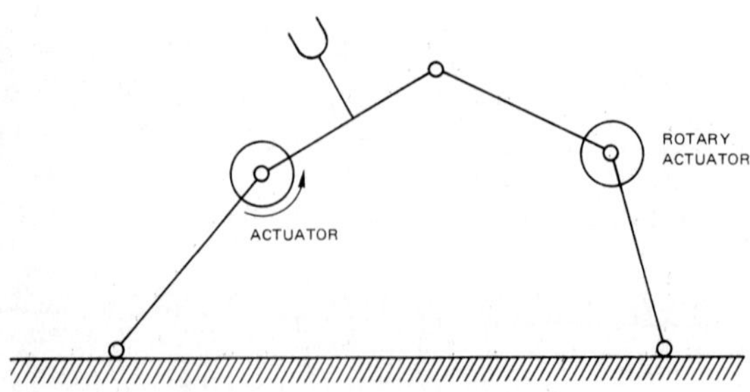

Figure 3.7 Parallel joints—five-bar linkage.

The most common linkage used in manipulators is the in-line chain. The linkages do not form closed loops. The Stanford Arm and all the designs shown in Figure 3.2 are examples of open-chain kinematics.

A common mechanism in manipulators is the three-roll wrist. This is simply three roll joints in a chain arranged so that the axes of rotation all pass through a single point (Figure 3.8). The common point simplifies coordinate transformations. The wrist can be viewed primarily as an orientation mechanism where the rotations do a minimum of positioning. The other three joints of the manipulator are then primarily for positioning.

3.2.3 Manipulator configurations

A common manipulator configuration is three chained, position-control joints and a three-roll wrist. Within this outline, there is great variety. In this section, we will examine some of the choices and discuss their advantages.

3.2.3.1 Coordinate system designs. Some manipulators are designed to produce motion in a well defined coordinate system. Prab, for example, has a series of robots where the first three joints are arranged to provide motion in a cartesian reference system. No coordinate conversions are required to position the manipulator at a choice of X, Y, and Z; the values become the settings for three prismatic joints. The advantage of the configuration, however, does not lie in eliminating coordinate transformations, but in the accuracy of motions that can be made in the direction of joint travel. If the designer wants to produce a particular shape of motion, he or she should consider a manipulator whose natural motion when under the control of a single actuator is the desired motion.

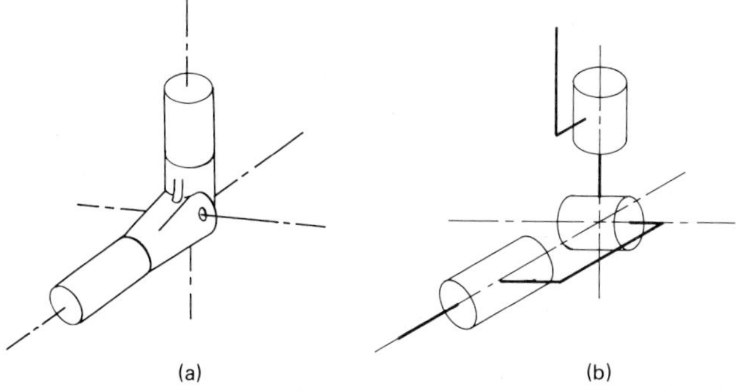

Figure 3.8 Three-roll wrist and schematic.

Figure 3.9 shows a Melfa RC-136P manipulator with cylindrical coordinates. The base has a rotational joint. The next two joints are lift and reach prismatic joints. The Stanford Arm and the Unimate 2000 (shown schematically in Figure 3.2c) are spherical-coordinate robots. The gantry type (Figure 3.2b) is a cartesian-coordinate robot.

3.2.3.2 Classical configurations.

The following configurations occur frequently enough to be given special attention. From an evolutionary viewpoint, they appear to meet needs. Paden (1986) has shown that the elbow manipulator or its dual maximizes the work volume for a six-revolute-joint robot when the forearm and upper arm are the same length and other links have negligible length. He has also shown that when one prismatic joint is used, the spherical-coordinate-system robot or its dual maximizes the work volume. Thus the popularity of these configurations has been explained by their utility.

The *dual* is the configuration when the base and the end effector are interchanged. The Intelledex 660 Robot shown in Figure 3.10 is an example of the dual of an elbow manipulator.

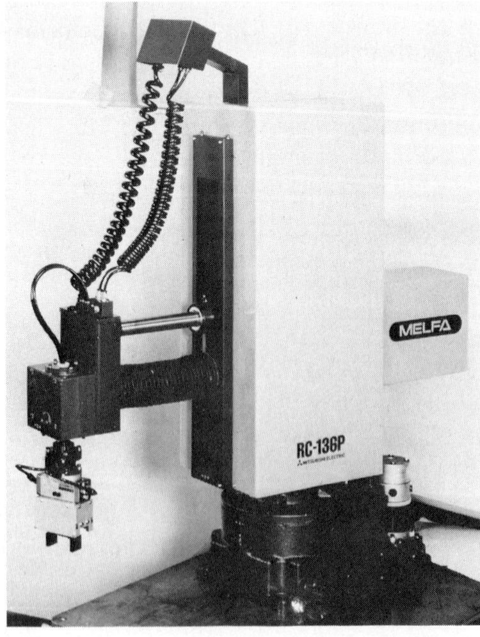

Figure 3.9 MELFA RC-136P cylindrical coordinate manipulator.

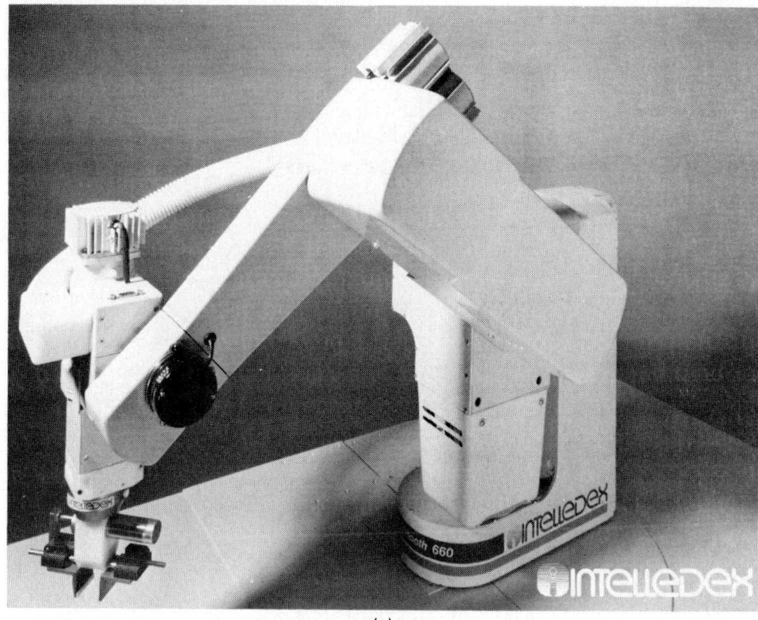

(a)

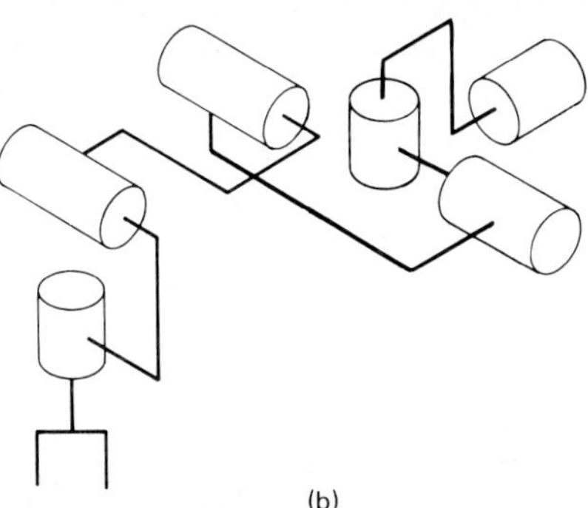

(b)

Figure 3.10   Intelledex 660 Robot with dual elbow kinematics.

3.2.3.2.1 Elbow manipulator. Figure 3.11 shows a Melfa RV-133P manipulator with the elbow configuration. The configuration is modeled on the human arm. There is a body rotational joint in the base. A shoulder rotational joint moves the upper arm up and down. The elbow joint connects to a forearm that is about the same length as the upper arm; the forearm then connects to a wrist and gripper.

3.2.3.2.2 One-prismatic joint. When one prismatic joint is used, it appears to be preferable to put the prismatic joint in the position that allows the most reach. This approach gives the configuration for the cylindrical-coordinate robot.

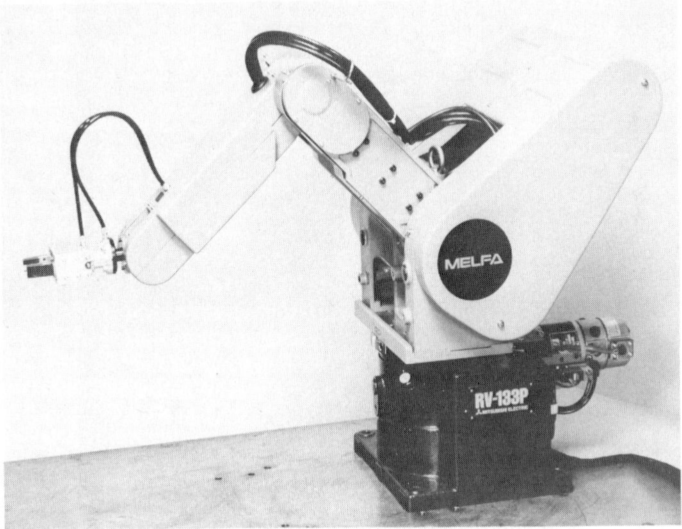

Figure 3.11   MELFA RV-133P with elbow kinematics.

3.2.3.2.3 Gantry. The gantry robot has already been noted as an example of cartesian-coordinate kinetics. The reason for its general popularity, however, appears to be its ability to position accurately within a large volume. The ability to measure an angle in a revolute joint is independent of the length of links used. As a result, the accuracy of a revolute manipulator varies inversely with size. On the other hand, the accuracy of positioning on a linear rail is more nearly independent of the length of the rail. Gantry robots have orthogonal rails to maximize the work volume.

3.2.3.3 Special linkages. Some linkages have found extensive use in robotics. The parallelogram linkage is an example. Figure 3.6 shows a parallelogram linkage and a kinematically equivalent chain linkage. The significant difference between the two is the location of the actuators. In the chain linkage, the actuator for the second joint must be located at the joint, or a transmission must be provided from the base. In the parallelogram linkage, both actuators can be located near the base. In a sense, the mechanism acts as a transmission. An additional advantage lies in the designer's ability to locate the actuators to achieve counterbalancing of the manipulator.

3.2.3.4 SCARA configuration. The Selective Compliance Assembly Robot Arm (SCARA) configuration has become popular in recent years (Makino and Furuya, 1980). As illustrated by the Melfa RH-211, Figure 3.12, the configuration is essentially the elbow robot lying on its side. The key feature is that the revolute joints have vertical axes so that gravity loads are not carried by the motors. Since the actuators do not handle

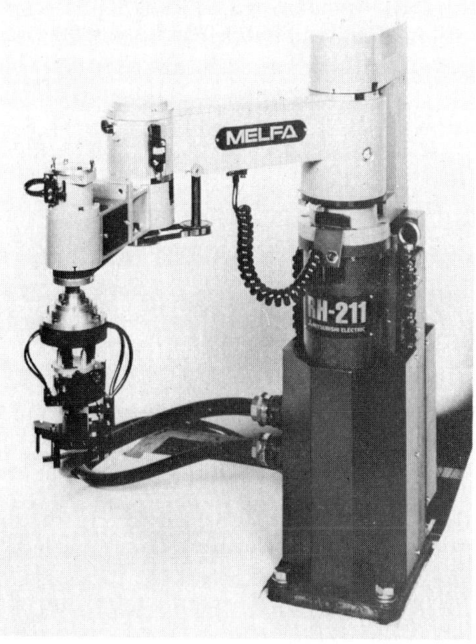

Figure **3.12** MELFA RH-211 SCARA configuration.

heavy loads, little or no gearing is required and the motor torque is directly related to the joint torque. This is to say that the motors are back-drivable. Compliance at the end effector, necessary for force control and needed for assembly operation, can now be built into the manipulator joints and controlled, hence the name.

Other features also recommend the SCARA configuration. The parallel hinged axes make simplified transmissions possible so that motors can be located one or more joints inboard of the joint they actuate. Power can be transmitted to the outer joints through systems of belts and pulleys. The ability to ignore gravity loads means that dynamic response is the only limit to speed. Inboard location of the motors, use of direct-drive torque motors, and emphasis on dynamics makes the SCARA configuration fast.

The SCARA manipulator with all vertical-axis joints illustrates the value of eliminating gravity loading of the actuators. When a vertical-axis joint is used, it should be the most heavily gravity loaded joint, the base joint in an open-chain kinematic design. Most manipulators have such a base joint.

3.3 Configuration

A few decisions in addition to the kinematic design selection will be needed for sizing. These decisions include a selection of actuator type, choices of actuator locations, and possible use of counterbalancing or spring loading. With this information, one can sketch the appearance of the manipulator, recognizing that exact size and other details may change.

3.3.1 Actuator kind

At this point we wish to select the kind of actuator, the type of motion produced and how it will couple to the manipulator (transmission), and become aware of the implications of this choice. In general, linear actuators are preferred for prismatic joints, while rotary actuators are preferred for rotary joints. Figure 3.3, however, illustrates the use of a linear actuator to control a rotary joint. The choice of actuators according to power source is shown in Table 3.1. The table notes advantages and disadvantages of each type.

Considerations relating to the choice of actuator are the robot's behavior when there is a loss of power and its power load when it is in a holding position. A dc direct-drive electric robot will deflect under applied loads when the power is lost and requires constant power to hold its position against loads. A brake is desired if a position is to be held for a long time or if alteration of position with loss of power must be prevented. Not all electrical drives respond the same way. Highly geared motors, especially worm gears, often cannot be driven backward and will hold their position when power is off. Stepping motors also hold their position. Hydraulic actuators in control systems are not back-drivable and need brakes only to prevent drift due to internal leakage (in the actuators or control valves). Unfortunately the same factors that eliminate the need for brakes make force control difficult.

TABLE 3.1   Kinds of Actuators

Power source	Motion	Comments
Electric	Rotary, can be linear, can be converted	Many types ranging from simple stepping motors to dc servos that provide excellent control. Sometimes preferred for simplicity. Sometimes preferred for control of force as well as of position. Generally has low starting torque, needs gearing
Hydraulic	Linear, can be rotary	High ratio of force and power to weight of actuator. Good position control possible. Makes for complex piping. Can be messy
Pneumatic	Linear, can be rotary	Difficult to control in position. Preferred for speed in simple applications, i.e., one-degree-of-freedom robots and grippers

3.3.2   Actuator location

Generally, one desires to locate actuators as close to the base of the manipulator as is reasonable. The actuators have a mass that must be supported by the structure of the robot and often by actuators on the joints between the base and the joint under consideration. This load may be gravitational, due to forces applied at the end effector or to dynamic loads. Inboard location therefore has the effect of reducing the structural size of the robot and the size of actuators.

It is difficult to locate motors several joints inboard from the joint to be controlled because the transmission must pass through intervening joints. At least, the motors should always be located on the supporting link rather than on the moved link.

3.3.3   Counterbalancing

One of the difficulties in the control of robots is that the torques required at joints depend on the configuration. The performance is therefore a strong function of the configuration, and often the control system is designed to accommodate the worst condition, thereby reducing performance at other positions. The problems of variable configuration and load have traditionally been dealt with in the control system.

Counterbalancing and springs may be used to reduce motor size and simplify control. Simple counterbalancing, however, increases inertia and is therefore counterproductive if rapid accelerations are desired. Youcef-Toumi and Asada (1986) have shown how to design a manipulator with a constant inertia tensor. Regardless of the position of the manipulator, the inertia seen by each actuator is constant.

Whether it is wiser to build manipulators which have constant properties or which accommodate changes in the control system is a subject of continual debate. A change in load will change the manipulator properties in any case, so the control system will have to do some accommodating. The answer to the question, however, is that the designer should use techniques that minimize the gravity loads and decrease the changes with position when it is convenient. Figure 3.6a shows the location of an actuator to counterbalance the limb.

3.4   Structural Design and Analysis

Structural design consists of planning the physical construction of the load-carrying parts of the robot. Specifically, this task includes specifying the shape, cross section, and materials used for each link of the arm and specifying the actuator sizes and transmissions.

Typically the designer begins at the end effector of the manipulator, designing the end link and its actuator-transmission to meet the applicable load, accuracy, speed, and other requirements. This link with its mass and reaction forces added to the applied load is used to design the next link, one step closer to the base. Subsequent links toward the base are designed until the design is complete.

This structural design which has proceeded one link at a time must now be subjected to an overall analysis to see whether it will perform as desired. Many performance features, such as the accuracy and overall stiffness, depend on all the joints and links and cannot be evaluated separately. Performance may also depend on the sensors and control system which are not yet specified. Representative values for unspecified quantities should be used.

There are two approaches to the overall structural design, *accuracy* and *frequency*. The accuracy approach is concerned with the structural stiffness and how it affects accuracy. The frequency approach is concerned with stiffness and inertia and how they affect the speed of the manipulator. Both approaches may be necessary to ensure that requirements have been met, although a single approach by itself is often sufficient since both approaches tend toward similar designs. Since the accuracy approach is simpler, it may be the only analysis used on simple manipulators, leaving performance verification to laboratory tests.

3.4.1 Accuracy analysis

Usually payload and accuracy are specified. The accuracy approach to the robot design assures that the endpoint of the manipulator is within the specified accuracy whether or not the robot has a payload. The analysis amounts to a determination of the spring constant of the endpoint of the manipulator in various configurations. The manipulator must deflect less than the specified accuracy under the specified payload, i.e., a minimum spring constant is required.

The deflection of the manipulator under load could be measured and compensated, but this is a more complicated approach. Allowing for compensation would lead to lighter-weight manipulators.

3.4.2 Frequency analysis

Manipulator motions involve starting and stopping. The motions can be decomposed into frequencies which are generally higher as the speed of the robot is increased. The ability of a controller to control typically falls off at higher frequencies. The frequency capability of a controller is referred to as its *bandwidth*. The higher the bandwidth, the faster the robot can be manipulated, or conversely, the more rapid the manipulation desired, the higher the control system bandwidth must be.

The relationship with the robot structure has to do with the ability of the control system to know where the manipulator is. Consider the simple example shown in Figure 3.13. The endpoint of the link may be in quite a different position and moving at a different velocity than indicated by the sensor measuring the joint angle. The control system may try to move the link to a desired set point and may deform the link still further. When it vibrates back, it will overshoot in the other direction. The danger is that the control system will compensate in an unstable fashion, pumping energy into the vibration. This can be avoided if the natural frequency of vibration is kept about 5 times higher than the bandwidth of the controller. The controller simply will not respond to the high frequency. The bottom line is that manipulation ability depends on the natural frequency of the manipulator.

The analysis required is modal analysis, the determination of the modes of vibration and especially the frequency under various conditions. The critical value is the lowest frequency. Modal analysis requires information on the inertia and stiffness of the links and joints. The frequency analysis method leads to a large, stiff manipulator, the same general trend as the accuracy analysis approach. Again, lighter links could be used if better control systems were used. In particular, the control system would need to have and use information on the state of the deformed link and is a topic of current research.

3.5 Levels of Control

Control is generally divided into two activities or functions that are labeled "Planning" and "Execution" in Figure 3.14 *Planning* is predeter-

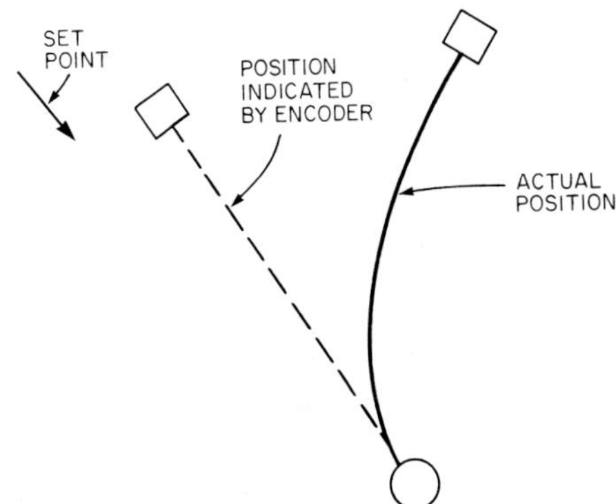

Figure 3.13 Difference between perceived and actual position due to bending.

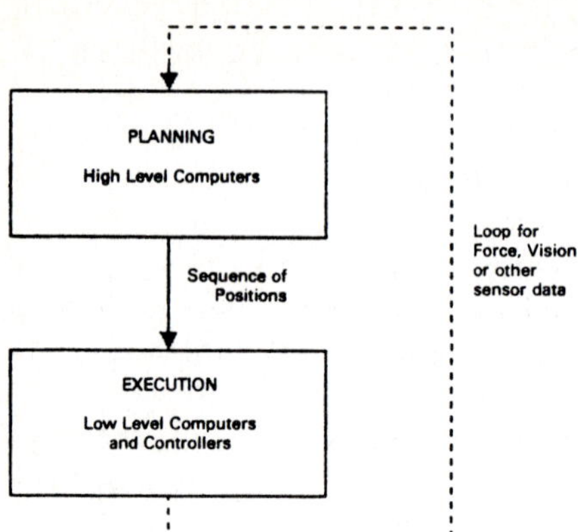

Figure 3.14 Standard control system structure.

mination of the motion, the selection of the path. It is usually done in a central computer prior to the execution. The planner sends a series of positions to the executor, often called the "controller." The controller is designed simply to be able to carry out these commands by sending proper signals to the actuators. Planning is done in a high-level computer while execution often requires only simple devices.

Most manipulators are position control devices. The execution system is designed to accept position information from the planner. A controller could be designed to apply an endpoint force, but this typically has not been done. Therefore, the planning computer must get into the loop in order to execute a control that requires force. The force is measured and sent to the planning computer. The planning computer works out a position that will provide the desired force and sends it to the controller. The loop through the planner could be eliminated by a different design of the controller unit.

The strategy of separating execution and control has the advantage of being compatible with interchangeability of hardware. One might like to generate the solution to an assembly problem and then have two different manipulators to do the job. The local controller would have the task of assigning the torques to provide the specified motions.

Separation of planning and execution is also compatible with the concept of offline programming. Much robot programming is done by leading the manipulator through a path (or a series of points); the operator thereby provides the high-level planning. It is imagined that in the future, computer-aided planning will decrease the time during which the manipulator is not actually doing a job. The operator would program the manipulator by means of a computer model, or perhaps the planning computer would eventually generate its own set of commands based on design information.

However, complete separation of planning and execution can never be achieved. Avoiding obstacles with the links of the manipulator, for example, requires that the planner have knowledge of the kinematics of the manipulator. Furthermore, optimal planning requires a complete model of the specific robot; an optimal path for one manipulator would not likely be optimal for another. A similar limitation is that as the speed increases, the dynamics of a robot become extremely important. One can argue that the dynamics should be built into the controller, but this is not always possible. In any case, the planner may specify motions that cannot possibly be executed. The planner must at least be aware of the kinematic and dynamic capabilities of the manipulator.

If a manipulator model is required, the question becomes whether the model is accurate enough. In some cases, particularly with increasing speed, we suspect that the answer is no. The foundation is extremely important in determining the lowest mode of manipulator resonance. Unfortunately, the foundation is difficult to model and is the one factor over which the manufacturer has no control.

3.6   Servo System Design and Analysis

3.6.1   Coordinates and data format

The format for communication between the robot and the system that commands the robot should be included in the performance requirements. The simplest (and most common) format for this communication is the joint coordinates. As robots become more sophisticated, they are being designed to respond to higher-level commands. For example, it is desirable for the robot to comprehend commands in "world" coordinates rather than in joint coordinates. Some robots can be instructed to follow a curved path at a specified velocity. Other desirable features include:

- Ability to freeze and unfreeze each joint.
- Ability to be compliant on command. In the compliant mode, the actuators exert sufficient torque to exactly cancel the effects of the weight of the robot. A compliant robot can be easily moved by hand for purposes of training.
- Ability to specify the torque (or force) exerted by each joint instead of by joint coordinates.

Other considerations relating to data format include:

- Type of data interface (such as bit parallel, bit serial, RS-232, IEEE-488, teaching pendant)
- Data rate of robot commands
- Requirements for internal storage of commands

3.6.2 Control strategy

The system servo selection is largely determined by the sophistication of motion specification desired. Manipulator systems may be asked to do several different kinds of tasks which may require different kinds of control. Tasks with some important subcategories are:

Go to another point. The manipulator is at rest at one point and is requested to go to another location and orientation. The task may be subjected to the following additional constraints:

No overshoot. A necessary characteristic when approaching a solid object where impact could cause damage

As fast as possible. Sometimes optimizing to the minimum time is valuable

In a specified time. Arrival must be coordinated with another event

Avoiding obstacles in the path. Either in the path of the end effector or of the limbs of the manipulator

Limiting accelerations. It may be necessary to limit accelerations on the payload to prevent damage or on joints of the manipulator itself

Pass through a series of points. Or at least close to them.

Follow a path. A path may be specified for any number of reasons. The path will certainly be in space and may be in time as well. We may desire the manipulator to be in certain locations at specified times.

Push with a specified force. Force control.

Provide a compliant behavior. Provide a force in relationship to a distance from a position.

Follow an edge or a surface. Move in one direction while providing a force in an orthogonal direction. There are several permutations of possibilities here.

The problem is to provide the robot with the ability to do these significantly different tasks.

A motion specification hierarchy, in order of decreasing difficulty, is given in Table 3.2.

The problem of control amounts to deciding on the torques to be applied to each joint based on the desired outcome and the present state of the manipulator. Conceptually the problem sounds simple. One simply measures the position of the robot, calculates the torques based on the desired function, and sends the values to the amplifiers. Certainly as computers increase in speed, this solution will be increasingly feasible. However, at present it is not possible. The calculations necessary to send the motor signals cannot be made rapidly enough by a single central com-

TABLE 3.2   Hierarchy of Motion Specification

Difficulty score*	Motion specified	Comments
10	Endpoints, factors to limit, optimization criterion	Limiting factors may be acceleration; approach vector; object to avoid. Optimization may involve time, energy, wear of parts
9	Endpoints, departure and arrival vectors, time, acceleration limits	
8	Endpoints, departure and approach vectors	Choice of paths is a complication
7	Sensor-controlled path	Current state of the art
6	Path as a function of time	Current state of the art
5	Path only	
4	Endpoints, time of travel, overshoot limitations via points	
3	Endpoints, via points	Via points may be used to avoid objects
2	Endpoints, overshoot limitations	
1	Endpoints only	Includes orientation. Allows overshoot that is unacceptable

*10 = most difficult

puter. By the time the information is calculated, the manipulator is too far from the correct position to make use of the calculations. In fact the manipulator may respond in a way that completely departs from the intended behavior. Consideration of obstacles in the path of the payload or limbs, optimization such as minimization of time, and path-planning activities make a complete real-time solution impractical.

Many practical control systems are unsophisticated and limited in their capabilities. The simplest control strategy is the "bang-bang," or motion-limit technique used in repetitive point-to-point robots. Usually these are pneumatic actuator devices for material handling. Adjustable limits are built into the joints, and the joints are driven to one limit or the other. The motion-limit strategy has only limited applicability to advanced robots, perhaps in open/close end effectors for holding tools.

Simple open-loop control of each joint is a strategy that has been used on stepper-motor-activated robots. Each joint is directed to a specified position without return information (feedback) on whether the robot has accomplished the objective.

The current state of the art is to use proportional, integral, differential (PID) controllers on each joint, using feedback of the joint position (and rate of change) for comparison with the command position. Joints may be commanded to move simultaneously, but each joint's controller is independent.

The current state of the art is an adaptation for each joint of conventional control technology. This is of limited value in robotics for the following reasons.

- The control strategy does not adapt to the changing inertia that is a function of manipulator position and payload. To ensure stability over all positions, the conventional PID controller will fall far short of the performance that is possible in other positions.

- The strategy does not take into account the simultaneous behavior of other joints that affect the performance of the PID control loop. Additional instabilities are possible over those available for a single joint. This is particularly true if the links and joints are more flexible than a single joint.

- PID control is not well suited to nonlinear components that robot designers tend to include for weight-saving.

Control strategies more elaborate than independent PID controllers have been proposed (Utkin, 1971, 1972; Raibert, 1976; Albus, 1975*a,b*; Blanchard, 1976). The more elaborate strategies will become more common in the future and should be considered in the design of new robots. The following advanced strategies entail modification of the controller and use advanced computer calculations, making the computer an integral part of the controller.

The most straightforward approach is to modify the gains of the PID controllers as a function of the arm position. This could be thought of as a step toward adaptive control. Adaptive control would include a means of measuring the payload and making gain modifications in response to the payload as well as to the arm position.

Learning could be used to speed up particular motions. Presumably, this would require a controller whose output could be preprogrammed for the basic motion with fine control superimposed. The robot would learn by doing, much as a child learns coordination through practice. This technology implies use in a frequently repeated motion such as on a production line. Because most robots will probably be used in repetitive tasks, the approach could be very important. The learning could be accomplished on simulations.

Use of faster computers and calculating procedures makes possible the advanced calculation of joint forces and torques for a given motion. In this strategy, the computer would deliver to the controller the basic joint torque and position, and the controller's function would be to make small modifications to account for wear or variable friction, for example. The strategy would be similar to learning except that the basic actuator signals would be computed on the basis of a path supplied. One has no guarantee that the given path would be optimal in any way.

Optimal control would entail calculations like those above but would also supply the path in order to minimize time or maximize some desired performance criterion. Optimal control is a complicated mathematical construction and may first be used to generate the signals for repeated motions offline. Online optimal robot control is probably decades away. These advanced control strategies are potentially applicable for use with flexible robots.

So far the discussion has focused on position control by feedback control of joints. Such control does not include feedback of the end effector of the robot and is in a sense open-loop control. The use of the robot for application of force or torque has not been discussed.

Knowledge of the endpoint could be improved if the deflection of the links were included. In existing robots, only the large deflection of the joints are measured. Measurement of the links will become more important as robots become more flexible.

Position or force (or torque) feedback from the end of the effector can currently be used as a feedback technique. The desired information can come from a sensor or a camera mounted either on the manipulator or on the workstage. Based on the known joint positions and assumed end position, and on the observed end position (or force), corrective action can be prescribed. End-effector feedback is usually used only for fine positioning, because it is slow and because fine measurements are possible only within restricted ranges.

3.6.3 Control system architecture

After the control strategy has been selected, the control system designers must make several decisions which, taken together, define the control system architecture. These include:

- Division of the control system into analog and digital subsystems

- Selection of the quantity and type of processor(s) used in the control system (for example, the Unimation PUMA robot uses seven processors in its control system)

- Selection of memory for the digital subsystem and provisions for shared memory if multiple processors are used

- Provisions for interprocessor communication in multiprocessor control systems

3.6.4 Servo system design and analysis

After the control data format, control strategy, and control system architecture have been established, those tasks normally associated with servo design (i.e., those topics covered in most servo system design textbooks) can be undertaken. This effort includes the selection of the PID coefficients, servo bandwith, and servo compensation.

At this time, it would be advisable to perform a mathematical analysis or numerical simulation of the servo system to verify that the performance requirements can be met. A thorough mathematical or numerical analysis of a typical robot servo system will be expensive and time-consuming, but the effort will probably be worthwhile, because it can result in huge savings in system debugging expense (described below). The servo system analysis should include the results of:

- The kinematic design
- The structural analysis
- The selection of sensors and actuators (ie.e, sensor and actuator input/ output relationships)

The servo system analysis will yield predictions of the robot performance in several areas including:

- Servo bandwidth
- Maximum slew rate and acceleration of the various joints
- Any servo instabilities
- Robot positioning accuracy
- Maximum payload capability

These results should be compared with the performance requirements. As shown in Figure 3.1, if the performance requirements are not met, it will be necessary to reconsider at least the selection of sensors and actuators and the servo system design. If the rule of thumb that the servo bandwidth is at least 5 times the lowest structural natural frequency with locked joints

$$\omega_1 \gtrsim 5\omega_s$$

has been met, it will usually not be necessary to reconsider the kinematic and structural design in order to achieve satisfactory servo system performance.

Many robots have been designed without mathematical analysis of the servo system. This approach typically results in a servo system that is, at best, marginally stable, which is manifest in the robot's shaking. If the servo system has not been analyzed, the servo instabilities are not evident until debugging begins. By then, it is usually too late (according to management) to consider a redesign of the structure, sensors, actuators, or servo system. The usual solution to the problem is to reduce servo band-width until stability is achieved. This expedience results in a product that fails to meet the performance requirements—a most unfortunate outcome.

3.7 Detailed Design

Once a satisfactory servo system design has been devised, the detailed design of the mechanical and electrical subsystems of the robot can begin. These tasks can be performed simultaneously, as shown in Figure 3.1. The lines in the figure represent the flow of information between the teams working on each subsystem.

The detailed mechanical design includes:

- Details of fabrication of the robot structure (design of castings and weldments)

- Selection of bearings for each joint

- Mounting of bearings, sensors, and actuators for each joint

- Devising a method and procedure for assembly of the robot

- Provision of cable raceways to permit routing of electrical, pneumatic, and hydraulic lines to the various joints

- Provision for mounting and cooling the electrical subsystem

- Provision for shear pins, break-away joints, and other mechanical features to minimize the robot's potential for self-destruction

The detailed electrical design includes:

- Design of the actuator drive circuits (servo amplifiers if servomotors are used)

- Design of the circuitry to implement the servo loops (PID controllers, for example)

- Design of the robot's interface and control system (interface transceivers, system bus, control computer or computers, and memory)

- Design of power supplies for all the above

- Design of systems for prevention of electromagnetic interference (EMI) between the subsystems in the robot and between the robot and its environment

- Writing software for computers (if used) in the robot

After the mechanical and electrical subsystems have been designed, they are built and tested. These two subsystems should be separately tested as much as possible before they are interconnected. It may be necessary to prepare some simple test fixtures and circuits to facilitate the independent testing of the electrical and mechanical subsystems. For example, a dynamometer and flywheel can be used as a simulated load for the joint actuators. Such an apparatus should be used to permit testing of the actuator drive circuits. Simple manually operated drive circuits for the joint can be used in testing to permit the mechanical design team to operate the robot at reduced speed without using the actuator drive circuits of the finished robot. This arrangement will allow the mechanical team to assess the robot's accuracy and stiffness before integration with the electrical subsystem.

Testing of the mechanical subsystem of the robot should include:

- Verification of kinematics (i.e., can the robot reach all points in the working volume?)

- Verification of stiffness

- Verification of the robot's natural frequencies

Testing of the electrical subsystem should include:

- Testing of servo drive circuits using simulated loads on the actuators

- Testing of the interface and control system

- Testing of power supplies with real and simulated loads

- Testing the electrical subsystem for both internal (i.e., between the various circuits of the electrical subsystem) and external (i.e., between the robot and its environment) EMI

- Testing of software using simulated sensor input

- Determination of temperature rise of highly stressed components such as transformers, electrolytic capacitors, and high-power semiconductors

Life tests of critical robot components should probably also begin at this time. There is no substitute for a life test of the entire robot, but schedule considerations will very likely preclude thorough life testing before the robot is in production. Some critical components can and should be separately life-tested, including:

- Bearings

- Actuators

- Gears, cables, flexible shafts, and other transmission components

- Power semiconductors

- Brushes and commutators, if any of the actuators use brush-type motors

3.8   System Integration and Debugging

After the mechanical and electrical subsystems have been tested separately, they are interconnected to create the complete robot. Despite the most careful design and testing of these subsystems, there will nearly always be unexpected problems that must be solved to achieve successful system integration. The solution of these unexpected problems is termed *debugging.*

It is difficult to provide debugging advice in this book because the problems encountered in integrating a specific system are unexpected. However, the following observations are offered. It is likely that one or more servo instabilities will be discovered during system integration. The first response to this problem should be to review the servo system analysis to determine why the physical system is not behaving as predicted by its mathematical model. Often, the physical system has certain aspects that were not modeled; backlash, friction, and excessive compliance in bearings and gears are frequently overlooked. Such problems can be handled in two ways. First, the model can be revised and then used to test various proposed solutions to the problem. Second, efforts can be made to minimize the backlash, friction, or compliance.

Despite efforts made during previous tasks, EMI between the different electronic subsystems in the robot can occur under certain unusual or unanticipated situations. Such problems are often intermittent, and thus very difficult to isolate.

Finally, the robot is a very complex system. A good knowledge of the disciplines of mechanical and electrical engineering is required to fully understand its operation. This broad knowledge base can be especially crucial during the debugging effort. Unfortunately, engineers with such comprehensive background knowledge are relatively rare. The wise robot manufacturer will cultivate engineers with this kind of broad background and assign them to oversee the task of robot debugging. Figure 3.15

Figure 3.15   The motion control engineer. A broader knowledge base: power electronics + electric motors + controls.

(*Drives & Controls International,* 1982) depicts the mix of talents needed for efficient debugging of complex robotic systems.

The debugging task is complete when the robot can meet all the performance requirements. As shown in the flowchart of Figure 3.1, partial or complete redesign of the robot will be required if debugging is unsuccessful.

Structural Design Methods

4.1 Introduction

This chapter describes some of the mathematical techniques used by designers of complex structures. Mathematical models and analyses are briefly described, and a detailed description is given of the finite-element method of structural analysis. Solution techniques are presented for static, dynamic, and modal analysis problems.

As part of the design procedure, the designer must analyze the entire structure and some of its components. To perform these analyses the designer will develop mathematical models of the structure that are approximations of the real structure; these models are used to determine the important parameters in the design.

The type of structural model the designer uses depends on the information that is needed and the type of analysis the designer can perform. Three types of structural models are:

- *Rigid members.* The entire structure or parts of the structure are considered to be rigid. Hence, no deformation can occur in these members.

- *Flexible members.* The entire structure or parts of the structure are modeled by members that can deform, but in limited ways. Examples of these members are trusses, beams, and plates.

- *Continuum.* A continuum model of a structure is the most general, since few, if any, mathematical assumptions about the behavior of the structure need to be made prior to making a continuum model. A continuum member is based on the full three-dimensional equations of continuum mechanics.

In selecting a model of the structure, the designer also must consider the type of analysis to be performed. Four typical analyses that designers perform are:

- *Static equilibrium.* In this analysis, the designer is trying to determine the overall forces and moments that the design will undergo. The analysis is usually done with a rigid member model of the structure and is the simplest analysis to perform.

- *Deformation.* This analysis is concerned with how much the structure will move when operating under the design loads. This analysis is usually done with flexible members.

- *Stress.* In this analysis the designer wants a very detailed picture of where and at what level the stresses are in the design. This analysis is usually done with continuum members.

- *Frequency.* This analysis is concerned with determining the natural frequencies and mode shapes of a structure. This analysis can be done with either flexible members or continuum members, but now the mass of the members is included in the analysis.

The following sections describe a unified mathematical method that includes all the above models and analyses.

4.2   Finite-Element Method

4.2.1   Introduction

The finite-element method was developed during the early 1960s in the aerospace industry. Aircraft designers needed a technique for analyzing complex structures and reanalyzing them when numerous design changes were necessary. Today the finite-element method is the most widely used structural analysis technique, and it is being applied in new areas such as fluid mechanics and biomechanics. The mathematical basis of the finite-element method is described below, with an emphasis on structural analysis. For a detailed account of the finite-element method, see Zienkiewicz (1977).

4.2.2   The method

The finite-element method is an approximation method of analyzing structures and continua. The approximations are:

- *Assumed displacements.*   In each finite element the mathematical form of the displacement field is assumed (displacement method).

- *Spatial discreteness.*   The continuum is approximated by a collection of finite regions (finite elements).

In dynamic analysis there is an additional approximation:

- *Temporal discreteness.*   The time behavior of some variables is approximated by a specified mathematical form (numerical integration).

In the next part of this chapter, we derive the stiffness matrices for the truss element, beam element, special joints, and read links.* Each stiffness formulation consists of the following steps:

1. The element displacement field is expressed in terms of the element nodal degrees of freedom (nodal displacements).

2. The strain field is expressed in terms of nodal displacements by applying the strain-displacement relations to the displacement field determined in step 1.

3. A constitutive law is introduced to relate stresses to strains.

4. Equations that describe nodal forces as functions of the stresses are constructed. Because equations for stress are available in terms of nodal displacements (step 3), it is now possible to relate nodal forces to nodal displacements, and the resulting equations are the element stiffness equations.

4.2.3   The truss element

A truss element (see Figure 4.1) is a one-dimensional structural element that can support only axial force (tension or compression), with no lateral forces or deformation allowed. We assume that the deformations in such

---

* The derivations are detailed mathematical presentations; readers who are not familiar with the finite-element method may prefer to study the examples at the end of each subsection before attempting to follow the mathematics.

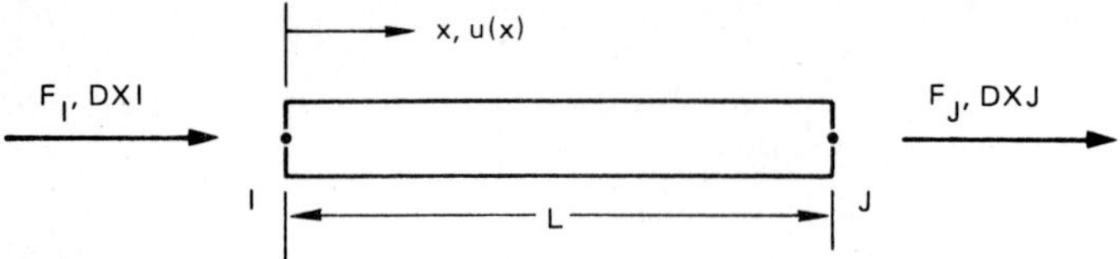

Figure 4.1   Truss-element nodal forces and displacements.

an axial member will be linear and of the form

$$u(X) = A + BX \qquad 0 < X < L \tag{4.1}$$

where $u(X)$ is the deformation field and L is the length of the element. Next we want to determine the two constants A and B in terms of nodal displacements such that $u(0) = DXI$ and $u(L) = DXJ$, where DXI and DXJ are the nodal displacement at the two ends of the element. Solving these two equations and two unknowns, we have

$$A = DXI$$

$$B = \frac{DXJ - DXI}{L}$$

or, rewriting Equation (4.1),

$$U(X) = DXI \frac{1 - X}{L} + DXJ \frac{X}{L}$$

We can write this equation in matrix form as

$$\mathbf{u}(X) = \phi \mathbf{D} \tag{4.2}$$

where $\mathbf{D}^T = [DXI \quad DXJ]$ are the unknown nodal displacements and $\phi = [(1 - X/L) \quad X/L]$. The ϕ matrix is called the "shape function matrix," because its elements determine the form or shape of the assumed displacement field. Once the relation between the displacement field $u(X)$ and the nodal displacements $\mathbf{D}$ has been defined in Equation (4.2), we can determine the strains in the element. Because we have only one nonzero component of displacement, $u(X)$, the strain displacement relation is

$$\epsilon_x = \frac{\partial u}{\partial X} \tag{4.3}$$

Substituting Equation (4.2) into the above relation, we have

$$\epsilon_x = \frac{\partial \phi}{\partial X} \mathbf{D} \tag{4.4}$$

and we define

$$\mathbf{B} = \frac{\partial \phi}{\partial X} = \frac{1}{L} \begin{bmatrix} -1 & 1 \end{bmatrix} \tag{4.5}$$

Thus the strain displacement relation becomes

$$\epsilon_x = \mathbf{BD} \tag{4.6}$$

Having defined the strain in terms of the nodal displacements we can now determine the stress. Because there is only one nonzero component of strain, Hooke's law reduces to

$$\sigma = E\epsilon$$

where E is the elastic modulus of the material. Substituting Equation (4.6) into the above relation and defining the constitutive matrix

$$\mathbf{C} = [E]$$

we have

$$\sigma = \mathbf{CBD} \tag{4.7}$$

This is the stress nodal displacement relation.

In the next step we make use of Castigliano's first theorem, which states that the set of admissible displacements that minimizes the potential energy of a system is also an equilibrium configuration for that system. Thus, we need to express the potential energy (or just the strain energy in the absence of forces) in terms of the nodal displacements. The general form of the potential energy is given by

$$U = \tfrac{1}{2} \int_{\nu} \sigma^T \epsilon \, d\nu - \int_{S} \mathbf{u}^T \mathbf{f} \, dS \tag{4.8}$$

where ν is the interior of the domain and S is the boundary where forces $\mathbf{f}$ are prescribed.* Substituting the above relations into Equation (4.8), we obtain

$$U = \tfrac{1}{2} \int_{\nu} \mathbf{D}^T \mathbf{B}^T \mathbf{CBD} \, d\nu - \int \mathbf{D}^T \phi^T \mathbf{f} \, dS$$

---

* U is the symbol for potential energy; it differs from $u(X)$, displacement.

Because the unknown nodal displacements $\mathbf{D}$ are independent of integration variables, we can move them outside of the integration

$$U = \tfrac{1}{2}\mathbf{D}^T \int_v \mathbf{B}^T \mathbf{C} \mathbf{B}\, dv\, \mathbf{D} - \mathbf{D}^T \int_S \boldsymbol{\phi}^T \mathbf{f}\, dS$$

Next we want to minimize this potential energy with respect to the nodal displacements

$$\frac{\partial U}{\partial \mathbf{D}} = 0 = \int_v \mathbf{B}^T \mathbf{C} \mathbf{B}\, dv\, \mathbf{D} - \int_S \boldsymbol{\phi}^T \mathbf{f}\, dS$$

and we define

$$\mathbf{K} = \int_v \mathbf{B}^T \mathbf{C} \mathbf{B}\, dv \text{ (element stiffness matrix)}$$

$$\mathbf{F} = \int_S \boldsymbol{\phi}^T \mathbf{f}\, dS \text{ (nodal forces)}$$

Then we can rewrite the above as

$$\mathbf{F} = \mathbf{K}\mathbf{D} \tag{4.9}$$

which is the classical statement of the finite-element equations of static equilibrium.

Returning to our example of the truss element, we can now easily evaluate the element stiffness matrix for an element of cross-sectional area A and length L:

$$\mathbf{K} = \int_v \mathbf{B}^T \mathbf{C} \mathbf{B}\, dv = A \int_0^L \mathbf{B}^T \mathbf{C} \mathbf{B}\, dX$$

$$= A \int_0^L \frac{1}{L} \begin{bmatrix} -1 \\ -1 \end{bmatrix} [E]\, [-1 \quad 1] \frac{1}{L}\, dX$$

or
$$\mathbf{K} = \frac{EA}{L} \begin{bmatrix} 1 & -1 \\ -1 & 1 \end{bmatrix} \tag{4.10}$$

The stiffness matrix for element 1 is Example 4-1 is

$$\mathbf{K}_1 = \frac{A_1 E_1}{L_1} \begin{bmatrix} 1 & -1 \\ -1 & 1 \end{bmatrix} = \begin{bmatrix} k_1 & -k \\ -k_1 & k_1 \end{bmatrix}$$

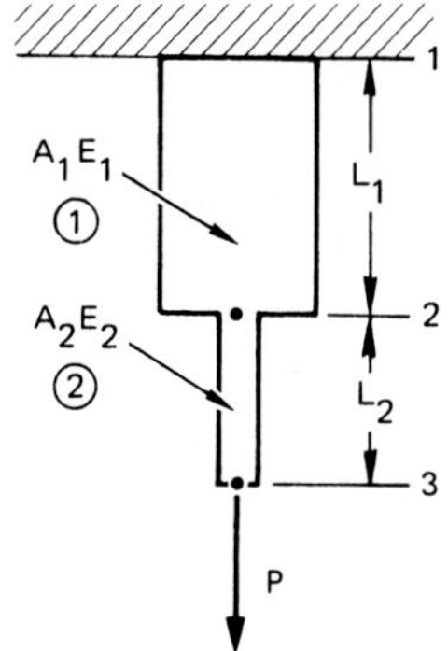

Example 4.1 Truss element.

and for element 2,

$$\mathbf{K}_2 = \frac{A_2 E_2}{L_2} \begin{bmatrix} 1 & -1 \\ -1 & 1 \end{bmatrix} = \begin{bmatrix} k_2 & -k_2 \\ -k_2 & k_2 \end{bmatrix}$$

By superposition (see the discussion of linear elastic analysis later in this section) we can add the two element stiffness matrices together; thus $\mathbf{K} = \mathbf{K}_1 + \mathbf{K}_2$ or

$$\mathbf{K} = \begin{bmatrix} k & -k_1 & 0 \\ -k_1 & k_1 + k_2 & -k_2 \\ 0 & -k_2 & k_2 \end{bmatrix}$$

And the equilibrium equations become

$$\begin{Bmatrix} F_1 = 0 \\ F_2 = 0 \\ F_3 = P \end{Bmatrix} = \begin{bmatrix} k_1 & -k_1 & 0 \\ -k_1 & k + k_2 & -k_2 \\ 0 & -k_2 & k_2 \end{bmatrix} \begin{Bmatrix} u_1 \\ u_2 \\ u_3 = 0 \end{Bmatrix}$$

which can be solved for the unknown displacements u_1 and u_2.

4.2.4 The beam element

A beam element (see Figure 4.2) is a one-dimensional structural element that can support lateral forces (shear forces and bending moments) with the corresponding shear displacements and bending rotations. For small strains, the axial behavior of the truss element can be superimposed with the beam element, but to keep this explanation simple, we derive the beam element without axial stiffness.

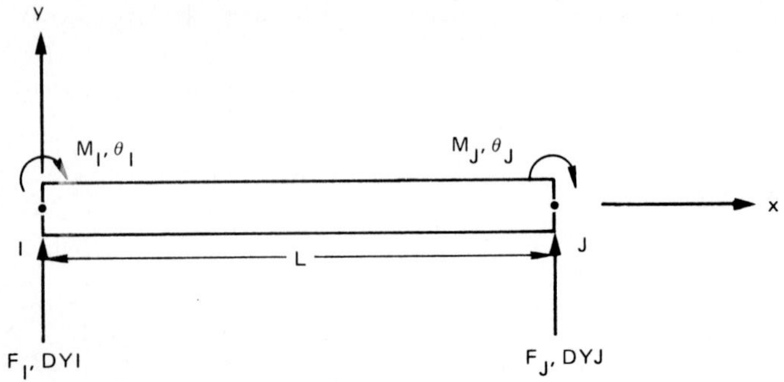

Figure 4.2 Beam-element nodal forces and displacements.

We assume that the transverse deformations in a beam element are cubic and of the form

$$v(x) = a_0 + a_1 x + a_2 x^2 + a_3 x^3 \qquad 0 < x < L \qquad (4.11)$$

where $v(x)$ is the deformation normal to the axis of the beam (transverse deformations). Again we want to determine the four constants a_0, a_1, a_2, and a_3 in terms of the unknown nodal displacements and rotations such that

$$v(0) = DYI$$

$$v(L) = DYJ$$

$$\left.\frac{\partial v}{\partial x}\right|_{x=0} = \Theta I$$

$$\left.\frac{\partial v}{\partial x}\right|_{x=L} = \Theta J$$

Here DYI and DYJ are the transverse nodal displacements at the two ends of the beam with ΘI and ΘJ the rotations of the beam cross section at each end. Solving for the constants a_0, a_1, a_2, and a_3 gives

$$a_0 = DYI$$

$$a_1 = \Theta I \qquad\qquad (4.12)$$

$$a_2 = \frac{3}{L^2}(DYJ - DYI) - \frac{1}{L}(2\Theta I + \Theta J)$$

$$a_3 = \frac{2}{L^3}(DYI - DYJ) + \frac{1}{L^2}(\Theta I + \Theta J)$$

We again want to make use of Castigliano's theorem in determining the stiffness matrix, but we will do so with a slightly different method. This method is especially helpful when dealing with the more complex structural elements.

Recall the form of the element stiffness in Equation (4.9). The strain energy can be written as

$$U = \tfrac{1}{2}\mathbf{F}^T\mathbf{D} \tag{4.13}$$

Here the strain energy is given by the average force acting over the total displacement. Substituting Equation (4.9) into Equation (4.13) gives

$$U = \tfrac{1}{2}\mathbf{D}^T\mathbf{K}\mathbf{D} \tag{4.14}$$

Equation (4.14) represents the quadratic form of the element strain energy. It can be shown (Martin, 1966) that the elements of the stiffness matrix $\mathbf{K}$ are given by

$$K_{ij} = \frac{\partial^2 U}{\partial D_i \partial D_j} \tag{4.15}$$

Next we need to express the strain energy in terms of the nodal displacements. From beam theory we know that the only nonzero strain is ϵ_x and the strain-displacement relation is given by

$$\epsilon_x = -y\,\frac{d^2v}{dx^2} \tag{4.16}$$

and from Hooke's law

$$\sigma_x = E\epsilon_x$$

Hence we can write the strain energy as

$$U = \tfrac{1}{2}\int_v \sigma_x\epsilon_x \, dv = \frac{E}{2}\int_v \left(-y\,\frac{d^2v}{dx^2}\right)^2 dv$$

$$= \frac{E}{2}\int_0^L \left[\int\int y^2 \, dy \, dz\right]\left(\frac{d^2v}{dx^2}\right)^2 dx$$

The term in brackets is called the "moment of inertia" of the beam cross section, which is designated by the symbol I. With this simplification, the strain energy becomes

$$U = \frac{EI}{2}\int_0^I \left(\frac{d^2v}{dx^2}\right)^2 dx \tag{4.17}$$

From Equation (4.11) we have

$$\frac{d^2v}{dx^2} = 2a_2 + 6a_3x$$

Substituting into Equation (4.17) and evaluating the integral gives

$$U = \frac{EI}{2}\,(4a_2^2L + 12a_2a_3L^2 + 12a_3^2L^3) \qquad (4.18)$$

Equation (4.18) provides the relation between the strain energy and the constants a_2 and a_3, which are known in terms of the nodal displacements. Next we apply Equation (4.15) to Equation (4.18) to obtain the desired stiffness matrix elements. As an example

$$K_{11} = \frac{\partial^2 U}{\partial DYI\,\partial DYI} = \frac{EI}{2}\frac{\partial}{\partial DYI}\left[8La_2\frac{\partial a_2}{\partial DYI}\right.$$

$$\left. + 12L^2\left(a_2\frac{\partial a_3}{\partial DYI} + a_3\frac{\partial a_2}{\partial DYI}\right) + 24L^3a_3\frac{\partial a_3}{\partial DYI}\right]$$

where Equations (4.12) are used to calculate the partial derivatives

$$\frac{\partial a_2}{\partial DYI} = -\frac{3}{L^2}\qquad \frac{\partial a_3}{\partial DYI} = \frac{2}{L^3}$$

As a result

$$K_{11} = \frac{12EI}{L^3}$$

If all the elements of the stiffness matrix are computed in this manner, we obtain

$$\mathbf{K} = EI
\begin{array}{cccc}
DYI & \Theta I & DYJ & \Theta J
\end{array}
\begin{bmatrix}
\dfrac{12}{L^3} & -\dfrac{6}{L^2} & -\dfrac{12}{L^3} & -\dfrac{6}{L^2} \\[2ex]
-\dfrac{6}{L^2} & \dfrac{4}{L} & -\dfrac{6}{L^2} & \dfrac{2}{L} \\[2ex]
-\dfrac{12}{L^3} & \dfrac{6}{L^2} & \dfrac{12}{L^3} & -\dfrac{6}{L^2} \\[2ex]
-\dfrac{6}{L^2} & \dfrac{2}{L} & \dfrac{6}{L^2} & \dfrac{4}{L}
\end{bmatrix}$$

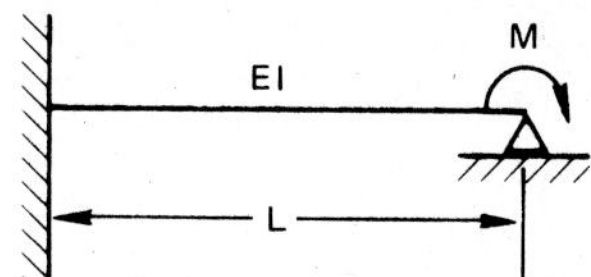

Example 4.2   Beam element.

The equilibrium equations for the element in Example 4-2 can be written as*

$$
\begin{Bmatrix} F_I = 0 \\ M_I = 0 \\ F_J = 0 \\ M_J = M \end{Bmatrix}
= EI
\begin{bmatrix}
\dfrac{12}{L^3} & \dfrac{6}{L^2} & \dfrac{12}{L^3} & \dfrac{6}{L^2} \\
 & \dfrac{4}{L} & \dfrac{-6}{L^2} & \dfrac{2}{L} \\
SYM & & \dfrac{12}{L^3} & \dfrac{-6}{L^2} \\
 & & & \dfrac{4}{L}
\end{bmatrix}
\begin{Bmatrix} DYI = 0 \\ \Theta_I = 0 \\ DYJ = 0 \\ \Theta_J \end{Bmatrix}
$$

which can be solved for the unknown nodal rotation Θ_J.

4.2.5 Special joints

Any mechanism that connects two structural elements is considered to be a joint. Although detailed methods for treating various mechanisms (extensional spring, torsional spring, bending connector, and shear connector) have been derived (see Highway Safety Research Institute, 1977), in many applications the joint can be modeled with a simple beam element representation. For this type of representation and in the more detailed derivation referenced, the stiffness of the member is usually determined by laboratory testing because an analytical determination would be very difficult or might not represent the true joint stiffness.

An idealization of the laboratory tests needed to determine the axial and bending stiffness of an idealized joint is shown in Figure 4.3. In this idealization, the complex joint is represented by a straight beam of length L. The axial stiffness (AE/L) and bending stiffness (EI/L) are estimated from the load-deflection curves measured in the laboratory tests. Once the joint stiffnesses—axial and bending—have been determined, the corre-

---

* SYM = symmetric.

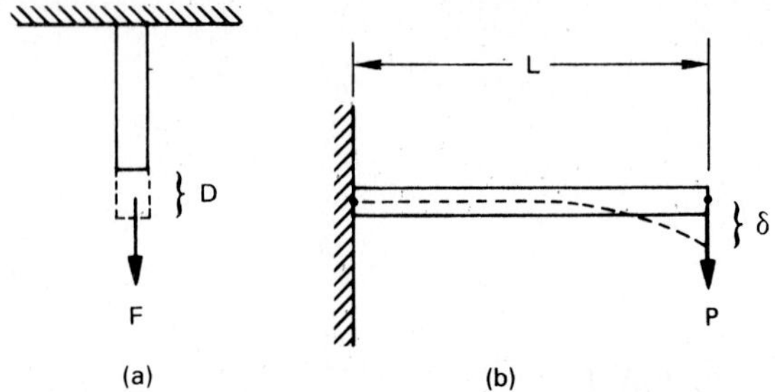

Figure 4.3 Tests to determine stiffness of an idealized joint. (*a*) Axial stiffness (AS) = AE/L = FD. (*b*) Bending stiffness (BS) = EI/L = $12L^2P/\delta$.

sponding element stiffness matrix is given by

$$
\mathbf{K} =
\begin{array}{cccccc}
DXI & DYI & \Theta I & DXJ & DYJ & \Theta J \\
\end{array}
$$

$$
\mathbf{K} =
\begin{bmatrix}
\text{AS} & 0 & 0 & -\text{AS} & 0 & 0 \\
0 & \dfrac{12\text{BS}}{L^2} & \dfrac{-6\text{BS}}{L} & 0 & \dfrac{-12\text{BS}}{L^2} & \dfrac{-6\text{BS}}{L} \\
0 & \dfrac{-6\text{BS}}{L} & 4\text{BS} & 0 & \dfrac{6\text{BS}}{L} & 2\text{BS} \\
-\text{AS} & 0 & 0 & \text{AS} & 0 & 0 \\
0 & \dfrac{-12\text{BS}}{L^2} & \dfrac{6\text{BS}}{L} & 0 & \dfrac{12\text{BS}}{L^2} & \dfrac{6\text{BS}}{L} \\
0 & \dfrac{-6\text{BS}}{L} & 2\text{BS} & 0 & \dfrac{6\text{BS}}{L} & 4\text{BS} \\
\end{bmatrix}
$$

where AS is the axial stiffness and BS is the bending stiffness.

4.2.6 Rigid links

Rigid links are elements that do not deform. In a structure, the rigid links are interconnected by flexible members such as the truss, beam, and joint elements described above. The geometry of a rigid link is determined by its connections to flexible members. These connection points are called "secondary" or "slave" nodes. The centroid of a rigid link is where we write the equations of motion and is called the "primary" or "master" node.

Consider a rigid link in the xy plane (Figure 4.4) with a local system of coordinates xy originating from the primary node (X_P, Y_P). The location

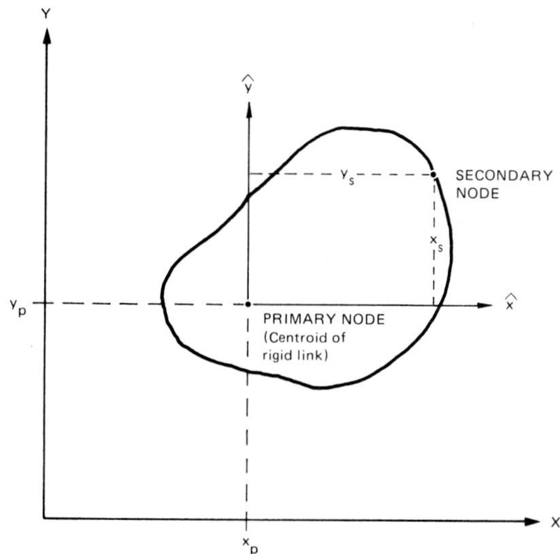

Figure 4.4 Rigid link notation.

of the secondary node (X_S, Y_S) in the local coordinates is given by

$$R_X = X_X - X_P$$

$$R_Y = Y_X - Y_P$$

If the secondary node is a point of connection for a flexible element, there will be three forces acting at the secondary node: F_{XS}, F_{YS}, and M_S. If we sum forces at the primary node, we obtain

$$F_{XP} = F_{XS}$$

$$F_{YP} = F_{YS}$$

$$M_P = M_S - F_{XS}R_Y + F_{YS}R_X$$

From this we see that inclusion of a rigid link only alters the moment equation.

We can write this in matrix form, assuming node I is where the rigid link connects to the flexible element

$$\left\{\begin{array}{c} F_{XI} \\ F_{YI} \\ M_I \\ F_{XJ} \\ F_{YJ} \\ M_J \end{array}\right\}_p = \left\{\begin{array}{c} F_{XI} \\ F_{YI} \\ M_I \\ F_{XJ} \\ F_{YJ} \\ M_J \end{array}\right\}_s + \left[\begin{array}{cccccc} 0 & 0 & 0 & 0 & 0 & 0 \\ 0 & 0 & 0 & 0 & 0 & 0 \\ -R_Y & R_X & 0 & 0 & 0 & 0 \\ 0 & 0 & 0 & 0 & 0 & 0 \\ 0 & 0 & 0 & 0 & 0 & 0 \\ 0 & 0 & 0 & 0 & 0 & 0 \end{array}\right] \left\{\begin{array}{c} F_{XI} \\ F_{YI} \\ M_I \\ F_{XJ} \\ F_{YJ} \\ M_J \end{array}\right\}_s$$

or
$$\mathbf{F}_p = \mathbf{F}_s + \mathbf{RF}_s = (\mathbf{I} + \mathbf{R})\mathbf{F}_s$$

where $\mathbf{I}$ is the identity matrix; further, since $\mathbf{F}_s = \mathbf{KD}$,

$$\mathbf{F}_p = (\mathbf{I} + \mathbf{R})\mathbf{KD} = \mathbf{K}_p\mathbf{D}$$

where $\mathbf{K}_p = (\mathbf{I} + \mathbf{R})\mathbf{K}$. Note that if the primary node and the secondary node are coincident ($R_X = R_Y = 0$), we have the standard form of equilibrium equations.

The rotation matrix for the element in Example 4-3 is

$$\mathbf{R} = \begin{bmatrix} 0 & 0 & 0 & 0 & 0 \\ 0 & 0 & 0 & 0 & 0 \\ -R_Y = 0 & R_X = L/2 & 0 & 0 & 0 \\ 0 & 0 & 0 & 0 & 0 \\ 0 & 0 & 0 & 0 & 0 \\ 0 & 0 & 0 & 0 & 0 \end{bmatrix}$$

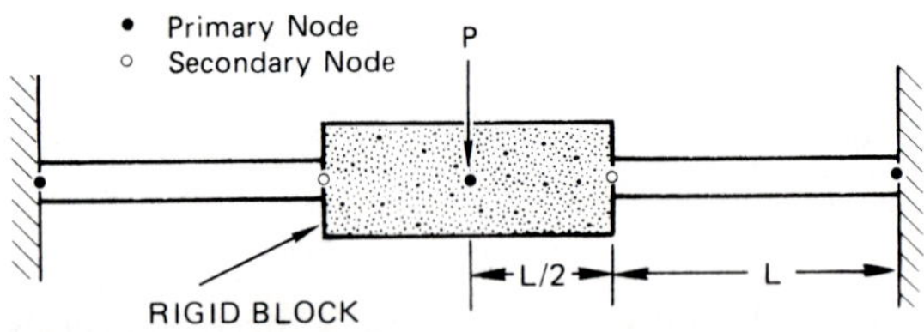

Example 4.3   Rigid link.

4.2.7   Linear elastic analysis

Assembling the stiffness matrix.   In the previous section the element stiffness matrices for several structural members were derived. Most structures of practical interest consist of several structural members fastened into an assemblage. In this section we show how the total structural stiffness of the assemblage is determined and how the solution is obtained.

Consider a structure consisting of two truss members with stiffness matrices given by

$$\mathbf{K}_A = k_A \begin{bmatrix} \overset{U_1}{1} & \overset{U_2}{-1} \\ -1 & 1 \end{bmatrix} \qquad \mathbf{K}_B = k_B \begin{bmatrix} \overset{U_2}{1} & \overset{U_3}{-1} \\ -1 & 1 \end{bmatrix}$$

These matrices cannot be superimposed because their columns are not compatible. To correct this, we will expand with 0s each element stiffness matrix to the order of the total stiffness matrix

$$\mathbf{K}_A = k_A \begin{array}{c} \begin{array}{ccc} U_1 & U_2 & U_3 \end{array} \\ \begin{bmatrix} 1 & -1 & 0 \\ -1 & 1 & 0 \\ 0 & 0 & 0 \end{bmatrix} \end{array} \qquad \mathbf{K}_B = k_B \begin{array}{c} \begin{array}{ccc} U_1 & U_2 & U_3 \end{array} \\ \begin{bmatrix} 0 & 0 & 0 \\ 0 & 1 & -1 \\ 0 & -1 & 1 \end{bmatrix} \end{array}$$

The two matrices are now compatible and can be added together to form the total stiffness matrix

$$\mathbf{K} = \mathbf{K}_A + \mathbf{K}_B = \begin{bmatrix} k_a & -k_a & 0 \\ -k_a & k_a + k_b & -k_b \\ 0 & -k_b & k_b \end{bmatrix}$$

This procedure for assembling the total stiffness matrix is based on equating all the internal and external forces at the nodes of the structure and is referred to as the "direct stiffness method."

Solving the stiffness matrix.   We can use the above total stiffness matrix to illustrate the solution procedure. First, we need to specify a boundary condition to assure a well-posed problem, assume $U_3 = 0$, and assume the applied loads are F_1, F_2, and F_3, with F_3 an unknown reaction (see Example 4.1). Thus

$$\begin{Bmatrix} F_1 \\ F_2 \\ F_3 \end{Bmatrix} = \begin{bmatrix} k_a & -k_a & 0 \\ -k_a & k_a + k_b & -k_b \\ 0 & -k_b & k_b \end{bmatrix} \begin{Bmatrix} U_1 \\ U_2 \\ U_3 = 0 \end{Bmatrix}$$

In practice, the reactions are seldom of interest, so the third equation can be ignored. This is done by setting equal to zero the elements of the row and column corresponding to the constrained degree of freedom U_3 and placing a 1 on the diagonal. This combined with zeroing the reaction force assures a zero value for U_3 during the solution; thus we have

$$\begin{Bmatrix} F_1 \\ F_2 \\ 0 \end{Bmatrix} = \begin{bmatrix} k_a & -k_a & 0 \\ -k_a & k_a + k_b & 0 \\ 0 & 0 & 1 \end{bmatrix} \begin{Bmatrix} U_1 \\ U_2 \\ U_3 \end{Bmatrix}$$

Inverting this matrix to solve for the displacements yields

$$\begin{Bmatrix} U_1 \\ U_2 \\ U_3 \end{Bmatrix} = \begin{bmatrix} \dfrac{1}{k_a} + \dfrac{1}{k_b} & \dfrac{1}{k_b} & 0 \\ \dfrac{1}{k_b} & \dfrac{1}{k_b} & 0 \\ 0 & 0 & 1 \end{bmatrix} \begin{Bmatrix} F_1 \\ F_2 \\ 0 \end{Bmatrix}$$

To determine the reaction at F_3, we have from the original matrix

$$F_3 = 0 \cdot U_1 - k_b \cdot U_2$$

The two steps described above—assembling the global stiffness matrix and solving for the displacements—form the general procedure for linear elastic static finite-element analysis.

4.2.8 Dynamic analysis

This section presents the dynamic equations of motion for a finite-element model and discusses several solution techniques for the dynamic equations.

When dynamic loading is applied to a structure, the displacements are functions of position and time. An infinitesimal element of volume dv is subjected to an inertia force $-\rho \ddot{u}\, dv$, where ρ is the density of the material and $\ddot{u}$ is the acceleration of the volume dv. In accordance with D'Alembert's principle, a corollary of Newton's second and third laws, the equations of motion are obtained from the condition of equilibrium of the element when the inertial forces are taken into account. Because the inertial forces are proportional to the volume of the element, they constitute body forces. The work done by these intertial body forces is added to the potential energy equation [Equation (4.8)] and gives

$$U = \int_v \rho \mathbf{u}^T \ddot{\mathbf{u}}^T \, dv + \tfrac{1}{2} \int_v \sigma^T \epsilon \, dv - \int_s \mathbf{u}^T \mathbf{f} \, dS \qquad (4.19)$$

Noting that the acceleration field is given by

$$\ddot{\mathbf{u}} = \phi \ddot{\mathbf{D}} \qquad (4.20)$$

where $\ddot{\mathbf{D}}$ is a vector of unknown nodal accelerations, the first integral in Equation (4.19) can be written as

$$\int_v \rho \mathbf{u}^T \ddot{\mathbf{u}} \, dv \int_v \rho \mathbf{D}^T \phi^T \phi \ddot{\mathbf{D}} \, dv = \mathbf{D}^T \left(\int_v \rho \phi^T \phi \, dv \right) \ddot{\mathbf{D}} = \mathbf{D}^T \mathbf{M} \ddot{\mathbf{D}}$$

where we define the consistent mass matrix as

$$\mathbf{M} = \int_v \rho \phi^T \phi \, dv$$

To find the minimum potential energy, we set the derivative of the energy with respect to the unknown nodal displacements equal to zero:

$$\frac{\partial U}{\partial \mathbf{D}} = 0 = \mathbf{M\ddot{D}} + \mathbf{KD} = \mathbf{F} \qquad (4.21)$$

which are the finite-element equations of dynamic equilibrium.

4.2.9 Solution techniques for the dynamic equations of motion

Two strategies are used in dealing with the transient problem: modal superposition and direct integration. The choice of the method depends on the frequency content of the load and the portion of the frequency response that is of interest. If only the lower spectrum of the discrete model's frequency domain is of interest, modal superposition techniques are of advantage in linear problems. When high frequencies are important in the response, as in wave propagation problems, direct integration methods are appropriate, because all the modes of the discrete model are treated.

Modal superposition. In modal superposition methods, the mass and stiffness equations are diagonalized by finding the eigenvalues and eigenvectors of the equation

$$\mathbf{Mx} = \lambda \mathbf{Kx}$$

These eigenvalues λ and eigenvectors $\mathbf{x}$ correspond to the natural frequencies and mode shapes of the discrete model. When these are known, the transient equations can be put in uncoupled form and easily solved. Knowing the natural frequencies and mode shapes of a structure can be very useful information for a robotics designer. (See Example 4.4.) How-

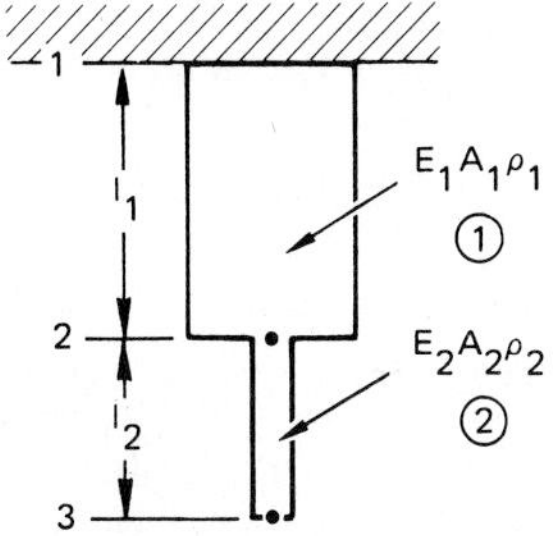

Example 4.4 Determination of natural frequencies.

ever, modal superposition techniques are not extensively used for transient problems, except for the analysis of axisymmetric structures under nonaxisymmetric loads. In the current state of robot design practice, transient analyses are not usually performed; a complete modal analysis is often, but not always, performed.

From Example 4.1, the total stiffness is

$$\mathbf{K} = \mathbf{K}_1 + \mathbf{K}_2 = \begin{bmatrix} k_1 & -k_1 & 0 \\ -k_1 & k_1 + k_2 & -k_2 \\ 0 & -k_2 & k_2 \end{bmatrix}$$

where
$$k_1 = \frac{E_1 A_1}{l_1} \quad \text{and} \quad k_2 = \frac{E_2 A_2}{l_2}$$

Because the mass matrix for an element is defined as

$$\mathbf{M} = \int_0^l \rho A \phi^T \phi \, dx = \frac{\rho A l}{6} \begin{bmatrix} 2 & 1 \\ 1 & 2 \end{bmatrix}$$

The two-element structure shown has a total mass matrix

$$\mathbf{M} = \mathbf{M}_1 + \mathbf{M}_2 = \frac{1}{6} \begin{bmatrix} 2m_1 & m_1 & 0 \\ m_1 & 2(m_1 + m_2) & m_2 \\ 0 & m_2 & 2m_2 \end{bmatrix}$$

where
$$m_1 = \rho_1 A_1 l_1$$

$$m_2 = \rho_2 A_2 l_2$$

The natural frequencies of a system are found by setting the forcing function equal to zero in the equations of motion; thus

$$\mathbf{M}\ddot{\mathbf{D}} + \mathbf{k}\mathbf{D} = 0$$

Assuming a periodic form for the displacements,

$$\mathbf{D}(t) = \mathbf{X}e^{i\omega t}$$

where $\mathbf{X}$ represents the nodal displacements independent of time. Using this assumed form, the equations of motion become

$$(\mathbf{K} - \omega^2 \mathbf{M})\,\mathbf{X} = 0$$

for which nontrivial solutions can be obtained only if

$$\det(\mathbf{K} - \omega^2 \mathbf{M}) = 0$$

where det is the determinant and ω TGT the natural frequencies of the system.

For the sample problem considered, noting that degree of freedom 1 should be constrained out of the problem, we have

$$\det\left\{\begin{bmatrix} k_1 + k_2 & -k_2 \\ -k_2 & k_2 \end{bmatrix} - \frac{1}{6}\begin{bmatrix} 2\omega^2(m_1 + m_2) & m_2 \\ m_2 & 2\omega^2 m_2 \end{bmatrix}\right\} = 0$$

To complete the example, let us assume $k_1 = 2k_2 = 2$ and $m_1 = 2m_2 = 12$, then

$$\det\begin{bmatrix} 3 - 3\omega^2 & -1 \\ -1 & 1 - 2\omega^2 \end{bmatrix} = 0$$

or
$$8\omega^4 - 9\omega^2 + 2 = 0$$

or
$$\omega_1 = 0.905 \quad \text{and} \quad \omega_2 = 0.552$$

If we consider degree of freedom 2 to be constrained, we have

$$\det\left\{\begin{bmatrix} k & 0 \\ 0 & k_2 \end{bmatrix} - \frac{1}{6}\begin{bmatrix} 2m_1\omega^2 & 0 \\ 0 & 2m_2\omega^2 \end{bmatrix}\right\} \leq 0$$

or, using the same numerical values as above,

$$\det\begin{bmatrix} 2 - 4\omega^2 & 0 \\ 0 & 1 - 2\omega^2 \end{bmatrix} \leq 0$$

Then, $(2 - 4\omega^2)(1 - 2\omega^2) = 0$, or $\omega_1 = \omega_2 = \sqrt{\tfrac{1}{2}} = 0.707$.

Direct integration. In direct integration methods, the equations of motion are integrated directly without any preliminary uncoupling by modes. There are two major classes of direct integration methods: implicit integration in time and explicit integration in time. Both the implicit and explicit methods are developed from difference formulas that relate the accelerations, velocities, and displacements. For example, the Newmark β method uses the following difference formulas.

$$\dot{D}(t + \Delta t) = \dot{D}(t) + \tfrac{1}{2}\Delta t\,[\ddot{D}(t) + \ddot{D}(t + \Delta t)] \qquad (4.22a)$$

$$\begin{aligned} D(t + \Delta t) = {}& D(t) + \Delta t\,\dot{D}(t) \\ &+ \Delta t^2[(\tfrac{1}{2} - \beta)\ddot{D}(t) + \beta\ddot{D}(t + \Delta t)] \end{aligned} \qquad (4.22b)$$

When $\beta = \frac{1}{4}$, these formulas may be derived by assuming that the acceleration between time t and $t + \Delta t$ is constant and equal to the average of the accelerations at the ends of the interval. For $\beta = 0$, they correspond to acceleration pulses at intervals of Δt. It can also be shown that the Newmark β method with $\beta = 0$ is equivalent to the central difference formulas (Belytschko, 1974).

Implicit integration. To develop the equations used in implicit integration, we consider the equations of motion [Equation (4.21)] for the time $t + \Delta t$ and eliminate any velocities or accelerations at $t + \Delta t$ by means of the integration equations [Equations (4.22)]. For the Newmark β method, this yields

$$\mathbf{K}^{\text{eff}}\mathbf{D}(t + \Delta t) = \mathbf{F}^{\text{eff}} \tag{4.23}$$

where

$$\mathbf{K}^{\text{eff}} = \mathbf{M} + \beta\, \Delta t^2\, \mathbf{K}$$

$$\mathbf{F}^{\text{eff}} = \beta\, \Delta t^2\, F(t + \Delta t) + \mathbf{M}[(\tfrac{1}{2} - \beta)\, \Delta t^2\, \ddot{D}(t) + \Delta t\, \dot{D}(t) + D(t)]$$

Thus the implicit method directly solves for the displacements at each time step, and the equation solver is an important component of implicit transient analysis.

Explicit integration. In explicit integration of linear problems, the accelerations are found at each time step by using an approximation of the equations of motion

$$\ddot{D}(t + \Delta t) = \mathbf{M}^{-1}[\mathbf{F}(t + \Delta t) - \mathbf{K}\mathbf{D}(t)] \tag{4.24}$$

The approximation in the above equations is that the previous nodal displacements $D(t)$ can be used to predict the current accelerations $\ddot{D}(t + \Delta t)$. This approximation is very reasonable if the displacements change slowly. To keep the displacements changing slowly the time step Δt must be small. Usually 50 to 80 percent of the wave transient time across the smallest element in the discrete model is used.

Sample problem. Two flexible members are interconnected by a flexible joint.

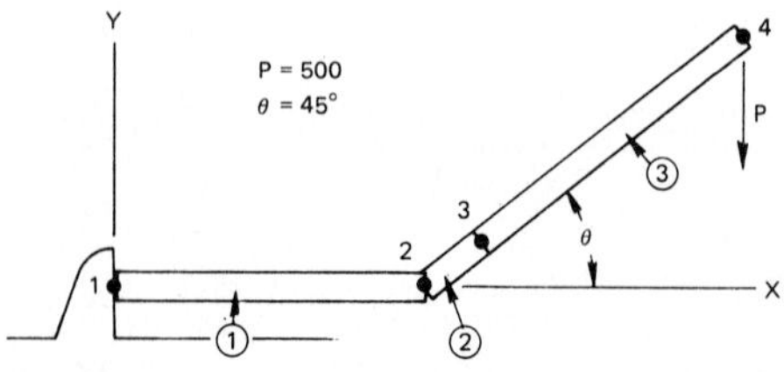

NODE	X	Y
1	0	0
2	10	0
3	10.7	0.7
4	17.77	7.77

ELEMENT	AE/L	EI/L	L	ρAL
①	10^6	8.3×10^5	10	2.6×10^{-3}
②	10^5	8.3×10^4	1	2.6×10^{-4}
③	10^6	8.3×10^5	10	2.6×10^{-3}

The local element stiffness for elements 1 and 3 can be written as

$$\overline{\mathbf{k}}_1 = \overline{\mathbf{k}}_3 = \begin{bmatrix} 10^6 & 0 & 0 & 10^6 & 0 & 0 \\ 0 & 10^5 & -5 \times 10^5 & 0 & -10^5 & -5 \times 10^5 \\ 0 & -5 \times 10^5 & 3.33 \times 10^6 & 0 & 5 \times 10^5 & 1.66 \times 10^6 \\ -10^6 & 0 & 0 & 10^6 & 0 & 0 \\ 0 & -10^5 & 5 \times 10^5 & 0 & 10^5 & 5 \times 10^5 \\ 0 & -5 \times 10^5 & 1.66 \times 10^6 & 0 & 5 \times 10^5 & 3.33 \times 10^6 \end{bmatrix}$$

and the local element stiffness for element 2 is

$$\overline{\mathbf{k}}_2 = \begin{bmatrix} 10^5 & 0 & 0 & -10^5 & 0 & 0 \\ 0 & 10^6 & -5 \times 10^5 & 0 & -10^6 & -5 \times 10^5 \\ 0 & -5 \times 10^5 & 3.33 \times 10^5 & 0 & 5 \times 10^5 & 1.66 \times 10^5 \\ -10^5 & 0 & 0 & 10^5 & 0 & 0 \\ 0 & -10^6 & 5 \times 10^5 & 0 & 10^6 & 5 \times 10^5 \\ 0 & -5 \times 10^5 & 1.66 \times 10^5 & 0 & 5 \times 10^5 & 3.33 \times 10^5 \end{bmatrix}$$

The term "local element stiffness" refers to the fact that all the stiffness matrices discussed this far have been derived in a local xy element system, with the x axis along the length of the element and the y axis perpendicular to the x axis. In the current example, the global xy axes do not correspond to the local axes of elements 2 and 3. Thus, it is necessary to transform the local element stiffness matrix to the global coordinate system before we can assemble the total global stiffness and solve for the nodal displacements. The transformation from local to global is that of a second-order tensor and is given by

$$\mathbf{K} = \mathbf{T}^T \overline{\mathbf{K}} \mathbf{T}$$

where $\mathbf{K}$ = global stiffness
$\quad\ \overline{\mathbf{K}}$ = local element stiffness
$\quad\ \mathbf{T}$ = transformation matrix

$\mathbf{T}$ is defined by

$$
\mathbf{T} =
\begin{bmatrix}
\lambda & \mu & 0 & 0 & 0 & 0 \\
-\mu & \lambda & 0 & 0 & 0 & 0 \\
0 & 0 & 1 & 0 & 0 & 0 \\
0 & 0 & 0 & \lambda & \mu & 0 \\
0 & 0 & 0 & -\mu & \lambda & 0 \\
0 & 0 & 0 & 0 & 0 & 1
\end{bmatrix}
$$

where $\lambda = \cos\theta$

$\mu = \sin\theta$

$\theta =$ angle between global and local coordinate system

If we perform the indicated transformation on the local stiffness matrix for an arbitrarily oriented beam element, we obtain

$$
\mathbf{K} = \frac{E}{L}
\begin{bmatrix}
 & DXI & DYI & \Theta I & \vline & DXJ & DYJ & \Theta J \\
A\lambda^2 + \dfrac{12I}{L^2}\mu^2 & & & & & & \\
\left(A - \dfrac{12I}{L^2}\right)\lambda\mu & A\mu^2 + \dfrac{12I}{L^2}\lambda^2 & & & & \text{SYM} & \\
\dfrac{6I}{L}\mu & -\dfrac{6I}{L}\lambda & 4I & & & & \\
\hline
-\left(A\lambda^2 + \dfrac{12I}{L^2}\mu^2\right) & -\left(A - \dfrac{12I}{L^2}\right)\lambda\mu & -\dfrac{6I}{L}\mu & A\lambda^2 + \dfrac{12I}{L^2}\mu^2 & & & \\
-\left(A - \dfrac{12I}{L^2}\right)\lambda\mu & -\left(A\mu^2 + \dfrac{12I}{L^2}\lambda^2\right) & \dfrac{6I}{L}\lambda & \left(A - \dfrac{12I}{L^2}\right)\lambda\mu & A\mu^2 + \dfrac{12I}{L^2}\lambda^2 & \\
\dfrac{6I}{L}\mu & -\dfrac{6I}{L}\lambda & 2I & -\dfrac{6I}{L}\mu & \dfrac{6I}{L}\lambda & 4I
\end{bmatrix}
$$

In our example problem $\Theta = 45°$ and the global stiffness for element 2 is

$$
\mathbf{K}_2 =
\begin{bmatrix}
5.5 \times 10^5 & & & & & \\
-4.5 \times 10^5 & 5.5 \times 10^5 & & & \text{SYM} & \\
3.53 \times 10^5 & -3.53 \times 10^5 & 3.33 \times 10^5 & & & \\
-5.5 \times 10^5 & 4.5 \times 10^5 & -3.53 \times 10^5 & 5.5 \times 10^5 & & \\
4.5 \times 10^5 & -5.5 \times 10^5 & 3.53 \times 10^5 & 4.5 \times 10^5 & 5.5 \times 10^5 & \\
3.53 \times 10^5 & -3.5 \times 10^5 & 1.66 \times 10^5 & -3.53 \times 10^5 & 3.53 \times 10^5 & 3.33 \times 10^5
\end{bmatrix}
$$

and for element 3 we have

$$\mathbf{K}_3 = \begin{bmatrix} 5.5 \times 10^5 & & & & & \\ 4.5 \times 10^5 & 5.5 \times 10^5 & & \text{SYM} & & \\ 3.53 \times 10^5 & -3.53 \times 10^5 & 3.33 \times 10^6 & & & \\ -5.5 \times 10^5 & -4.5 \times 10^5 & -3.53 \times 10^5 & 5.5 \times 10^5 & & \\ -4.5 \times 10^5 & -5.5 \times 10^5 & 3.53 \times 10^5 & 4.5 \times 10^5 & 5.5 \times 10^5 & \\ 3.53 \times 10^5 & -3.53 \times 10^5 & 1.66 \times 10^6 & -3.35 \times 10^5 & 3.35 \times 10^5 & 3.33 \times 10^6 \end{bmatrix}$$

The next step is to form the global stiffness matrix by adding together the individual element stiffnesses. The addition of element stiffness matrices is done on the basis of nodal connectivity, i.e., which nodes are connected by the element. Because nodes 1 and 2 are related by $\mathbf{K}_1$, nodes 2 and 3 by $\mathbf{K}_2$, and nodes 3 and 4 by $\mathbf{K}_3$, we can illustrate the matrix addition by superimposing the element stiffness matrices as shown by

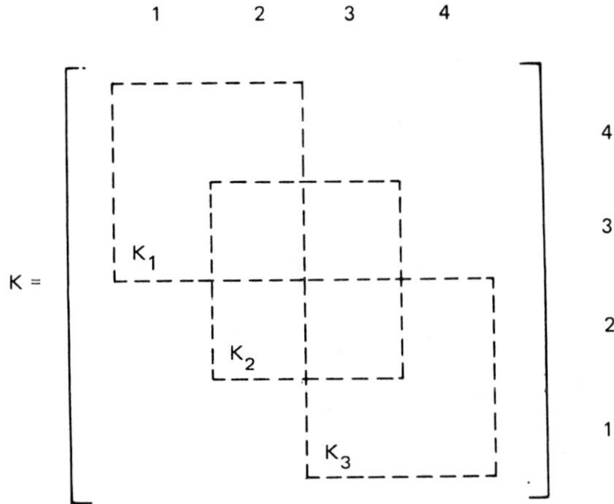

where an overlap of element stiffness matrices indicates addition of the matrices. Thus for our numerical example, the global stiffness matrix is shown on p. 4.24.

The next step is to apply the constraints to the global stiffness matrix by eliminating the rows and columns of the stiffness matrix corresponding to constrained degrees of freedom. In our sample problem the constraints are that all degrees of freedom at node 1 are constrained. This means we

$$
\mathbf{K} =
$$

	DX1	DY1	Θ1	DX2	DY2	Θ2	DX3	DY3	Θ3	DX3	DY4	Θ4
	10^6											
	0	10^5										
	0	-5×10^5	3.3×10^6						SYM			
	10^6	0	0	1.55×10^6								
	0	-10^5	5×10^5	-4.5×10^5	6.5×10^5							
	0	-5×10^5	1.66×10^6	3.53×10^5	1.47×10^5	3.66×10^6						
				-5.5×10^5	4.5×10^5	-3.53×10^5	1.1×10^6					
				4.5×10^5	-5.5×10^5	3.53×10^5	9×10^5	1.1×10^6				
				3.53×10^5	-3.5×10^5	1.66×10^5	0	0	3.66×10^6			
			0				-5.5×10^5	-4.5×10^5	-3.53×10^5	5.5×10^5		
							-4.5×10^5	-5.5×10^5	3.53×10^5	4.5×10^5	5.5×10^5	
							3.53×10^5	-3.53×10^5	1.66×10^6	-3.3×10^5	3.35×10^5	3.33×10^6

4.24

can eliminate the first three rows and columns of our stiffness matrix. The resulting 9×9 matrix can then be inverted, yielding $\mathbf{K}^{-1}$.

The next step is to construct the external force vector, which for our sample problem is given by

$$
\mathbf{F} = \left\{
\begin{array}{c}
0 \\
0 \\
0 \\
\hline
0 \\
0 \\
0 \\
\hline
0 \\
0 \\
0 \\
\hline
0 \\
-500 \\
0
\end{array}
\right\}
\begin{array}{c}
1 \\ \\ \\
2 \\ \\ \\
3 \\ \\ \\
4 \\ \\
\end{array}
$$

Note that the reaction forces at node 1 are not included in the external force vector.

The final step in solving for the displacements is to multiply the external force vector by the inverted stiffness matrix, thus $\mathbf{D} = \mathbf{K}^{-1}\mathbf{F}$. For the sample problem considered we obtain the following results

$$
\mathbf{D} = \left\{
\begin{array}{rcl}
DX1 &=& 0 \\
DY1 &=& 0 \\
\Theta1 &=& 0 \\
DX2 &=& -9.37 \times 10^{-5} \\
DY2 &=& -4.33 \times 10^{-2} \\
\Theta2 &=& -7.66 \times 10^{-3} \\
DX3 &=& 6.77 \times 10^{-3} \\
DY3 &=& -5.03 \times 10^{-2} \\
\Theta3 &=& -1.20 \times 10^{-2} \\
DX4 &=& 0.101 \\
DY4 &=& -0.146 \\
\Theta4 &=& -1.42 \times 10^{-2}
\end{array}
\right\}
$$

If the reactions at node 1 are required, they can be obtained by multiplying the displacement vector by the original unconstrained stiffness matrix; thus $\mathbf{F} = \mathbf{KD}$.

In the last part of this example we outline how to determine the natural frequencies of the sample problem under consideration. The determination of natural frequencies requires that a global mass matrix be assembled in the same manner as the global stiffness matrix was assembled.

The consistent mass matrix for a beam element in local coordinates is given by

$$
\overline{\mathbf{M}} = \frac{\rho A L}{420}
\begin{bmatrix}
140 & & & & & \\
0 & 156 & & & & \\
0 & 22L & 4L^2 & & & \\
70 & 0 & 0 & 140 & & \\
0 & 54 & 13L & 0 & 156 & \\
0 & -13L & -3L^2 & 0 & -22L & 4L^2
\end{bmatrix}
$$

Another type of mass matrix that is often used, a diagonal matrix that is easy to calculate and thus saves computer time, is the lumped mass matrix, which is given by

$$
\overline{\mathbf{M}} = \frac{\rho A L}{2}
\begin{bmatrix}
1 & & & & & \\
& 1 & & & \mathbf{0} & \\
& & \dfrac{L^2}{12} & & & \\
& & & 1 & & \\
& \mathbf{0} & & & 1 & \\
& & & & & \dfrac{L^2}{12}
\end{bmatrix}
$$

The local element mass matrix transforms to the global coordinate system in the same way the stiffness transforms, $\mathbf{M} = \mathbf{T}^T \overline{\mathbf{M}} \mathbf{T}$, and the assembly of the global mass matrix is performed in the same manner as is the local-to-global stiffness assembly.

The natural frequencies of the system are now determined by solving the eigenvalue problem

$$
\det(\mathbf{K} - \omega^2 \mathbf{M}) = 0
$$

where $\mathbf{K}$ and $\mathbf{M}$ are taken to be the global stiffness and mass matrices of the system. It should be noted that the natural frequencies can be determined for an unconstrained system (i.e., no constraints are imposed on $\mathbf{K}$

and **M**), or the constraints may be applied by eliminating the constrained equations from the system before solving for the natural frequencies.

For the sample problem being considered, a lumped mass matrix was used to determine the natural frequencies of the constrained system with the following results for the lowest three natural frequencies:

$$\omega_1 = 374.4 \text{ rad/s}$$

$$\omega_2 = 1222.1 \text{ rad/s}$$

$$\omega_3 = 4206.4 \text{ rad/s}$$

Mechanical Design Methods

5.1 Considerations and Alternatives

Detailed mechanical design of a robot entails more than simply selecting the bolts, brackets, and bearings. Many of the detailed mechanical design decisions can have unexpected and far-reaching consequences. As an extreme example, consider a robot with excellent performance—large working volume, high speed, and large payload capability—that also has an attractive price. Unfortunately, the robot's mechanical design requires that it be disassembled when the brushes in one of the joint actuator motors must be replaced (an otherwise simple maintenance function). After initial acceptance, the robot would probably become very unpopular with users because of the inevitable maintenance complexity. In the final analysis, this robot would lose money for the manufacturer because of poor user acceptance and lack of repeat purchases.

The above example should help demonstrate the crucial importance of design decisions to a robot's ultimate success. During the mechanical design effort, designers must always consider the following factors:

- Manufacturability (production cost and ease of assembly of the robot *by robots*).

- Ease of installation.

- Ease of modification or reconfiguration to adapt the robot to specific tasks.

- Ease of adjustment and calibration.

- Ease of maintenance, including maintenance intervals, materials, and labor costs.

- Ease of diagnosis and repair in the event of a failure in one of the robot subsystems.

- Availability of spare and replacement parts. (Standard or readily available components—actuators, sensors, bearings, and gears—should be used whenever possible.)

- Compatibility with equipment from other vendors (e.g., grippers, wrist sensors, welding torches, riveters, and paint sprayers).

- Provisions for safe robot behavior in the event of component malfunctions.

- Likelihood that the robot will damage or destroy itself as a result of faulty electronic hardware or software.

5.2   Mechanical Design Decisions

The mechanical design of a robot is guided by a number of decisions that are usually made in an iterative fashion; the ordering below is offered to help robot designers perform the first one or two iterations in a reasonably efficient manner:

- Select transmission components (gear trains, tendon linkages, and drive chains and belts).

- Select bearings for each joint.

- Locate the actuators and transmission components within (or on) the robot structure.

- Route the necessary cables and hoses through the robot structure.

- Design the overall robot structure; select parts that will be weldments, castings, and composites. Consider methods of assembly.

- Complete detailed mechanical design; design brackets, castings, and weldments; choose fasteners; generate manufacturing drawings.

The following sections discuss some of these processes in more detail.

5.2.1   Selecting joint bearings

To achieve good positioning accuracy, the accuracy requirements on joint bearings can be quite severe. This is especially true for the shoulder joint(s) because the length of an outstretched arm can magnify errors.

Most of the joint bearings will probably require preloading to eliminate running clearance and associated errors. Despite proper preloading, the "run-out" of joint bearings may be large enough to cause errors in robot motions. (In other words, a shaft supported by a bearing will not revolve precisely about a single axis.) Run-out is customarily measured with a dial indicator resting on a precisely round shaft supported by the bearing. As the bearing is rotated, variations in the dial-indicator reading correspond to bearing run-out. Rolling-element (i.e., ball, needle, or roller) bearings are often used for robot joint bearings. Run-out of these bearings often exceeds 0.05 mm (0.002 in). Such error may not at first seem significant when compared with a robot's positioning accuracy, which might be 1 mm. However, a 0.05-mm run-out in a shoulder joint bearing can cause an error of perhaps 0.001 rad in the encoder associated with that joint. This angular error corresponds to a positioning error of 1 mm at the end of a 1-m-long arm.

Run-out can be minimized by using special high-precision rolling-element bearings. However, even the finest rolling-element bearings have run-outs of about 0.01 mm (0.0004 in). Journal bearings can be used to achieve less run-out, but journal bearings are not well suited to robot applications. For example, journal bearings tend to have a very limited life when used in intermittent-motion applications, and robot joint bearings are always subjected to intermittent motion.

Two other types of bearings that achieve excellent run-out are hydrostatic and air bearings. These are not likely to be used in most robots because of their extreme cost and certain reliability problems.

5.2.2   Locating actuators and transmission components

The first consideration in locating components should be to place the actuators as far inboard (i.e., toward the base and away from the gripper) as is practical. Such a design will minimize the inertia of the robot and thereby maximize the robot's speed and payload. Designers of some robots, such as the Microbot MiniMover-5, have gone so far as to mount most or all actuators in the base.

Transmission components are needed if actuators are mounted far inboard of their respective joints. The clever robot designer will attempt to use the same components to provide mechanical advantage and to transmit motion from actuators to their respective joints.

For ease of maintenance and service, actuators and transmission components should be mounted where they are accessible. Ideally, it should be possible to replace any actuator or transmission in the robot without disassembling any other parts of the robot.

5.2.3 Routing cables and hoses

Modern robots require many electric cables and pneumatic and hydraulic hoses, most of which must pass through (or around) one or more of the joints. Providing sufficient room for the cables and hoses in the robot and careful treatment of their passage through the joints are very important parts of the mechanical design. Unfortunately, these aspects of the design often receive less attention than they deserve.

There are usually several mechanical devices that must be mounted in or near the joint. These include actuators, transmissions, brakes, bearings, and position and velocity sensors. Finding additional room for cables and hoses can be difficult. For many reasons, it is preferable to route cables and hoses within the joint, but if this is impossible, they can be formed into loops external to the robot structure to pass around a joint.

Routing cables and hoses at rotary joints is difficult, but the design problems presented by linear (sliding) joints are even more formidable. It is normally necessary to increase the cable length by an amount equal to the stroke of the linear joint. Then some means must be provided for reliably coiling and uncoiling the cable as the linear joint moves. Ideally, the entire cable-coiling mechanism should be located within the robot structure rather than mounted on the outside in a vulnerable "outrigger" fashion. These requirements can be met but normally require elaborate mechanisms that may jam or tangle the cables. These problems may explain why linear joints are used much less frequently than rotary joints in modern industrial robots.

Cable-routing problems can be alleviated if the electrical system of the robot is designed so that the various sensors and actuators can communicate with the control system via a shared "bus." Power for the various sensors and actuators can also be distributed on a bus. Using this approach, the number of cables that pass through each joint can be reduced from 50 or 100 to less than 10.

As grippers and their associated sensors become increasingly complex, it will be necessary to use some form of bus to carry data to and from the gripper. The data bus may use unshielded metallic conductors, coaxial cables, or optical fibers for data transmission. The design considerations relating to robot data buses are discussed in detail in Chapter 8.

Robot designers should appreciate that the eventual users of the robot may have in mind diverse applications unforeseen by the designers. Wise designers will provide room in their designs for future addition of cables and hoses by the user, because some of these applications will inevitably require that additional cables or hoses be routed to the gripper.

5.2.4   Overall mechanical design

The usual considerations of strength, weight, and manufacturing cost will govern many of the decisions relating to the design of the robot structure. If strength and weight were the only considerations, robot structures would use techniques common to aircraft: stressed-skin and space-frame construction using materials like high-strength aluminum, titanium, and steel alloys. Advanced aircraft are also made of glass- and carbon-fiber composites. Although cost and maintenance considerations preclude the use of aerospace-type construction for most robots, designers should carefully consider the use of this construction for the more outboard parts of their robots. (Low weight is especially important for the outboard parts of *high-speed* robots.)

Assembly of the robot should be considered from at least two perspectives:

- Ease of maintenance and service
- Compatibility with automatic assembly

For reasons discussed above, robots tend to have their actuators located inboard of the associated joint. Power transmission components, such as belts, cables, or shafts, transmit power from the actuators to the joints. Robots also tend to have many cables and hoses which pass through several joints. Disassembling the robot for maintenance or repair can therefore be very complex and time-consuming. It is not unusual that removal of a defective sensor in, for example, joint 2, will require removal of joint 5; pulling several cables out of joints 5, 4, 3, and 2; removal of the joint 3 actuator; and removal of the joint 2 transmission. Careful consideration of these problems during the early phases of the mechanical design effort can result in a final design that is reasonably easy to service. A few additional connectors in long cables and maintenance access ports in the robot structure may greatly simplify servicing the robot.

One factor in the manufacturing cost of a robot is the cost of labor required to assemble the robot. It should be apparent from the discussion above that assembly labor can become a significant part of the total cost. Therefore, automatic assembly of robots can be quite desirable. The designers should consider this factor during the mechanical design process. Designing for automatic assembly is a complex topic beyond the scope of this book, but several guidelines can be offered:

- The structures requiring assembly should be as open as possible to allow a robot manipulator to reach easily the areas requiring assembly operations.

- The ends of shafts and other parts requiring insertion should be chamfered.

- Ideally, all assembly operations should be performed from one side of the structure.

- Use of limp and flexible parts such as belts, cables, hoses, and fabric should be avoided. Automatic handling of limp materials is very difficult.

5.3 Gripper Interface

The diverse uses to which any given robot design may be applied require that different types of grippers be used on the robot. The robot industry is beginning to recognize the need for one or more standardized robot/gripper interfaces. Such a standard interface would facilitate the use of grippers with robots made by different vendors—not to mention the interchange of grippers among robots of the same vendor.

A standard for the robot/gripper interface should specify:

- The mechanical interface, including location and thread size of bolts; location and size of locating pins or bosses; and tolerance for these parameters and for flatness, perpendicularity, and concentricity of mounting surfaces

- Connectors and mounting provisions for electric conductors that may pass through the interface

- Signal levels, power capability, and frequency response of the above electric conductors

- Connections and mounting provisions for pneumatic or hydraulic lines that may pass through the interface

- Maximum loads (forces and moments) that the mechanical interface is required to support

As robots become more sophisticated and pervasive in the factory, there will be an increasing need for robots that can interchange their grippers without human assistance. This exchange could be accomplished with a suitable robot/gripper interface that did not rely on bolts for mechanical connection. Several kinds of electrically, hydraulically, or pneumatically actuated latches could be used for such a gripper interface.

Design of a "quick-change" robot/gripper interface is not a simple problem, however. Inherent positioning inaccuracy of the robot will prevent the robot from mating with a gripper (held in a fixture or carousel, for example) unless the robot/gripper interface incorporates suitable tapered or conical surfaces and compliance to permit the gripper to move as much as a few millimeters while mating with the arm.

Careful design of the connectors, electric, optical, pneumatic, and hydraulic, will be required to ensure reliability in the presence of dirt, and to ensure leak-free connections. Useful ideas may be provided from quick-change mechanisms used for numerical control machines and docking mechanisms for spacecraft. Several quick-change interfaces are commercially available. Autonetics, Kennametal, and Mecanotron are companies that provide such equipment.

Servo Design Methods

6.1 Introduction

Servomechanism technology is crucial in industrial robotics, because it allows a manipulator to perform predictably in the presence of unpredictable influences. In a modern, sensor-equipped robot, the concept of feedback is applied at many levels. For example, a low-level feedback loop might ensure that the torque exerted by a motor that moves an arm joint is not influenced by changes in the motor's resistance that are caused by heating. A high-level feedback loop might use sophisticated image-analysis techniques to allow the robot to adapt to variations in the position of workpieces.

This chapter considers those feedback loops that control the position of the individual joints of the manipulator, without regard to the specific task being performed or to the external information from visual or tactile sensors that is used to guide arm motion at a higher level. The following topics are relevant and are treated in the order mentioned:

- Task requirements for assembly

- Control problems

- Characteristics of the devices used to move arm joints

- Solutions to the control problems

- Methods for verifying the correctness of a servo design

6.2   Task Requirements for Assembly

Assembly is basically a select-and-place activity in which the placement action may involve fitting objects together, fastening them in place, or both. The precise trajectory that the arm follows when moving from the selection and grasping point to the placement point is usually unimportant, assuming that the arm does not collide with something along the way. Some manipulator characteristics that are *less important* in assembly than in other tasks (such as spray painting) include:

- Accurate dynamic positioning
- Long reach
- Accurate positioning under large loads

The performance characteristics that are *most important* in designing a manipulator that is to be used for assembly include:

- Accurate static positioning
- Ability to make small, controlled motions
- Rapid fine motions
- Rapid gross motions
- Stability
- No overshoot

The following sections discuss the need for these important characteristics in assembly robots.

6.2.1   Need for accurate static positioning

The essence of assembly is bringing parts together. If workpieces are supplied in stationary positions, rather than on moving conveyors, the robot does not have to *track their motion* accurately; it only needs to be able to *go to their positions* accurately. This greatly simplifies the design of the servo systems in the arm. Minimizing the need for robots to handle moving objects should not be difficult in most assembly situations.

In many cases, the position in which a part should be placed and the tolerance on its position for successful part mating will be known. In these cases, if the manipulator can position the part within the allowed tolerances, there will be no need for the manipulator to be equipped with special compliance mechanisms (such as a remote center compliance tool), force feedback equipment, or vision systems. This can reduce the equipment cost, simplify the control software requirements, and reduce assembly time.

Unfortunately, high accuracy is expensive to achieve and difficult to maintain when the manipulator is heavily used. Therefore, without detailed consideration of the uses to which a manipulator will be put, it is impossible to predict whether an expensive, accurate, sensorless manipulator will be a better choice than a low-cost, inaccurate manipulator equipped with external sensors. Both may be capable of performing an assembly task, but they may have different throughput rates, different downtime characteristics, different life-cycle costs, and different tooling and jigging requirements. One manipulator may be adaptable to more assembly tasks than the other, or it may adapt to them more easily; these factors can affect the minimum economic batch size for a production run.

6.2.2 Need for smooth small motions

To make effective use of tactile sensing—especially sensing of forces and torques exerted by or on the end effector—a manipulator should be able to make very small motions. Most tactile sensors have a very limited range of compliance; if the end effector moves more than this distance in one sampling interval, damage will result. Preferably, the end effector should be able to move controllably a small fraction—say one-tenth, or better, one-hundredth—of that distance.

Although such small movements need not be made at a high *velocity,* they should be carried out in a short *time.* That is, there should be a minimum of delay between the time the manipulator receives a command to move and the start of the motion, and a minimum of delay after the cessation of motion before the manipulator reports that the motion is completed and is ready to accept the next motion command. This allows a high sampling rate, and therefore rapid, but highly sensitive, tactile servo-controlled motions.

6.2.3 Need for rapid gross motions

Assembly activity usually entails gross motions over distances from several centimeters to a meter interspersed with fine motions on the order of a millimeter. These motions are characteristics of such activities as part acquisition and part mating, respectively. To maximize cost-effectiveness, both sorts of motions should be as rapid as possible. Because of the limitations of existing external sensors, the fine motions now take most of the time in assembly. As sensing improves, increasing the speed of gross motions (beyond the 1-m/s average velocity at which they commonly take place today) will become more important.

6.2.4 Need for stability

Stability means that bounded input to a control system yields bounded output. The servomechanisms in a robot manipulator can easily become unstable and go into violent oscillation. When this happens, the arm shakes back and forth at high speed until something breaks. Instability is both costly and dangerous and absolutely must be prevented.

Sensory feedback can make a stable manipulator unstable if it is improperly applied. Tactile feedback can cause oscillations at frequencies of tens of cycles per second. Image-processing delays in a visual feedback path can cause oscillation at lower frequencies, which can be equally hazardous.

6.2.5 Need to prevent overshoot

Assembly, like many manipulation tasks, requires that objects come into contact or close proximity with each other. If the arm as a whole tends to overshoot its target position in such situations, a damaging collision can occur. This must be prevented. It is usually a sufficient precaution to prevent the individual joints from overshooting their commanded positions.

6.3 Control Problems

A servomechanism that operates the joint of a robot arm must overcome problems created by five major influences, which are discussed individually below:

- Friction

- Gravity

- Inertia variation

- Dynamic coupling between joints

- Structural resonances

With the exception of friction, the strength of these influences on any joint usually depends strongly on the position, velocity, and acceleration of the other joints, and on the load being carried by the arm. This is particularly true of conventional serial-link manipulators whose joints are all, or mostly all, rotary. One particular manipulator design, however, manages to reduce many of these undesirable influences almost to zero. This is the so-called overhead crane design such as is used in the Olivetti Sigma robot and the IBM RS1 robot. In this design, the first three joints are prismatic and orthogonal to one another, and the first two joints are horizontal. The first two do not have to support the weight of the other joints; instead, the weight is borne by a fixed framework. Furthermore, because the axes of the first three joints are rotary, they do not interact dynamically in the ways described below. The arm solution (i.e., computing the joint angles for a given hand position) is thus greatly simplified.

Large stiff manipulators such as the Unimation model 2000 and the Cincinnati Milacron T746 have natural frequencies on the order of 5 Hz. They are normally controlled adequately with oversized actuators and simple feedback circuits. However, their response is more sluggish than necessary considering the power of the actuators.

Some manipulators of this type use one classical PID controller for each joint, uncoupled to the other controllers, and it is not even compensated for gravity loading. Others engage a brake when a joint reaches its commanded position in order to resist any forces coupled into that joint from other joints that are still moving.

One commercial manipulator was stabilized by attaching a large lead weight in a box of oil to one link of its arm. Early models of another manipulator had three different servo circuits, each tuned for a different load to be carried. It was the responsibility of the operator to select which circuit was to be used for each motion in a task. If that arm accidentally dropped a heavy load that it was carrying, however, its servos were grossly mistuned for the arm's inertial characteristics and the arm whipped around violently. Such control methods will probably be inadequate for operating high-performance assembly robots.

6.3.1 Frictional forces

Three kinds of frictional effects are important in joint operation: viscous friction, static Coulomb friction, and dynamic Coulomb friction. We will discuss these effects in terms of friction-induced torques, rather than forces, because rotary joints are the kind most commonly used in assembly robots.

Viscous friction always acts to oppose joint motion, and its magnitude is roughly proportional to the velocity of the joint motion. This viscosity of the lubricants in gearboxes and motor bearings is the major source of viscous friction; the friction in a joint's main bearing is usually negligible in comparison. Viscous friction tends to cause steady-state position error in a joint moving at a constant velocity.

Static Coulomb friction acts to oppose motion in a joint when the joint is *not* moving. Dynamic Coulomb friction does the same when the joint *is* moving. Unlike viscous friction, however, the dynamic Coulomb force is approximately the same for any joint velocity. It arises from microscopic surface-surface interactions in the joint and is always *smaller* than the static friction force. Static Coulomb friction makes it more difficult to make small motions with the end effector, and dynamic Coulomb friction tends to cause errors in positioning the joint.

6.3.2   Gravity loads

Gravity imposes large loads on the joints of a manipulator even if the manipulator is not carrying an object because the weight of the manipulator itself from the "shoulder" out is typically 100 to 400 lb. Gravity loads are typically larger on the inboard joints (except for cartesian, or "X-Y-Z," arms).

In a conventional serial-link manipulator design, each joint carries all the links outboard from it, so it must support their combined weight as well as its own. The gravitational load on a joint usually depends not just upon the joint's position but upon the positions of some of the other joints as well, both outboard and inboard from it.

Manipulators are often hung upside down, which reverses the direction of gravity loading. The servomechanisms should be able to accommodate this. If the manipulator is mounted at some angle other than right-side up or upside down, however, compensating for gravitational loading becomes much more complicated.

6.3.3   Variations in inertial loads

In a conventional serial-link manipulator design, each joint moves all the links outboard from it so that it sees their combined inertial load. The positions of the inboard joints have no effect, but those of the outboard joints have a dramatic effect. The inertial load seen by inboard joints can easily vary by a factor of 10:1. This cannot be ignored in servo design, because it has a major impact on the natural frequency of the overall servo loop and the damping factor, and hence on the speed of response and the possibility of overshoot.

The cartesian, or X-Y-Z, joint configuration tends to minimize the variation in inertial loading because the motions of the first three joints are linear; there is no variable lever arm to consider, as with a rotary joint.

6.3.4   Dynamic coupling between joints

Dynamic coupling exerts forces and torques at a joint that depend upon the motions of one or more of the other joints. This happens because

- Each link of the arm has a mass and moments of inertia
- Motion of one joint can accelerate links other than the two to which it is attached
- The inertial forces and torques that arise from acceleration of any link can be transmitted to any joint through the intervening joints and links

Dynamic coupling effects are usually significant only for rapid joint motions.

There are three coupling effects. They produce, respectively, a force or torque at a joint i that is proportional to one of the following:

- The acceleration of joint j (cross coupling)
- The square of the velocity of joint j (centripetal coupling)
- The product of the velocities of joints j and k (Coriolis coupling)

where i, j, and k are all different.

Cross coupling predominates over the other two for arm motions of moderate speeds—say below 1 m/s. The various coupling proportionality factors, however, are usually extremely complicated functions of the shape, mass, and moments of inertia of many links, as well as of instantaneous joint angles. There are several ways to deal with this complexity.

1. Ignore it completely. This approach is favored by many robot manufacturers, who so far have been able to develop servo designs (by trial and error and ad hoc methods) that are fast enough for most purposes and stable under light and heavy loads and in all arm positions. This approach may not be good enough to produce arms that are an order of magnitude faster and more accurate, however.

2. Determine which coupling effects predominate and design the servo to compensate for them only. This requires extensive mathematical analysis based on reasonably accurate estimates of link properties, physical reasoning, experience with servo design, and some intuition. Experimental verification of the predictions is advisable. This method will probably become increasingly important for the design of high-performance robots.

3. Evaluate the coupling effects numerically in real time and compensate for them with feedforward terms in the servo. (The servo still requires *feedback* paths in order to respond to unpredictable disturbances.) A computational algorithm that is fast enough for this purpose was developed at Purdue University (Luh et al., 1980).

Luh's method, an advanced, yet practical real-time algorithm for dynamic control of a manipulator with six or more joints, computes the force or torque that each joint must exert, assuming frictionless joints and rigid arm links that have constant weights and moments of inertia. Additional simple computations can correct for frictional effects at each joint on the basis of the joint velocities. Additional feedback is required to sense perturbations from the nominal payload trajectory and guide the joint back in place.

The Luh method is based on a recursive Newton-Euler formulation of the dynamics problem, rather than on the complicated classical lagrangian formulation used by previous authors. Although the details of the method are too lengthy to present here, we can give a brief summary that shows the elegance and simplicity of the reasoning behind it. The method is as follows. For each instantaneous position of the payload along the desired trajectory,

1. Compute the positions, velocities, and accelerations of each joint. These depend directly upon the position and velocity of the payload and upon the acceleration that it is supposed to experience at that instant. These can be the nominal values along some specified trajectory if one is precomputing joint efforts offline. Alternatively, the position and velocity can be the actual values measured in real time and the acceleration can be specified arbitrarily by some higher-level feedback control algorithm.

2. From the results of step 1, compute the positions, velocities, and accelerations of each link and the payload.

3. From the accelerations and the corresponding inertias, compute the net force and torque that must act at the center of gravity of the payload and of each link.

4. By subtracting the net force and torque on each pair of links from that of the successive pair, obtain the force f and torque n that one link exerts on the other through the intervening joint. If the joint is a sliding joint, then it is rigid in all directions except along its axis. Therefore it can transmit any torque n without actuator effort. Similarly, it can transmit the component of force f at right angles to its axis without actuator effort. So the actuator of a sliding joint only has to "resist" the component of f along its axis, and that gives the actuator effort required. If the joint is a rotary joint, then it is rigid except around its axis. Then it can transmit any force f without actuator effort. Similarly, it can transmit the component of torque n normal to its axis. Therefore, for a rotary joint the actuator effort must be equal to the component of n along its axis.

The computation time obviously increases only linearly with the number of joints in the arm. This is much better than the classical lagrangian approach, in which the computation time increases as the fourth power of the number of joints. Furthermore, the computations in steps 1 and 2 can be performed simultaneously for each joint, starting from the most inboard joint and working out to the wrist. Steps 3 and 4 can also be performed together for each joint, working from the wrist back to the base. The method is described as *recursive,* because the computation of any joint effort depends upon having previously obtained the efforts at all the joints outboard from it. One of the secrets of this method is that it makes frequent use of the 4 × 4 coordinate transformation matrices described later to transform forces and torques into the frame of reference of each joint in turn. All four computations can be performed for a six-joint arm in 4.5 ms on a PDP-11/45 computer, when the algorithm is coded efficiently in assembly language and arithmetic is performed in floating-point form.

Hollerbach (1980) developed a similar calculation method. Hollerbach's method is based on a recursive lagrangian analysis of the dynamics problem and is almost as fast. Either method by itself is not sufficient for manipulator control, because errors in the arm model and external influences can cause the payload to deviate in unpredictable ways from the trajectory that it is supposed to follow. A higher-level control algorithm is needed to sense such deviations and adjust the desired acceleration of the payload used in step 1 so as to bring it back on course. Both methods are exact, that is, they do not ignore any coupling effects even if they are negligible or too complex to compute.

6.3.5 Structural resonances

The links of a manipulator are not perfectly rigid. The Space Shuttle teleoperator arm—which has two 30-ft-long links—is an extreme example. Although conventional manipulators are much stiffer, they still have low enough fundamental frequencies of vibration to be worth consideration when designing servos for them. Paul (1979) recommends that the bandwidth of a joint servo be limited to one-half of the natural frequency of the joint or less. Resonant frequencies as low as 4 to 5 Hz are sometimes encountered in the inboard link of an arm that is suitable for assembly tasks. This would require bandwidths of 2.0 to 2.5 Hz in the inboard joint servos. Higher frequencies could induce vibrations in the link and send the arm into uncontrollable oscillations.

The outer links of an arm are usually smaller and have a much higher natural frequency—perhaps 5 times higher. The servos for the joints in these links can operate safely at correspondingly higher bandwidths. They may well excite vibrations in the larger inboard links, but the servos that drive those links will be too sluggish to respond to such rapid disturbances in the joint angles they control, and the vibrations will die out.

6.4 Equipment

The servo designer works with a variety of equipment, including actuators, position feedback transducers, rate feedback transducers (tachometers), analog circuitry, digital computers, and power amplifiers.

6.4.1 Actuators

The actuators in common use in joint servos today are dc electric servo-motors and hydraulic actuators. A few robots use stepper motors, ac servomotors, or exotic combination actuators (such as an hydraulic actuator whose servo valve is driven by a stepper motor).

6.4.2 Position transducers

Position transducers are used to measure the angle of rotation of a rotary joint or the distance that a prismatic joint is extended. The transducers in common use today for this purpose include potentiometers, optical encoders (single or multichannel, Grey or binary code), and resolvers.

It is difficult to manufacture such transducers with enough linearity and resolution to meet the needs of joint position servoing. A solution that many manufacturers adopt is to use a transducer with high linearity but with a resolution that is $1/n$th of what is required. They connect it to a joint through a $1:n$ gear train so that the transducer travels through its entire range approximately n times as the joint rotates (or extends) once through its range of motion. Resolvers and optical encoders are particularly suited to this method, but ordinary potentiometers are not, because they have a dead band over some fraction of their rotation. Two potentiometers or a single potentiometer with dual wipers are sometimes used to overcome this problem.

Gearing the transducer in this way makes the transducer reading ambiguous, of course, because the same reading appears at n different positions throughout the range of motion of the joint. Some manufacturers, including Cincinnati Milacron, make it the responsibility of the operator to resolve this ambiguity once, at start-up time. After that, the control system keeps track of which "turn" the transducer is on. The operator of a T3 manipulator, for example, uses the remote-control box to position each joint to a reference position as indicated by index marks on the links on either side of each joint. Other manufacturers provide a secondary transducer system, with much coarser resolution and linearity, that provides the control system with just enough information to allow it to resolve the ambiguity automatically. In the PUMA manipulator, for example, a potentiometer coupled 1:1 to the joint serves as the coarse-resolution position transducer. The fine-resolution system is probably a single-channel optical encoder with an index mark, judging by the behavior of the arm during its automatic calibration procedure.

The transducer system should indicate the absolute angle or extension of each joint. This provides the necessary information to solve the kinematic equations of the arm. That, in turn, allows the control system to know the arm's position when motion is initiated so that it has complete control immediately. This allows the arm to recover from accidents, malfunctions, or programming errors much more quickly and easily, whether the recovery is carried out automatically or with the assistance of the operator.

6.4.3   Rate transducer (tachometer)

Tachometers are not strictly necessary in a joint servo, because theoretically the velocity of the joint may be derived by differentiating the position of the joint with respect to time. Either analog or digital methods may be used. In practice, such differentiation introduces noise into the velocity signal, and most manufacturers use some sort of tachometer as well as a position feedback transducer at each joint.

A variety of tachometers are suitable for use in joint servos. Two important characteristics are bidirectionality and linear response through zero rate. Many simple transducers, such as dc generators, produce discrete pulses at low rates; these pulses must be smoothed by filters, with a consequent loss in frequency response. Other tachometer designs provide true proportional response to rate. One such design is based on the differential transformer, in which shaft rotation turns an eddy-current disk that unbalances the field distributions in each output winding. However, it requires ac excitation and phase-sensitive demodulation of the output, which increases the cost.

6.4.4   Analog versus digital methods

In general, the servo designer will use analog circuitry where it provides a simpler, cheaper, or more convenient solution than digital signal-processing methods. Often, the need for an extremely wide bandwidth will dictate analog methods. High bandwidths are required, for example, in controlling the torque in a stiff system. Small, precise, force-controlled motions may require great stiffness when the tool and workpiece are both rigid.

6.4.5   Power amplifiers

Power amplifiers are a major cost factor in a joint servo. They must meet a number of demanding requirements, including power level, duty cycle, linear amplitude and phase response, flat frequency response, and controlled roll-off above the cutoff frequency. Power amplifiers are usually designed especially to suit the requirements of individual joints or groups of joints with similar power requirements.

Usually the limitation on servo performance imposed by the power amplifier is negligible compared to physical effects such as Coulomb friction, interjoint coupling effects, and the characteristics of the actuators and feedback transducers.

6.5 Servo Design Methods

Servo design encompasses the following steps, which need not be carried out in the order shown:

1. Modeling physical effects in the manipulator
2. Feedforward compensation for physical effects
3. Feedback compensation to tailor dynamic response characteristics
4. Feedforward compensation to tailor steady-state response

The following sections discuss each of these steps in more detail.

6.5.1 Modeling physical effects in the manipulator

If the manipulator were ideal, the acceleration of any joint would be exactly proportional to the actuator effort commanded by the control system. Then we could make a joint servo as shown in Figure 6.1. Clearly, the actual joint position in this system must exactly follow the desired joint position command.

In any real manipulator a number of physical effects, shown in Figure 6.2, would defeat this highly idealized design. In order of increasing complexity, they are

1. Viscous damping
2. Inertial loading
3. Coulomb friction (static and dynamic)
4. Gravity loading
5. Interjoint coupling effects

The following sections describe the origins and characteristics of the first four effects and their impact on manipulator performance.

6.5.1.1 Viscous damping. Viscous friction in a servo usually arises from the drag imposed on moving parts by the lubricants used on them; for example, a shaft turning in a greased journal bearing or the seals in a hydraulic piston sliding on the inside walls of its cylinder.

The characteristics of viscous friction are that it exerts a force in the direction opposite to any motion and that the force is usually proportional to the velocity of the motion. Viscous friction can be reduced by using ball bearings, for example. The major source of friction in an electric servo is usually the small amount in the motor and gearing, because it is amplified by the gear ratio.

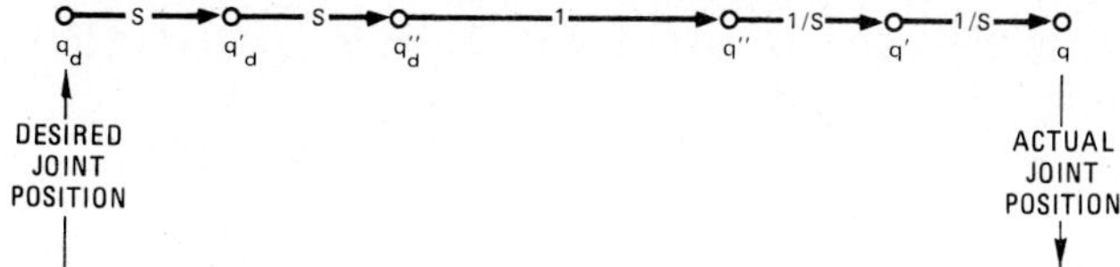

Figure 6.1 Ideal servo.

The effect of viscous friction can be beneficial, because it tends to make a servo more stable. It can also be detrimental; because it tends to make response more sluggish, it causes a position error proportional to steady-state velocity and tends to increase energy consumption.

6.5.1.2 Inertial loading.

The origin of inertial loading is the mass of the moving parts of the manipulator and the mass of the load carried by the manipulator.

Inertial loads are proportional to the inertia of the parts and their accelerations. They act in the opposite direction to the acceleration. Inertial loads can appear as forces and torques; they exist in all joint servos. The inertia of an electric drive motor reflected through a high gear ratio can have as much effect as the inertias of the links and payload themselves.

The effect of an inertial load on a servo is to destabilize the servo and to increase the drive requirements for a given acceleration.

6.5.1.3 Coulomb friction (static and dynamic).

Coulomb friction arises from contact between parts. It depends upon the nature of the surfaces in contact.

When two parts are stationary with respect to one another, a certain minimum force or torque is required to make them start moving. This is the static friction force. The dynamic friction force develops after the two parts are moving. It is usually smaller than the static friction force and is independent of velocity.

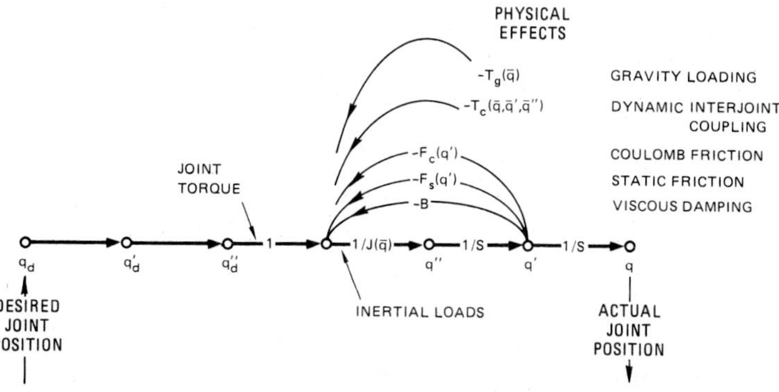

Figure 6.2 Nonideal servo showing physical effects that degrade performance.

Coulomb friction usually degrades servo performance by preventing the servo from responding to small inputs. In the presence of some compliance in the servo drive, it can also cause "stick-slip" oscillation.

6.5.1.4 Gravity loading. Gravity loading arises from the force of gravity acting on the links of the arm and on the payload. Gravity loads appear as forces or torques at the joints of the arm. They are usually complex functions of the posture of the arm, except in cartesian arms. Gravity loads tend to introduce errors in servo positioning. Compensating for them can be complex.

6.5.2 Feedforward compensation for physical effects

Each of the physical effects described in the preceding section can be compensated for by applying equal and opposite efforts to the joint. These efforts can all be precisely computed in real time, although some are best computed digitally. Figure 6.3 shows where these feedforward signals appear in the flow diagram of the resulting joint servo.

6.5.3 Feedback for dynamic response

The joint servo system in Figure 6.3 would be adequate only for an ideal manipulator in which there were no unpredictable disturbances. In any real system, numerous disturbances would prevent the manipulator from moving as commanded. Some appear as acceleration errors at the joint; these include wind resistance, cogging in gear trains, reaction forces exerted by a spray gun, and contact forces exerted on a tool by external objects. Others appear as rate errors; they can be caused by a slight imbalance in the analog amplifiers in the motor drive circuitry or by changes in viscous friction as lubricants heat or cool. Finally, some appear as position errors; static and dynamic Coulomb friction, the weight of an object in the

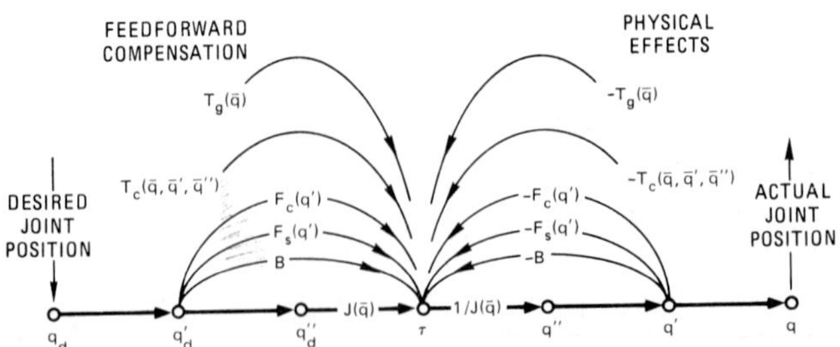

Figure 6.3 Ideal servo with feedforward compensation for physical effects.

hand, and the variation in gravity loading that results from the changing positions of the outboard joints are the most common causes of position errors.

Feedback provides a method for reducing the magnitudes of these errors by changing the dynamic response of the system. In Figure 6.4, the three feedback paths most commonly used in joint servos today have been added to the servo. Velocity feedback $(-K_v)$ reduces the rate at which the joint moves in response to a given disturbing effort. Position feedback $(-K_p)$ makes the joint stiffer by increasing its resistance to displacement from its correct position. This reduces position errors, but it cannot eliminate them completely. Integral feedback $(-K_i)$ does reduce position errors to zero, but it takes time to do so.

Feedback paths such as these are a potential source of instability and damaging oscillations, and the feedback gains in each path must be adjusted properly to avoid this. (We discuss servo tuning below.)

6.5.4 Feedforward for steady-state response

If the feedback paths in Figure 6.4 are applied only by themselves, as is, they will tend to resist valid joint motion commands from the control system—they would resist any joint motion, resulting in large steady-state *rate* errors, and would always push it back toward its zero position, resulting in large steady-state *position* errors. In order to prevent this, we must apply matching feedforward signals, as shown in Figure 6.5.

In Figure 6.5, the feedforward path K_p provides the "position set point" for the joint, which feedback path $-K_p$ strives to balance. This prevents steady-state position errors. Similarly, feedforward path K_d supplies the "rate set point" that feedback path $-K_v$ matches. This prevents steady-

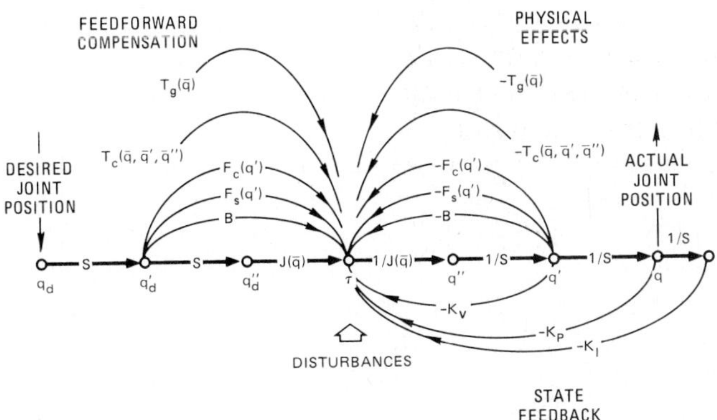

Figure 6.4 Servo with state feedback to resist disturbances.

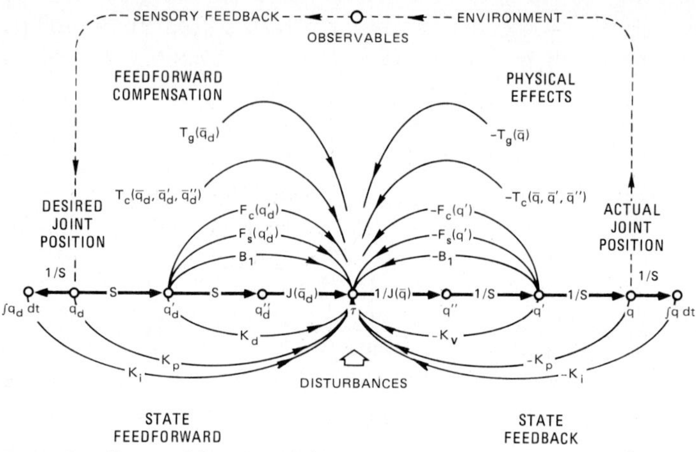

Figure 6.5 Servo with state feedforward paths to compensate for resistance to commands with feedback paths.

state rate errors caused by the sluggish response that feedback path $-K_p$ produces. Finally, feedforward path K_i balances the growth of the integrated position that is fed back along path $-K_i$ for nonzero joint set points. The effect of these three feedforward paths is to remove the feedback paths from the system for control input but not for external disturbances.

6.5.5 Practical servo implementations

Figure 6.5 is only a theoretical design for a joint servo. Its value is that it makes explicit the symmetry between the various contending influences in the servo and the central importance of the net effort applied to the joint. The major causal influences in the system flow along the horizontal centerline of the figure in the directions shown by the arrows. The vertical centerline of the figure divides the desired state of the system (on the left) from its actual state (on the right). The effort on the joint is easily seen to be, first, the point at which all influences in the servo come together, and second, the only channel through which the control system can affect the behavior of the controlled system.

It would not be practical to implement a joint servo exactly as shown in Figure 6.5. For example, the signals on the integral feedforward and feedback paths would grow indefinitely whenever the joint was held at a nonzero angle, leading to overflow in a digital system or saturation in an analog system. The characteristics of the actuator and power amplifier are not shown either, although they can be important factors in designing a high-performance manipulator. Finally, the close relationship of this design to that of the classical controllers used widely in industry for process control is not readily apparent.

Finally, in practice it may not be necessary to include every type of feedforward compensation shown in Figure 6.5. For example, the inter-joint coupling effects may be small enough to be considered random disturbances for which the feedback paths can adequately correct. In that case the feedforward path T_c can be omitted. Similarly, the integral feedback will tend to correct automatically for static Coulomb friction and gravity loading. The feedforward path F_s can then be omitted, and the calculation for the T_g path can be simplified to a rough approximation of the gravity load based on the positions of only the heaviest links. If accurate tracking is not required, the velocity feedforward term K_p can also be omitted. The result will be a somewhat larger position error, but only while the arm is moving rapidly. A very common design decision today is to approximate the inertial load on a joint by a constant worst-case (i.e., maximum) value and to tune the feedback gains for that load. The only disadvantage is that when it is carrying a lighter load, the arm moves more slowly than the speed at which it could actually move. More precisely, the motion is overdamped, or more sluggish than necessary, and the arm's motors are not used as effectively as they could be.

In Figure 6.6 we have redrawn the feedback paths and their compensating feedforward paths to emphasize the conventional aspects of this design. The three parallel vertical paths in the lower half of the figure comprise a classical proportional controller with rate and reset adjustments. These are used widely in many factories and are also called PID controllers.

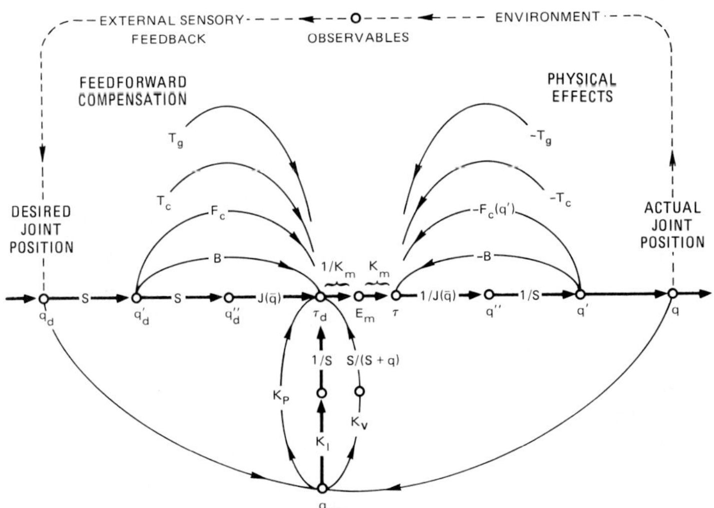

Figure 6.6 Servo redrawn to show proportional, integral, differential (PID) controller.

In Figure 6.6, also, the effort node has been split into two nodes that represent desired effort on the left and actual effort on the right, separated by arcs representing the power amplifier and actuator. In practice, those two devices will have leads, lags, and resonances that may be potential sources of instability when the feedback gains are adjusted for rapid response.

Finally, in Figure 6.6, we have indicated schematically the external feedback paths from the external robotic sensors such as vision systems and tactile fingers that control the overall movements of the manipulator. We can indicate these influences only schematically because any sensor may affect the motion of each joint in a different and extremely complex way, depending upon the algorithms in the high-level task control software.

The robot system designer should be aware that even if all the joint servos are stable, both individually and working together, the feedback from the external sensors can still make the overall system unstable, causing violent oscillations. Thus, the engineer responsible for providing sensory feedback capabilities should coordinate with the servo designer to ensure success. They should agree on performance characteristics for the manipulator itself that will make stable sensor-controlled motion possible.

6.6   Servo Design Again

This section essentially repeats the information of the last section but from a slightly different perspective. For example, the last section used signal flow diagrams, while this section uses block diagrams. This second perspective is presented to help clarify ideas for the reader who is not already familiar with servo controls. Another recommended perspective is provided in a survey article by Luh (1983).

6.6.1   Basics

The concept of feedback control is illustrated in Figure 6.7. The system is designed to respond to the difference between a desired and an actual value. A desired value is fed to the system. This value is compared to a measurement of the actual value by a transducer. The difference tells the amplifier which direction and how forcefully to respond while driving the device.

Disturbances impact the system. They are usually added as inputs to the device being controlled. Motor torques will have the same kind of effect on the manipulator as someone pushing on it. A feature of the servo system is that disturbances produce errors which in turn cause action. The effect of disturbances is thus removed.

The concept is adaptable to any kind of value, for example, tempera-

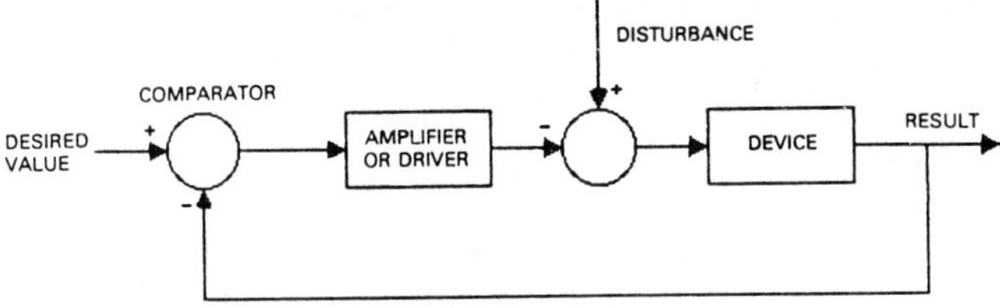

Figure 6.7 Feedback control concept.

ture, velocity, force, and position. As has been noted, in most robotic controllers the controlled variable is position.

The simple concept of responding to a position difference always produces a corrective force, but this may not be sufficient. When dynamics are considered, the corrective force may give a response that causes overshoot of the desired value; the system may come to the correct position but be traveling at such a high speed that it overshoots the position. The classical solution has been the proportional, differential (PD) controller shown in Figure 6.8. The system in this case is considered to be a dynamic body with a moment of inertia I. A torque results in an acceleration. The acceleration is integrated to give the velocity, and integrated again to give the position. (Multiplication by $1/s$ in Laplace-transform notation produces integration; in the block diagram the $1/s$ therefore indicates integration.) In addition to feeding back the actual position, the velocity is also fed back multiplied by a constant that determines the amount of damping. By adjusting the amplifier gain and the damping multiplier, the designer can establish a motion without overshoot for a given inertia.

Often an additional integration is added in the system giving a PID controller. The integration accumulates errors so that small but persistent errors can be overcome. For example, the integration will accumulate a small error until the force produced is large enough to overcome the coefficient of friction and cause motion. Integration will also correct for the following error produced by a constantly increasing, or ramp, command.

A simple manipulator control system consists of a PID controller for each joint. The controller can be constructed as an analog or digital system. In digital systems, several joints may be controlled by the same pro-

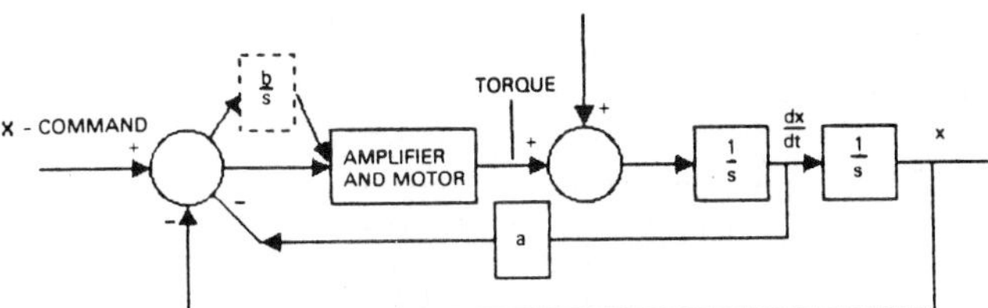

Figure 6.8 Proportional, differential (PD) controller (PID controller with integrator).

cessor. The controller for a joint can be located anywhere, but usually the power conditioning equipment (amplifier) is located in the base or in a separate rack. One of the problems of manipulators is the difficulty in running the power conduits (wiring for electric robots) for the joints from the power conditioning, through other joints, to the actuator.

The disadvantage of the PID controller is its one set of values. When the load changes, the inertia changes and the performance is altered. This is particularly disadvantageous in a manipulator where the locations and motions of the other joints constitute changes in inertias or disturbances. These disadvantages are particularly important when large changes in step positions are used. Depending on the load, the manipulator may take very different paths between points. Nevertheless, the PID controller is the basis of many robot control systems where small steps are given and planned trajectories are followed.

6.6.2 PID improvements

Figure 6.9 illustrates two improvements to a PID control system for a joint. The first is the forward compensation, which accommodates for lags in the control system for changing command levels. It recognizes that the control system will not follow exactly and creates a modified signal that will more nearly follow the input.

The second and more important modification is a consideration of the effects of load, and the positions and velocities of the other limbs. This compensation involves calculations based on a model of the manipulator. It may be considered an intermediate-level device.

Figure 6.10 shows a similar adjustment world model using inputs of the payload and the other manipulator joints to adjust the PID control parameters. The idea is to tune the PID control system, which is then called an *adaptive control system,* for the inertia involved. Note that using the load as an input may be difficult, but measurement of the load

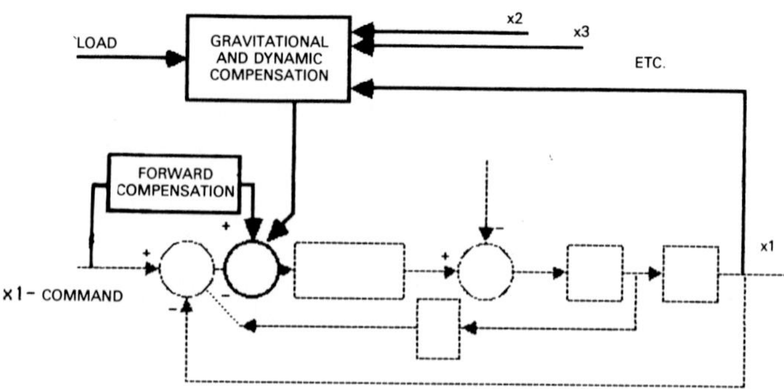

Figure 6.9 Compensation for other joints.

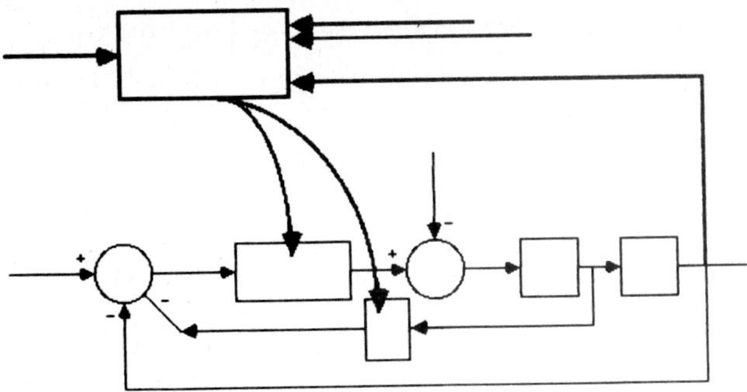

Figure 6.10 Adaptive control.

is not easy. One way of inferring the load is to know how the manipulator
would respond for different loads and make adjustments based on this
response. This is known as *model adaptive control* (Kornbluh).

Finally, all kinds of compensations can be combined into a single control system as shown in Figure 6.11. Robot control systems may be said
to be built upon existing technology to accomplish what is conceptually
not too difficult but which is time-intensive. Robot control systems may
be viewed as distributed computing machines.

6.7 Summary of Servo Design Pitfalls and Solutions

Table 6.1 is a chart of the more common problems that may be encountered in developing servos for a manipulator: An asterisk in the chart indicates that the problem in that column may result from the cause in that
row.

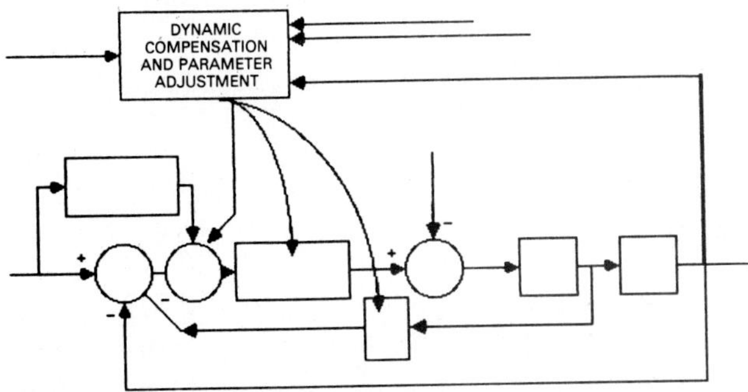

Figure 6.11 Compensation and adaptive control.

TABLE 6.1   **Common Causes of Servo Design Problems and Their Results**

Result	Cause*											
	1	2	3	4	5	6	7	8	9	10	11	12
Slow response		*	*							*		
Inaccurate positioning		*			*	*		*			*	*
Instability	*			*	*	*	*	*	*	*		*
Underdamped response	*			*			*					
Jerky short moves					*	*						*
Inaccurate force control					*						*	*
Inadequate stiffness		*						*				

*Position feedback gain too high (1) or too low (2); velocity feedback gain too high (3) or too low (4); static friction (5); cogging (6); velocity feedback saturation (7); coarse feedback position (8) or velocity (9) quantization; inaccurate dynamic (10) or kinematic (11) model; amplifier (12).

- *Slow response.*   The arm takes too long to reach a commanded position. It may move rapidly at first, but begins to slow down too soon and makes its final approach too slowly.

- *Inaccurate positioning.*   When the arm stops moving, it is not close enough to the commanded position.

- *Instability.*   Instead of standing still or moving smoothly, the arm shakes back and forth.

- *Underdamped response.*   The end effector overshoots the target position.

- *Jerky short motions.*   Instead of accelerating and decelerating smoothly when commanded to move, the arm may do nothing for a brief time, then start moving with a high acceleration, then stop abruptly. It may overshoot the target position and not return to it.

- *Inaccurate force control.*   The magnitude or direction of the force or torque exerted by the end effector is inaccurate.

- *Inadequate stiffness.*   The end effector moves too much in response to forces and torques exerted on it.

The solutions to these problems are:

- *Position feedback gain.*

    Too high: Reduce the gain.

    Too low: Increase the gain.

- *Velocity feedback gain.*

    Too high: Reduce the gain or reduce viscous damping in drive train.

    Too low: Increase the gain or increase viscous damping in drive train.

- *Static friction too large.*   Reduce friction by lubrication or use of low-friction materials such as Teflon, redesign drive train, or introduce "dither" into the control system.

- *Cogging (in the electric motors or gear train).*   Reduce by changing the motor or gears or by adjusting the amount of mesh between gears.

- *Velocity feedback saturation.*   Shift gains forward around velocity feedback loop, change tachometer sensitivity, increase voltage range of analog circuitry, rescale all loop gains. In a digital servo, rescale integer arithmetic, convert to floating-point computations, or increase number of bits used to represent quantities.

- *Coarse quantization.*

  Position: Use a digital position feedback transducer with more resolution, use an analog-to-digital converter with more bits of resolution if the transducer is analog, rescale the signal from the transducer (e.g., by changing gear ratios), or use a dual-rate transducer (i.e., use both a coarse and a fine position transducer on each joint).

  Velocity: Rescale by changing gear ratio or convert from a pulse-generator type of transducer to an analog transducer (e.g., a homopolar generator) with an analog-to-digital converter.

- *Inaccurate dynamic model.*   Increase accuracy by including effects such as centripetal and Coriolis forces; check inertias, weights, link dimensions; check assumptions such as perfect rigidity of links; check for compliance in base mounting; check for assumptions about which dynamic effects can be ignored in normal use.

- *Inaccurate kinematic model.*   Increase accuracy by checking assumptions such as that an offset in a joint is negligibly small or that the axes of two consecutive joints are exactly at right angles or parallel; verify link lengths; check for excessive compliance in links, drive trains, wrist socket, and base mounting.

- *Noise in analog circuitry.*   Reduce noise level by using low-noise preamplifiers (especially on strain gauges) and coaxial cables, by separating signal cables from motor-drive cables, by checking for ground loops or inadequate grounding, and by verifying adequate resolution in analog-to-digital and digital-to-analog converters.

Manipulator Simulation

A *manipulator simulation* is the solution of the equations of motion of the manipulator system that yields the positions, velocities, and accelerations of the system elements as functions of time. In some cases, these simulations also yield the drive forces and torques required to produce this motion and the resulting internal forces in the system's mechanical elements. The drive loads are important to the design of the system's actuators and their circuits. The internal forces are required for the design of the mechanical components of the manipulator.

Simulations are typically done offline, often during the system design stage. (Simulations should not be confused with the computations performed by the manipulator's control computer.) The simulations, in general, are detailed and include the properties of the manipulator's payloads, drive elements and control systems, and the manipulator's mechanical configuration. In some cases, manipulators working together may be modeled in the same simulation. The simulation may also include descriptions of the surrounding work space and its machine system, in order to study the interaction of the robot with its environment.

To describe the behavior of such complex systems, the analyst must formulate the system's equations of motion and then solve them. These equations are usually large sets of mixed, highly nonlinear differential and difference equations whose closed-form analytical solution is not possible. Some information may be obtained from analytical solutions of their linear form; however, to predict the manipulator's performance accurately and more completely, the full nonlinear equations of motion must be solved numerically on large digital computers. Such solutions are often called *digital simulations.* Using the graphics capabilities found on many digital computers, it is possible to generate displays of the robotic manipulator motions to aid in evaluation and interpretation of the manipulator system performance which was predicted by the simulation.

The major disadvantage of simulations is that if they become very large, their cost grows rapidly. Large simulations can require an analyst to spend a substantial amount of time formulating manipulator equations and writing and debugging simulation programs. Large, complex simulation programs can also require a great deal of computer time to produce the required results. However, the costs of simulation are far less than the cost of evaluating various designs experimentally, where design changes can only be studied by the expensive and time-consuming method of manufacturing new hardware. So, in most cases, simulations are the best design tool for initial design studies. Simulation costs can also be greatly reduced if the analyst uses several smaller simulations, each tailored to investigate a particular aspect of the design. For example, a simulation of the manipulator's interactions and phasing with surrounding machines might use only a simple model of the manipulator dynamics and control systems' characteristics, while the detailed dynamic analysis of the manipulator could be done with detailed models of the supporting machines.

7.1   Levels of Simulation

In most cases, three levels of simulation are used to study a manipulator design. The first is a *kinematic simulation.* In this simulation the dynamics of the manipulator and its control systems are neglected. The motion of the manipulator arm is assumed to be determined by the commanded joint displacements: angles for revolute joints or distances for sliding joints. The positions and velocities of the manipulator and associated equipment are calculated using standard kinematic methods, such as 4×4 matrix techniques. These equations are algebraic and are easily solved. The usefulness of this approach is discussed later in this section.

The second and more complex form of simulation uses a *rigid link dynamic model* of the arm. At this level of simulation the dynamic characteristics of the control systems, drive actuators and transmission, and the discrete-time control computer are considered. Here the equations of motion are coupled sets of nonlinear, algebraic, differential, and difference equations. Their solution requires the use of numerical forward-integration techniques. This form of simulation is most useful in designing and evaluating the manipulator's control systems, which include its control computer and its drive units.

In the third and most complex simulation form, the *distributed mass*

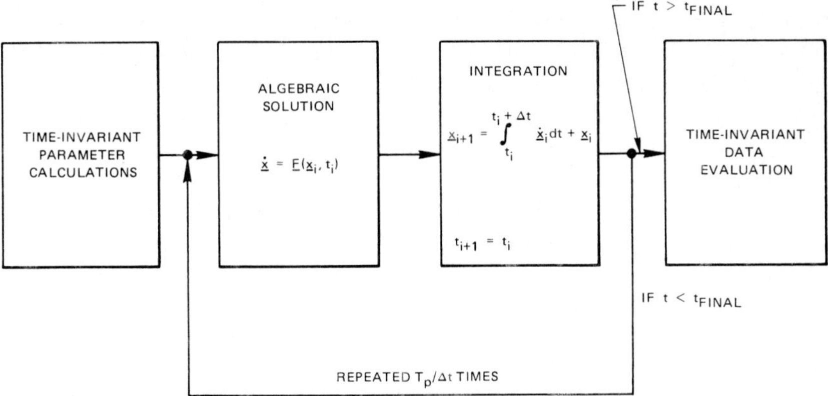

Figure 7.1   General simulation block diagram.

and flexibility of the manipulator's mechanical elements are included in the model, in addition to the control and drive properties contained in the rigid link analysis discussed above. In order to practically consider the links of the geometric complexity found in industrial manipulators, finite element methods must be used. However, standard finite-element software packages will not handle the nonlinear dynamic behavior of manipulators, and, as discussed later, special methods are required.

7.1.1 The general simulation structure

The basic structure of all three levels of simulation described above is shown in Figure 7.1. In this structure the data describing the system parameters and commands are input to the simulation program and all calculations that are not time-dependent are performed. In advanced simulation packages, these data are obtained by the computer directly from computer-aided design (CAD) models of the manipulator's elements. Then, with the user-furnished or default initial conditions represented by a vector $\mathbf{x}(t_0)$, the equations of motion of the system are evaluated. These equations are usually written in the well-known state variable form—typically, the joint-position state variables and those associated with the control systems. The equations of motion may also include algebraic relationships, because of the manipulator kinematics that must be solved before integration. The equations are usually highly nonlinear, because of the manipulator geometry and the drive actuator saturation characteristics.

The vector $\dot{\mathbf{x}}$ given by the state equation is then integrated using standard numerical forward-integration algorithms to obtain the state of the system at $t_0 + \Delta t$. This provides the new initial conditions to repeat the process, so the solution to the equations of motion advances in time (Korn, 1978) in the manner of classical forward interaction. At certain times during the solution the analyst may choose to have certain states displayed and to have the calculated program display critical parameters of the manipulator system, such as the error in the position and orientation of the end effort. Using computer graphics capabilities, the analyst may also see the position of the manipulator with respect to its environment.

Using the simulation, the analyst is able to evaluate the performance of the manipulator as a function of its design. In order to change a manipulator property the analyst usually need only change a number in the original simulation data. Hence, the simulation method provides a very effective technique with which to optimize the design.

7.1.2 Kinematic simulations

Kinematic simulations are simpler than dynamic simulations. However, they do not provide information about such factors as control system interactions, actuator requirements, or dynamic transients. Their main use is outlined below.

Kinematic simulations can be used for kinematic evaluation of the robot and its work-space environment. For example, it is possible, using a kinematic simulation, to determine the best locations within the work cell for the manipulator and its surrounding equipment to achieve maximum productivity. Figure 7.2 is a representation of a robotic work cell for painting aircraft. Using the computer to simulate this work cell, the analyst is able to determine the dimensions of the transportable robots, which will assure that every part of the aircraft is accessible to at least one of the arms.

Kinematic simulation can also be used to program manipulators offline. These simulations can be written to accept the manipulator program instructions and then demonstrate the motion that the manipulators will

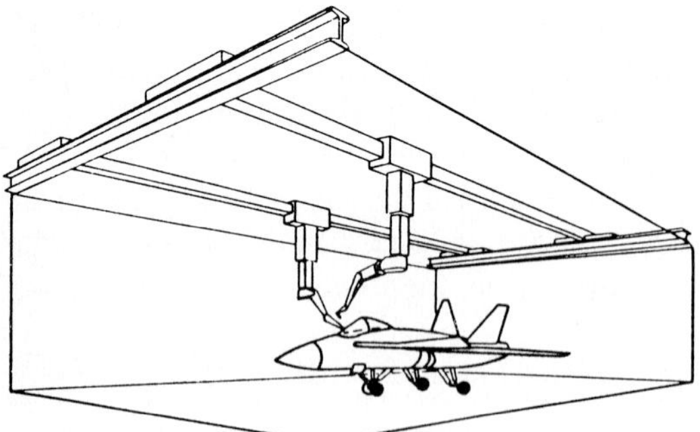

Figure 7.2 Manipulator work cell design based on kinematic considerations.

have when controlled by these instructions. By such means, the programmer can safely correct errors in the manipulator programs without damaging the arm, the work cell fixtures, or the machines and without endangering workers. By using the simulation program, the analyst does not need to remove the robotic work cell and its manipulator from production in order to test the programs for the next assignment. Such simulation is not a practical design tool, but it is useful in factories during production.

Kinematic simulation programs also permit the user to evaluate different robots for a given task and to design associated transfer machinery and fixtures. Such simulations probably will be the simulation used in computer-aided machining (CAM) systems to schedule work steps.

The kinematic positions and velocities of the manipulator can be obtained directly from the commanded joint angles using either classical geometry of 4×4 matrices, as discussed in Chapter 14 of this book and in Paul (1981). The robot kinematics can be described by a vector position equation that contains all the kinematic properties of the system. This equation has the form

$$\mathbf{W} = [\mathbf{T}_1] \, [\mathbf{T}_2] \, [\mathbf{T}_3] \, [\mathbf{T}_k]\mathbf{M} \tag{7.1}$$

where $\mathbf{W}$ = position vector in world coordinates of any point M fixed to the k link of the manipulator
$[\mathbf{T}_i]$ = homogeneous transformation matrices for each link in the manipulator
$\mathbf{M}$ = vector position of point M with respect to the manipulator $K + n$ link

The homogeneous coordinate transformations are the same 4×4 matrices used to control the manipulator. Given the commanded joint angles, the vector position equation can easily be solved for location of the manipulator end effector and for location of the manipulator links. The graphic displays can be generated using standard graphics software packages or specialized procedures for robotic manipulators. Examples of kinematic simulation results and procedures can be found in Soroka (1980) and Heginbotham et al. (1979).

7.1.3 Rigid-body dynamic simulations

In the kinematic simulations described above, the manipulator is assumed to faithfully follow its commanded trajectory. This form of analysis has value, but it is not the whole picture of manipulator performance. As manipulators move faster to achieve higher productivity, dynamic effects begin to dominate their performance. These effects can saturate the manipulator drives. Nonlinear dynamic coupling and inertial effects can also make the manipulator deviate significantly from the design trajectory, degrade transient performance with excessive overshoot, and destabilize the system. Finally, the discrete-time and time-delay characteristics of the manipulator control computer sometimes interact with the manipulator dynamics to further degrade and destabilize the system's performance.

These problems must be considered in the manipulator design. The drive motors and circuits must be sized to produce the desired performance. The links must be constructed to minimize the dynamic effects, and the control systems and the control computer must be designed to ensure dynamic performance over the entire work profile of the manipulator. In order to accomplish these goals, good analytical dynamic models or simulations of the manipulator are required.

The first step in developing the dynamic simulations for manipulators whose links may be considered rigid is to write the equation of motion of the system. Techniques for developing efficient analytical models are available. Methods for dealing with rigid systems have originated in the study of both mechanisms (Uicker, 1969; Woo and Freudenstein, 1971) and spacecraft dynamics (Hooker and Marguillies, 1965). The use of these methods results in essentially the same set of highly nonlinear, coupled, ordinary differential equations. Here the development of these equations will be outlined for a typical manipulator configuration using a classical lagrangian approach.

Figure 7.3 shows a kinematic configuration typical of a large class of manipulators. The system elements are assumed to be rigid and effects such as connection clearances and motor backlash are neglected. The joint angles and angular velocities are assumed measurable for control purposes.

The equations describing the motion of the manipulator are in terms of Lagrange's equation:

$$\frac{d\left(\dfrac{\partial T}{\partial \dot{\phi}_i}\right)}{dt} - \frac{\partial T}{\partial \phi_i} + \frac{\partial V}{\partial \phi_i} = Q_i \qquad i = 1, 2, \ldots, 6 \qquad (7.2)$$

where T and V are the kinetic and potential energies of the system and Q_i are the generalized forces. The kinetic energy of the manipulator may be written as

$$T = \tfrac{1}{2}m_1(\dot{\mathbf{R}}_1 \cdot \dot{\mathbf{R}}_1) + \tfrac{1}{2}m_2(\dot{\mathbf{R}}_2 \cdot \dot{\mathbf{R}}_2) + \tfrac{1}{2}m_p(\dot{\mathbf{R}}_p \cdot \dot{\mathbf{R}}_p)$$

$$+ \tfrac{1}{2}\omega_1^T \cdot \mathbf{I}_1 \cdot \omega_1 + \tfrac{1}{2}\omega_2^T \cdot \mathbf{I}_2 \cdot \omega_2 + \tfrac{1}{2}\omega_3^T \cdot \mathbf{I}_3 \cdot \omega_3 \qquad (7.3)$$

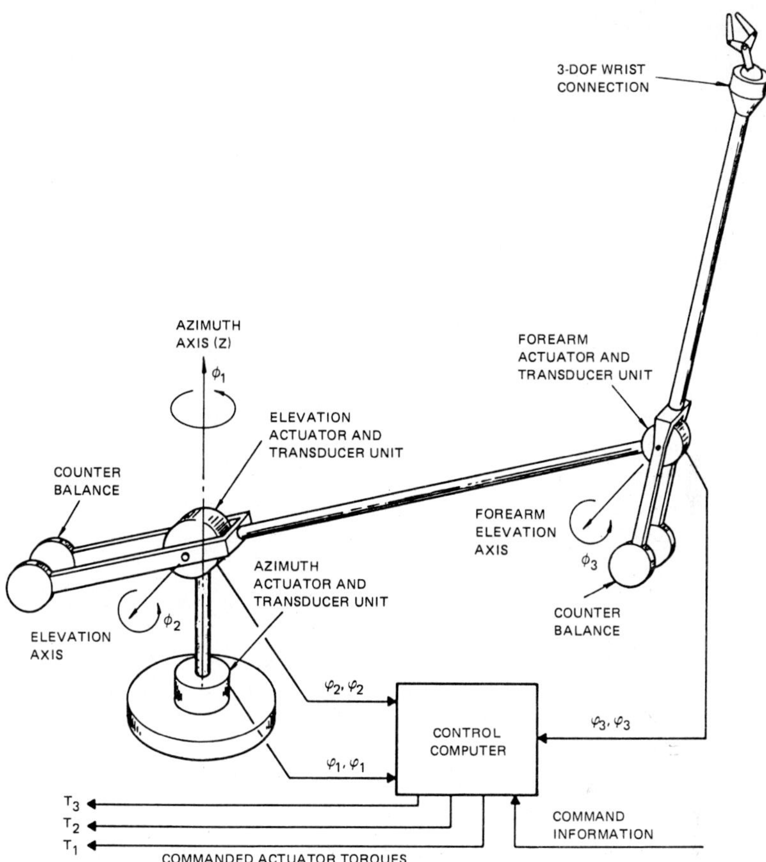

Figure 7.3 Typical manipulation configuration.

where $\mathbf{R}_1, \mathbf{R}_2, \mathbf{R}_p$ = position vectors to centers of mass of upper arm assembly (m_1), forearm assembly (m_2), and combined wrist and payload assembly (m_p)

$\mathbf{I}_1, \mathbf{I}_2, \mathbf{I}_3$ = inertia dynamics of assemblies

$\omega_1, \omega_2, \omega_3$ = inertial angular velocity vectors of assemblies

For this analysis the payload is assumed to be firmly grasped by the manipulator end effector, so the mass properties of the payload and wrist may be combined. It is also assumed that the wrist and payload dimensions are small compared to the other system-link lengths. Applying these assumptions results in $\mathbf{I}_3$ being small and the system dynamics being functions only of the generalized coordinates ϕ_1, ϕ_2, and ϕ_3, and not of the wrist joint angles. The angular velocities of the two major links can be written in terms of the generalized coordinates ϕ_1, ϕ_2, and ϕ_3 and their derivatives.

Using simple coordinate transformations, $\dot{\mathbf{R}}_1$, $\dot{\mathbf{R}}_2$, and $\dot{\mathbf{R}}_p$ can also be written in terms of selected generalized coordinates as

$$\dot{\mathbf{R}}_1 = \dot{\phi}_1 l_1^* \cos \phi_2 \, \hat{\mathbf{a}}_1 - \dot{\phi}_2 l_1^* \hat{\mathbf{a}}_3 \qquad \dot{\mathbf{R}}_2 = l_1(-\dot{\phi}_1 \cos \phi_2 \, \hat{\mathbf{a}}_1 + \dot{\phi}_2 \hat{\mathbf{a}}_3)$$

$$\dot{\mathbf{R}}_p = -\dot{\phi}_1[l_2 \cos \phi_2 + l_2 \cos (\phi_2 + \phi_3)]\hat{\mathbf{a}}_1 - [l_2(\dot{\phi}_2 + \dot{\phi}_3) \sin \phi_3]\hat{\mathbf{a}}_2$$
$$+ [l_2\dot{\phi}_2 + l_2(\dot{\phi}_2 + \dot{\phi}_3) \cos \phi_3]\hat{\mathbf{a}}_3 \qquad (7.4)$$

Substituting the expressions for ω_1, ω_2, and ω_3 and Equation (7.4) into (7.3) yields a rather lengthy expression for T that cannot be presented here due to space limitations. The potential energy term V in Equation (7.2) is due to the effects of gravity and can be written in terms of the generalized coordinates.

After computing the derivatives of T and V, as indicated in Equation (7.2), the resulting equations of motion of the manipulator system can be written in the form

$$\begin{bmatrix} m_{11} & m_{12} & m_{13} \\ m_{21} & m_{22} & m_{23} \\ m_{31} & m_{32} & m_{33} \end{bmatrix} \begin{bmatrix} \ddot{\phi}_1 \\ \ddot{\phi}_2 \\ \ddot{\phi}_3 \end{bmatrix} + \begin{bmatrix} g_{11} \dot{\phi}_1 & g_{12} \dot{\phi}_1 & g_{13} \dot{\phi}_1 \\ g_{21} \dot{\phi}_2 & g_{22} \dot{\phi}_2 & g_{23} \dot{\phi}_2 \\ g_{31} \dot{\phi}_3 & g_{32} \dot{\phi}_3 & g_{33} \dot{\phi}_3 \end{bmatrix} \begin{bmatrix} \dot{\phi}_1 \\ \dot{\phi}_2 \\ \dot{\phi}_3 \end{bmatrix}$$

$$= \begin{bmatrix} \overline{Q}_1 \\ \overline{Q}_2 \\ \overline{Q}_3 \end{bmatrix} \qquad (7.5)$$

where m_{ij} are the elements of the effective mass matrix and nonlinear functions of the generalized coordinates. This matrix is symmetrical. A typical term is

$$m_{12} = m_{21}I_{13}^1 \sin \phi_2 + I_{23}^1 \cos \phi_2 + I_{13}^2 \sin (\phi_2 + \phi_3)$$
$$+ I_{23}^2 \cos (\phi_2 + \phi_3) \qquad (7.6)$$

where I_{ij}^k is the i,j element of the inertia dyadic for the kth link. The elements g_{ij} are not symmetric and are nonlinear functions of the system generalized coordinates. For example,

$$g_{12} = -2(m_1 l_1^{*2} + m_2 l_1^2) \sin \phi_2 \cos \phi_2 - 2m_p[l_1 \cos \phi_2$$

$$+ l_2 \cos (\phi_2 + \phi_3)][l_1 \sin \phi_2 + l_2 \sin (\phi_2 + \phi_3)]$$

$$+ 2(I_{11}^1 - I_{22}^1) \sin \phi_2 \cos \phi_2 - 2(I_{12}^1)(\sin^2 \phi_2 - \cos^2 \phi_2)$$

$$+ 2(I_{11}^2 - I_{22}^2) \sin (\phi_2 + \phi_3) \cos (\phi_2 + \phi_3)$$

$$- 2(I_{12}^2)[\sin^2 (\phi_2 + \phi_3) - \cos^2 (\phi_2 + \phi_3)] \qquad (7.7)$$

and

$$g_{21} = 0 \tag{7.8}$$

The $\overline{Q}_i$ terms arise from the differentiation of T and V and from the generalized forces. The generalized forces Q_i are functions of the command motor torques applied at the joints, as well as any nonconservative torques and forces that may arise from friction. The generalized forces are obtained from the relationship

$$Q_i = \sum_{i=1}^{n} \mathbf{T}_i \cdot \frac{\partial \omega_i^!}{\partial \dot{\phi}_i} \tag{7.9}$$

where $\omega_i^!$ is the inertial angular velocity of the system at the point of application of the torque $\mathbf{T}_i$. The motor torques are generally applied at the connection between two links and hence act on two of the manipulator links.

A typical $\overline{Q}$ term is

$$\overline{Q}_1 = T_1(t) - (I_{12}^1 \cos \phi_2 - I_{23}^1 \sin \phi_2)\dot{\phi}_2^2$$
$$- [I_{13}^2 \cos (\phi_2 + \phi_3) - I_{23}^2 \sin (\phi_2 + \phi_3)](\dot{\phi}_2 + \dot{\phi}_3)^2 \tag{7.10}$$

where $T_1(t)$ is the combined servomotor torque and friction acting at the shoulder joint about the azimuth (ϕ_1) axis. This torque may be a complex nonlinear function including such factors as motor saturation effects. The discrete-time and time-delay characteristics of the control computer would enter the analysis through the T_i terms.

The terms given above exemplify the highly nonlinear, complex, and coupled nature of the manipulator dynamics. Equation (7.5) is the basic dynamic equation of motion for this example and can be transformed to the state variable form required for the simulation structure shown in Figure 7.1.

An example of the results that can be obtained using this model is shown in Figure 7.4. The simulation results clearly demonstrate the significant effect the mass properties of the manipulated object can have on the dynamic performance of the system. In this figure the arm, without any payload, is first commanded to swing to the ϕ_1 axis from 0° to 22.5° in 1 s at a constant velocity. The arm remains in that position for 3 s and then returns to its original station at a constant velocity. This simple maneuver is typical of many industrial manipulator tasks. During this action ϕ_2 and ϕ_3 remain at 0° and the arm moves in the xy plane as shown in Figure 7.5. The motion of the arm is well behaved: It closely follows the desired behavior given by a design model during this first cycle. Just before the second cycle, the arm grasps a substantial payload, with dra-

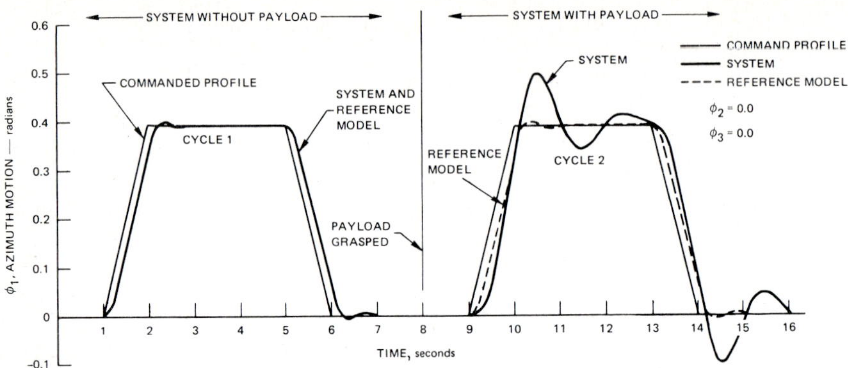

Figure 7.4 Manipulator performance as predicted by rigid link simulation.

matic effect; during the second cycle the arm performance is seriously degraded and deviates substantially from the design model. During the dwell when useful work might be performed on the manipulated object, the manipulator continues to swing about the ϕ_1 axis. Using modern adaptive control techniques and this simulation, it has been possible to develop a control algorithm that will compensate for the undesirable effects of unknown payloads and will produce uniformly good dynamic performance over the entire work space (Dubowsky and DesForges, 1979; Dubowsky, 1981).

The results obtained for this system, which are discussed above, could not be generated using analytical methods and clearly show the importance of dynamic simulations in the design of robotic manipulators.

7.1.4 Simulations of manipulators with flexible links

The method of dynamic analysis discussed above is for systems whose links are assumed to be ideally rigid. The problems associated with manipulator dynamics are compounded by flexibility in the manipulator links, connections, and drive trains, which can result in high-frequency resonances. These resonances have been found to interact with the discrete-time control computers to degrade system performance and increase system instability (Luh and Kinne, 1965). This problem is especially critical for large manipulators and for manipulators that operate at high speeds. The inertial forces acting on large or high-speed manipulators can also cause deformations that result in significant position errors at the end effector and large inertial stress within the mechanical components.

It might appear that the solution to the problem is simple: Make manipulators more rigid. If we make the manipulator's elements more massive, however, we produce much greater internal loads for the manip-

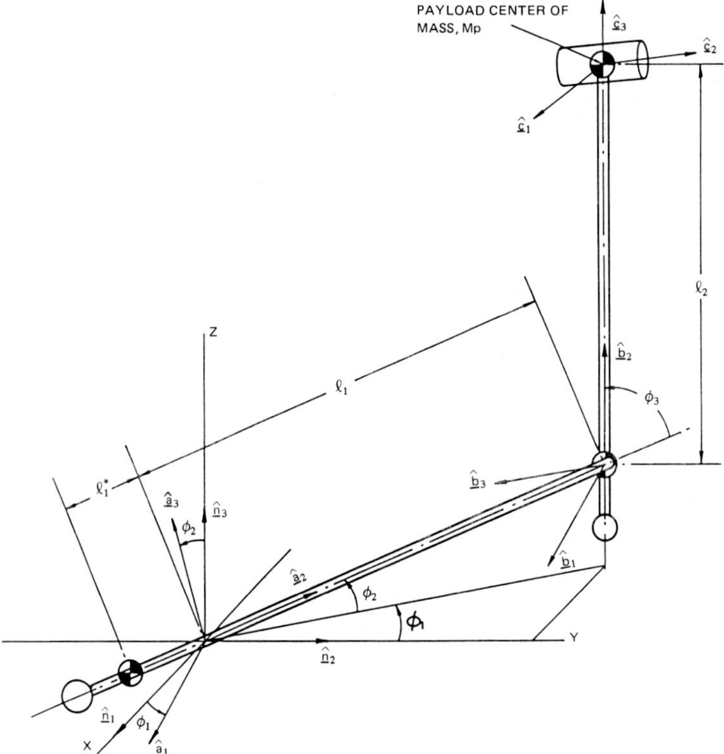

Figure 7.5   Manipulator dynamic schematic.

ulator's actuators, thus requiring larger and more massive actuators. These more massive elements reduce the structural frequencies, in part defeating the purpose for which the massive robot manipulator links were intended. One possible solution is to develop new manipulator materials that are both rigid and lightweight, such as fiber composites, but significant research is required to make such materials practical for industrial robotic manipulators because of

- The current high cost of the materials
- The problem of mounting end connections and actuators on fiber composite links, without significantly reducing their strength

Another possible approach is to exploit flexibility for high-speed motion by designing ultralightweight manipulators, much as a fly angler uses the flexibility of the fishing rod to extend the distance of the cast. However, such high-speed flexible manipulators would require control algorithms that are more advanced than any in use today. Methods to control highly flexible systems have been developed for spacecraft control (Balas, 1978;

Sesak et al., 1979), but these systems are far easier to control than structurally flexible manipulators, which exhibit very complex, nonlinear behavior. To develop such control methods for robots, good analytical dynamic models (simulations) of flexible manipulators will be required. In the near term, the best manipulator designs will be constructions that are rigid and yet lightweight. To design such structures requires accurate simulations that can predict the dynamic performance of the manipulator, including the mechanical flexibility.

7.2   A Method for Modeling Dynamic Behavior

Outlined below is a method for modeling the dynamic behavior of manipulators, including the effects of distributed mass, flexibility, and nonlinear dynamic nature. This method also considers the characteristics of the manipulator's drive system actuators and control policies and is formulated so as to be applicable to manipulators with the complex-shaped links commonly used in industry. The method's computation costs are relatively low, and the method's application can be sufficiently automated, making it a practical design tool. In this method, the individual link dynamic models are obtained from standard finite-element analysis. These link models are combined with both the descriptions of the manipulator joints, expressed by 4 × 4 matrices, and the system control policies into a complete dynamic description of a flexible manipulator system. By using standard finite-element procedures, the analyst is relieved of the very time-consuming task of three-dimensional geometric modeling of the links.

Existing finite-element programs will not, by themselves, handle the nonlinear dynamic nature of robot manipulators. A computer software package called FLEX-ARM (*FLEX*ible *A*nalysis of *R*obotic *M*anipulators) has been developed to mechanize this analytical method (Sunada and Dubowsky, 1983); see Figure 7.6. The method is developed by first writing Lagrange's equations for the ith manipulator link as

$$\frac{d}{dt}\left[\frac{\partial(T_i)}{\partial \dot{p}_{i\alpha}}\right] - \frac{\partial(T_i)}{\partial p_{i\alpha}} + \frac{\partial(V_i)}{\partial p_{i\alpha}} = \tilde{f}_{i\alpha} \qquad \alpha = 1, \ldots, NP(i) \qquad (7.11)$$

where $\tilde{f}_{i\alpha}$ = generalized forces (forces and torques external to the link
$\quad T_i, V_i$ = ith link kinetic and potential energies
$\quad\quad p_{i\alpha}$ = generalized coordinates of the ith link
$\quad NP(i)$ = total number of degrees of freedom of the ith link

To represent the manipulator's spatial motion, 4 × 4 matrices are used (Uicker, 1969). In this well-known method, the transformation between

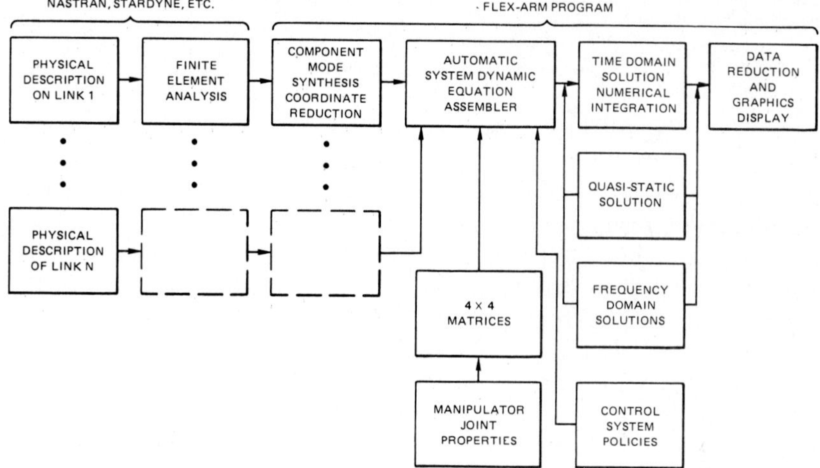

Figure 7.6. Block diagram of flexible manipulator simulation program.

any two coordinate systems, $\mathbf{T}^i_{i-1}$, can be obtained by the multiplication of a set of appropriate 4×4 transformation matrices. In particular, the transformation matrix from the ith frame to the inertial frame will be called $\mathbf{T}^i_0$.

The kinetic energy term in Equation (7.11) requires the inertial velocity of any point P in the manipulator. This can be written

$$\mathbf{v}_i = \sum_{j=1}^{i} \mathbf{U}^\Theta_{ij}\mathbf{r}_i\dot{\Theta}_j + \sum_{j=1}^{i} \mathbf{U}^H_{ij}\mathbf{r}_i H_j + \mathbf{T}^i_0\dot{\mathbf{r}}_i \qquad (7.12)$$

where

$$\mathbf{U}^\Theta_{ij} = \frac{\partial \mathbf{T}^i_0}{\partial \Theta_j}$$

$$\mathbf{U}^H_{ij} = \frac{\partial \mathbf{T}^i_0}{\partial H_j} \qquad (7.13)$$

where Θ is the angle of revolute joints, and H is the displacement for sliding joints. The first two terms in Equation (7.12) are functions of the manipulator joint's relative velocities $\dot{\Theta}_j$ and H_j. The third term is the point's velocity relative to the ith coordinate frame as observed from inertial space. The last term is not present in the rigid link case. The partial derivatives of the matrix $\mathbf{T}^i_0$ in Equation (7.13) can be replaced by simple and convenient matrix multiplication operations. For example:

$$\mathbf{U}^\Theta_{ij} = \begin{cases} \mathbf{T}^{j-1}_0\mathbf{Q}^\Theta\mathbf{T}^i_{j-1} & j \le i \\ 0 & j > i \end{cases} \qquad (7.14)$$

where

$$\mathbf{Q}^{\Theta} = \begin{bmatrix} 0 & 0 & 0 & 0 \\ 0 & 0 & -1 & 0 \\ 0 & 1 & 0 & 0 \\ 0 & 0 & 0 & 0 \end{bmatrix} \tag{7.15}$$

A similar expression for $\mathbf{U}_{ij}^{H}$ can be written. For simplicity, it is assumed here that the manipulator is composed of only revolute connections; the Θ superscript will be dropped. Higher-order partial derivatives of the $\mathbf{T}_{i-1}^{i}$ matrices, with respect to Θ_i and H_i, can also be written in terms of the $\mathbf{Q}$ matrices. These terms will appear in the final equations.

The elastic and mass properties of complex-shaped links are most easily described by finite-element procedures. In these methods each link is divided into a set of elements connected at various grid points. It has been found that the numerical computations are most efficient if their motion is expressed as small perturbations about some known nominal motion of the system, such as those obtained from the commanded motion assuming a rigid system. For spatial systems there are six scalar perturbation coordinates (three translation and three rotation) for each grid point on the ith link. If a link has $NG(i)$ grid points, then there will be six $NG(i)$ or $NP(i)$ scalar coordinates called $p_{i\beta}$. The first six coordinates (p_{i1} to p_{i6}) are the three elastic translational displacements and the three elastic rotations of the first grid point with respect to the nominal motion of the ith link. Similarly, all $NP(i)$ coordinates are defined. These $p_{i\beta}$ variables will be generalized coordinates used in Lagrange's equation (7.11) for the ith link.

The 4×1 deformed position vector of the gth grid point on the ith link with respect to the local coordinate system of the link is given by

$$\mathbf{r}_{ig} = \sum_{\beta=1}^{NP(i)} \phi_{ig\beta} p_{i\beta} + \mathbf{b}_{ig} \tag{7.16}$$

The constant 4×1 vectors $\phi_{ig\beta}$ are defined by

$$\phi_{ig\beta} = \begin{cases} [0 \quad 1 \quad 0 \quad 0]^T & \text{for } \beta = 1 + 6(g-1) \\ [0 \quad 0 \quad 1 \quad 0]^T & \text{for } \beta = 2 + 6(g-1) \\ [0 \quad 0 \quad 0 \quad 1]^T & \text{for } \beta = 3 + 6(g-1) \\ [0 \quad 0 \quad 0 \quad 0]^T & \text{for all other } \beta \end{cases} \tag{7.17}$$

$\mathbf{b}_{ig}$ is the undeformed position vector of the gth grid point.

Differentiating Equation (7.16) and substituting it into Equation (7.12) yields the inertial velocity of the gth grid point required by T_i in Equation (7.11). The kinetic energy for all grid points on the ith link T_i is defined by

$$T_i = \sum_{g=1}^{NG(i)} \tfrac{1}{2} m_{ig} Tr(\mathbf{v}_{ig} \mathbf{v}_{ig}^T) \tag{7.18}$$

Note that m_{ig} is the mass of the gth grid point and Tr indicates the trace operation. Details of these operations are presented in Sunada and Dubowsky (1983), where it is also shown that the potential energy v_i for the ith link associated with structural elasticity, assuming small elastic deformations $p_{i\beta}$, has the linear form

$$V_i = \sum_{\beta=1}^{NP(i)} \sum_{\gamma=1}^{NP(i)} \tilde{k}_{i\beta\gamma} p_{i\beta} p_{i\gamma} \tag{7.19}$$

where $\tilde{k}_{i\beta\gamma}$ is the β,γ element of the structural stiffness matrix for the ith link $\mathbf{k}_i$.

The dynamic equations of motion for the ith link, obtained from Equation (7.11), can then be written in scalar form as

$$\sum_{\beta=1}^{NP(i)} m_{i\alpha\beta}\ddot{p}_{i\beta} + \sum_{\beta=1}^{NP(i)} g_{i\alpha\beta}\dot{p}_{i\beta} + \sum_{\beta=1}^{NP(i)} k_{i\alpha\beta} p_{i\beta} = f_{i\alpha}$$
$$\alpha = 1, \ldots, NP(i) \tag{7.20}$$

In their general form, coefficients in Equation (7.20) are time-dependent functions. However, a careful analysis of these terms shows they can be obtained from the sum of constant matrices multiplied by a few time-dependent scalar factors (Sunada and Dubowsky 1983). This property is the key to the practical use of the method.

Equation (7.20) represents a very large set of equations, six for each grid point in the link. The numerical solution to such a large set would be prohibitively expensive. To solve this problem, a method called *component mode synthesis* (CMS) is applied (Hurty, 1965). CMS allows a richly detailed finite-element model of the link that gives good dynamic accuracy, yet is represented by a small number of equations.

In the CMS method, Equation (7.20) is rewritten in matrix form as

$$\mathbf{m}_i\ddot{\mathbf{p}}_i + \mathbf{g}_i\dot{\mathbf{p}}_i + \mathbf{k}_i\mathbf{p}_i = \mathbf{f}_i \tag{7.21}$$

and a transformation matrix $\mathbf{A}_i$ is developed so that

$$\mathbf{p}_i = \mathbf{A}_i\mathbf{a}_i \tag{7.22}$$

where $\mathbf{a}_i$ is the reduced set of coordinates describing the motion of the ith link (Sunada and Dubowsky, 1983). Applying this transformation to Equation (7.21) results in a much smaller set of dynamic equations for the ith link, which can be written in matrix form as

$$\mathbf{M}_i\ddot{\mathbf{a}}_i + \mathbf{G}_i\dot{\mathbf{a}}_i + \mathbf{K}_i\mathbf{a}_i + \mathbf{f}_i^a \tag{7.23}$$

where $\mathbf{M}_i$, $\mathbf{G}_i$, and $\mathbf{K}_i$ are obtained by premultiplying $\mathbf{m}_i$, $\mathbf{g}_i$, and $\mathbf{k}_i$, respectively, by $\mathbf{A}_i^T$ and postmultiplying them by $\mathbf{A}_i$. The $\mathbf{f}_i^a$ term is obtained by premultiplying $\mathbf{f}_i$ by $\mathbf{A}_i^T$. Using this method it has been shown that for a typical manipulator link, the number of required equations can be reduced from 264 to 12 without a significant loss in dynamic accuracy.

The last step in the formulation of the equations of motion is the assembly of the reduced individual link equations (7.23) into a set of system equations. In this process the interlink constraint forces are eliminated from the force vector $\mathbf{f}_1^a$. This step is performed using compatibility matrices based on the kinematic nature of the manipulator's connections. The matrices are used to transform the reduced link coordinates $\mathbf{a}_i$ into a set of independent system (global) coordinates $\mathbf{q}$; the $\mathbf{q}_i$ can be chosen to be an independent set of coordinates for which displacement and slope compatibility are always maintained between links.

The relationship between the system coordinates and the link-reduced coordinates can be written from simple geometry in terms of the manipulator joint angles. Assume that the joint angles can be approximated by their normal values. These relationships can be written in matrix form as

$$\mathbf{a}_i = \mathbf{B}_i(\Theta_j)\mathbf{q} = \mathbf{B}_i\mathbf{q} \tag{7.24}$$

$\mathbf{B}_i$ is called the compatibility matrix for the ith link and is, in general, a time-varying function of the nominal joint angles Θ_j. Conventional finite-element applications for stationary structures do not require time-varying compatibility matrices. Taking the first and second time derivatives of $\mathbf{a}_i$, substituting them into Equation (7.23), applying the principle of virtual work, and summing over all the manipulator links yields

$$\mathbf{M}\ddot{\mathbf{q}} + \mathbf{G}\dot{\mathbf{q}} + \mathbf{K}\mathbf{q} = \mathbf{Q} \tag{7.25}$$

where

$$\mathbf{M} = \sum_{i=1}^{NL} \mathbf{B}_i^T \mathbf{M}_i \mathbf{B}_i \tag{7.26}$$

$$\mathbf{G} = \sum_{i=1}^{NL} \mathbf{B}_i^T \mathbf{G}_i \mathbf{B}_i + \sum_{i=1}^{NL}\sum_{j=1}^{NL} 2\mathbf{B}_i^T \mathbf{M}_i \mathbf{B}_{ij}\dot{\Theta}_j + \mathbf{G}_d \tag{7.27}$$

$$\mathbf{K} = \sum_{i=1}^{NL} \mathbf{B}_i^T \mathbf{K}_i \mathbf{B}_i + \sum_{i=1}^{NL}\sum_{j=1}^{NL} \mathbf{B}_i^T \mathbf{G}_i \mathbf{B}_{ij}\dot{\Theta}_j \tag{7.28}$$

$$+ \sum_{i=1}^{NL}\sum_{j=1}^{NL}\sum_{k=1}^{NL} \mathbf{B}_i^T \mathbf{M}_i (\mathbf{B}_{ijk}\dot{\Theta}_j\dot{\Theta}_k + \mathbf{B}_{ij}\ddot{\Theta}_j)$$

$$\mathbf{Q} = \sum_{i=1}^{NL} \mathbf{B}_i^T \mathbf{f}_i^a \tag{7.29}$$

$$\mathbf{B}_{ij} = \frac{\partial \mathbf{B}_i}{\partial \Theta_j} \qquad \mathbf{B}_{ijk} = \frac{\partial \mathbf{B}_{ij}}{\partial \Theta_k} \tag{7.30}$$

and NL = number of links in the manipulator.

Equations (7.25) to (7.30) describe the complete dynamic behavior of the elastic manipulator. The above assembly of these equations may be done in symbolic form, using the computer because of the special form of matrix terms (see Figure 7.6).

The influences of the gravity loading on the system, the control system torques, and the joint friction enter the formulation through the generalized forces $\mathbf{Q}$. It is a relatively straightforward task to construct subroutines that provide these external forces as a function of time and the system state. Such effects as actuator nonlinearities and the discrete-time nature of a control computer are also easily included in the analysis in this way. The resulting dynamic equation of motion may be used in a number of different ways (Sunada and Dubowsky, 1983). An important use is the simulation structure shown in Figure 7.1, after transformation to state-space form.

An example of an industrial manipulator to which this analysis has been applied is shown in Figure 7.7. The manipulator is a six-degree-of-free-

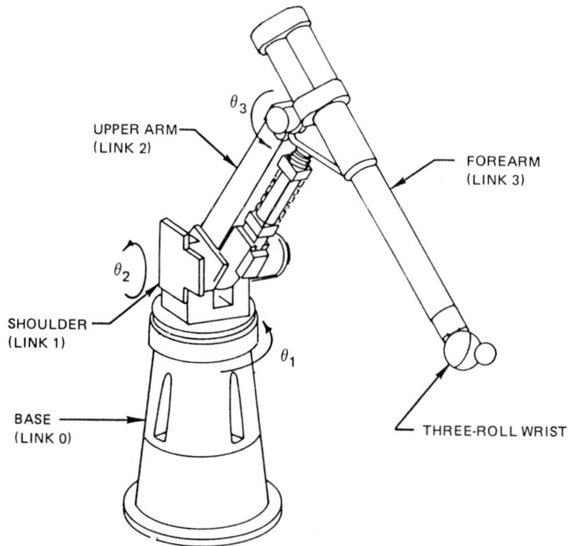

Figure 7.7 An industrial manipulator with significant mechanical flexibility.

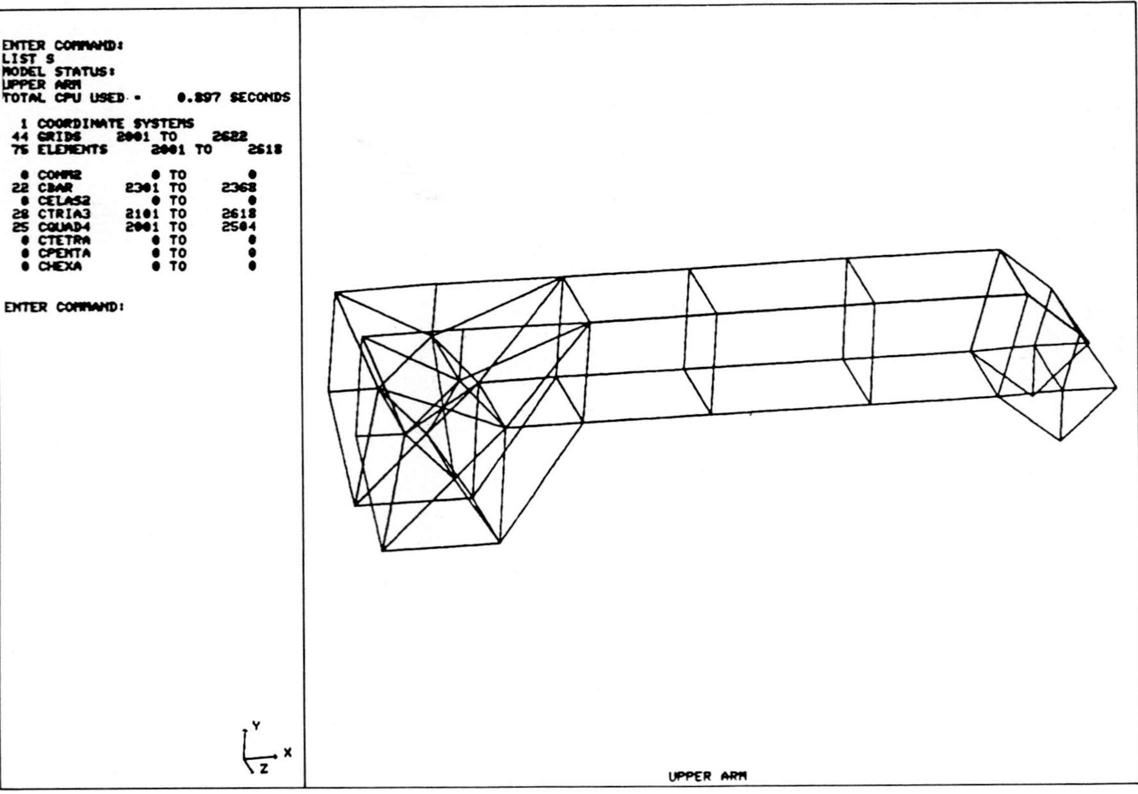

Figure 7.8   A detailed finite element model of an individual manipulator link.

dom (DOF) system. It has a reach of over 2.75 m (9 ft), weighs more than 1800 kg/m (4000 lb/m) and can carry a 23-kg (50-lb) payload to a maximum velocity of 1.27 m/s (50 in/s). For the analysis, the manipulator was assumed to be attached to a compliant, 20-cm (8-in) thick concrete floor. (The thickness of the concrete floor had a significant effect on the manipulator's behavior.) All the links, including the base, were assumed to be flexible, as were the bearings, which included both thrust and radial compliances. The analysis also included detailed models of the dynamic characteristics of the drive motors and the control system's filters and compensation networks.

Following the procedure shown in Figure 7.6, a finite-element model of each of the individual links was developed using NASTRAN (the finite-element model of the upper arm is shown in Figure 7.8). The combined finite-element model for the entire system is shown in Figure 7.9. The links contain a total of 155 grid points and 272 elements, which correspond to 930 mechanical DOFs, in addition to those associated with the drive and control systems. These DOFs were reduced to approximately 45 by applying the CMS reduction procedure to each of the individual links.

The compatibility matrices for the system were written directly from the manipulator geometry. The interface generalized coordinates for this system are shown in Figure 7.10. Descriptions of the drive actuator dynamics and control circuits were automatically integrated into the equations of motion by the software.

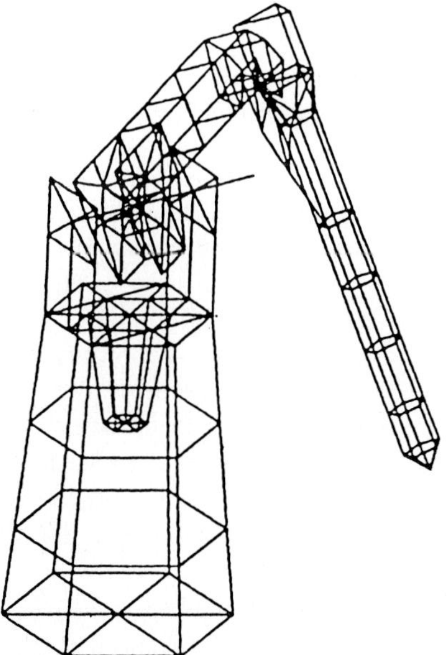

Figure 7.9   Finite element model of complete manipulator.

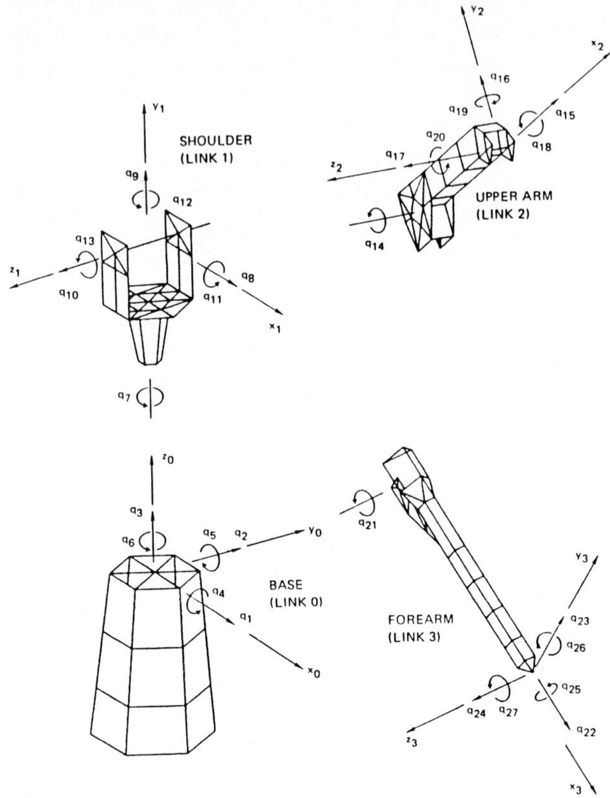

Figure 7.10   Interface global coordinate for manipulator system.

The final nonlinear dynamic equations representing the complete system were integrated numerically to obtain the time-domain response as required by Figure 7.1. It was also relatively easy to generate the natural frequencies (eigenvalues) and mode shapes (eigenvectors) for a linearized system model for small motions of the manipulator about any position in its working range (see Figure 7.6). This information proves valuable for comparison with experimental data to verify the analytical model.

The simulation developed using this method can be used to study the effects of flexibility during large high-speed motions of the arm. In these studies the initial and final positions of the tip manipulator are specified in three-dimensional space and the tip is commanded to follow a straight line between these points. An example is shown in Figure 7.11. The move shown corresponds to a maneuver in which the manipulator picks up an object on the floor and moves it to a high position on its left.

Figure 7.11 is an example of computer graphics system output, which is useful in these simulation programs. Such output has been used to produce animated displays that can give the designer insights into the flexible

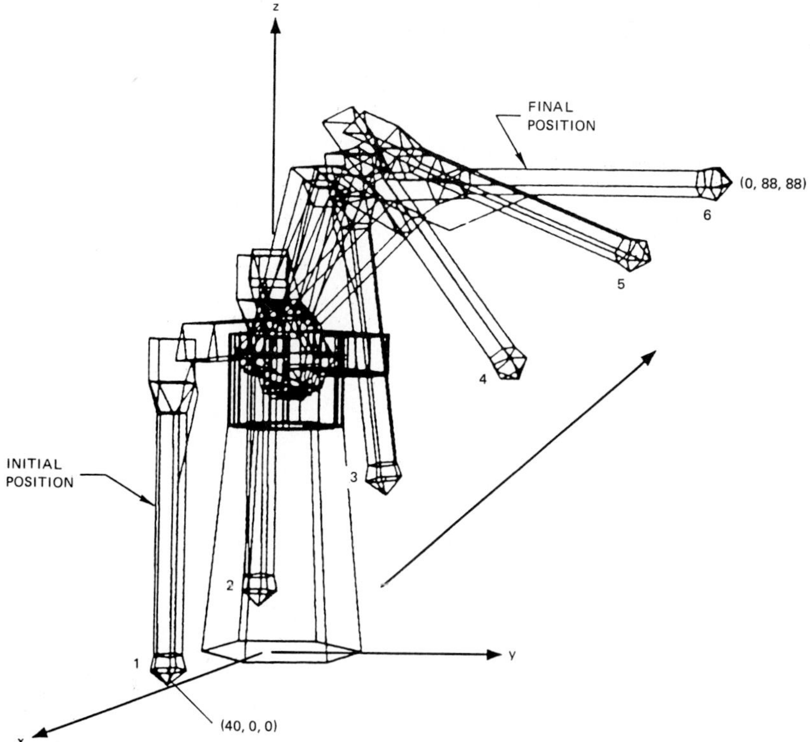

Figure 7.11   Computer graphics representation of manipulator maneuver.

motion of the manipulator. For the motion shown in Figure 7.11, the tip
has a constant acceleration for 0.25 s up to a constant velocity of 1.27
m/s (50 in/s). The tip then moves at this selected constant velocity along
its straight-line path and decelerates in 0.25 s to a stop at the final point.
Because of the complex manipulator geometry, the individual joints
undergo complex motions for the simple straight-line movement of the
manipulator end.

The simulation results show the importance of manipulator flexibility
(see Figure 7.12). The tip error for the maneuver shown in Figure 7.11 is
plotted in Figure 7.12 as a function of time for the manipulator with rigid
and flexible links. Relatively large starting and stopping transients are
seen in both cases. For the rigid system, these transients are due to actuator and control system dynamics. In the flexible case, these transients are
substantially larger, because of the addition of link elasticity. Although
the error during the straight-line motion may not be critical, the time
required for errors to die out at the end of a maneuver may be important.
A goal for this manipulator was to have a repeatability of ±0.5 mm (0.20
in). Taking one-half of this value as a goal for dynamic errors, the rigid

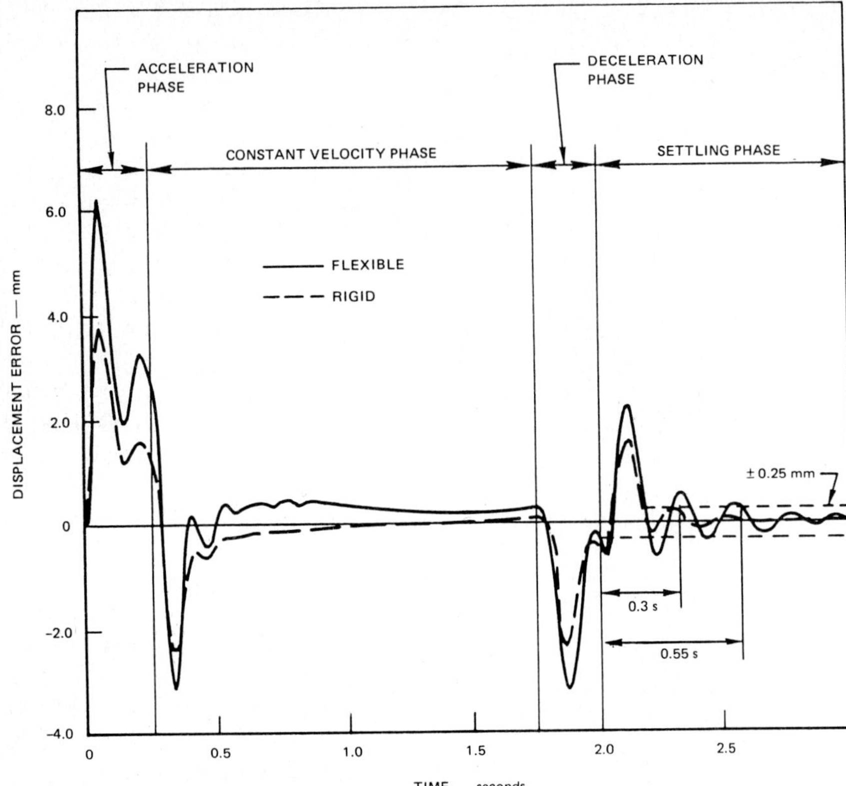

Figure 7.12 Simulation-predicted tip error for rigid and flexible manipulator models.

arm requires 0.30 s to achieve this tip error and the flexible arm requires 0.55 s, or nearly twice the time. This difference is important for applications where speed and performance are critical. It should be noted that this system is rugged—and thus relatively stiff. One would expect even larger flexible effects in high-speed, lightweight manipulators. In addition to such displacements as the tip error, the analysis also gives such useful design information as the stresses in the links and bearings.

Digital Electronics and Electrical Design Methods

8.1 Considerations and Alternatives

8.1.1 Robot control using microcomputers

The trend in robot control is toward distributed microcomputer systems. For example, in the Unimation PUMA assembly robots, the top-level microcomputer is the LSI-11/2 with up to 16K bytes of RAM; each of the five servo arm joints is controlled by a 6205 microcomputer. The VAL language for programming the LSI-11 is stored in EPROM.

Intelledex offers a robot controller using the Intel microprocessor pair 8086/8087 plus two microprocessors per motor (joint). The robot controller also utilizes an Intel 8088 to process sensor data. Other manufacturers using multiple processors for robot control include Adept Technology and Automatix. It is customary to use one microprocessor for overall surveillance and safety sensor alarming. The coming generation of controllers will make use of 32-bit processors to further improve the speed of operations of the robot.

A few robot manufacturers (Westinghouse, Olivetti, and Bendix) supply a digital controller for their robots that is adapted from a standard numerically controlled machine tool control unit.

TABLE 8.1 Some Control Systems

System	CPU	RAM	ROM	Programming language	Input/output	Peripherals
Adept One	Microprocessor M68000 (32 bit)	1000K bytes	10,700K bytes	VAL II	Parallel computer interface, 8-bits wide, IEEE 488, 16 binary, RS232 serial (5)	CRT terminal, floppy disk (250K bytes), vision system (2)
Automatix AI32	M68000 (2) (32 bit)	152K bytes	NA*	RAIL	RS232C (2) 16 input/output lines	Tape cartridge, vision system
Cincinnati-Milacron (Acramatic, version 4)	Minicomputer (16 bit with 8-bit data paths)	48K bytes 1.1-μs cycle	NA	NA	RS232C RS422	CRT terminal, cassette tape
Cybotech (RC5)	Microprocessor	4000 points (256 trajectories)	NA	LPR	16 input, 16 output	Tape cassette, "syntaxer"
General Electric	Minicomputer (LSI-11/2, 16 bit + 8080)	NA	NA	HELP (Pascal-type)	16 sensor inputs	Keyboard printer, Terminet 2030
IBM (7552)	Intel 80286 (16 bit)	512K bytes	NA	MAP	NA	Disk(s), vision system
Intelledex	8086/8087 + 8088	896K bytes	128K bytes	Robot Basic	RS232C (2)	Teach pendant, vision system
Prab (Model 700)	Microprocessor (16 bit)	32K bytes (128 programs)	NA	NA	32 input, 32 output	Teach unit, vision system
Unimation (Puma)	Microcomputer (LSI-11/2, 16 bit)	16K bytes (32K bytes CMOS)	16K bytes (battery backup CMOS)	VAL or VAL II	10 input, 8 output	Manual control ("teach box"), CRT or printing terminal, floppy disk
Motoman (RX CNC)	Microcomputer	64K bytes standard; 128K bytes optional	NA	MAP	48 input, 24 output RS232C	Teach pendant, printer cassette, CRT
Nachi Uniman 8000 (AK)	NA	1000 points in wire memory	NA	NA	16 input, 16 output	Teach box
Reis Robot Star II	Microprocessor M68000	7000 points	NA	NA	48 input, 32 output	Floppy disk, vision system

*NA = Information not available.

IBM's 7552 Industrial Computer using the Intel 80286 processor is physically rugged enough to operate in a factory environment (without refrigeration).

Prab Robotics robots have several different digital controllers. Their Model 700 controller uses an LSI-11/2 microcomputer to control up to seven servo axes at once on the Model F robots. It can handle up to 128 programs. It also provides 32 input and 32 output lines to interface with other devices in the automation system. The controller can be programmed with a handheld teaching unit or directly through a keyboard mounted on the main printed circuit board.

In Japan, there is a distinct trend toward the use of 16-bit microcomputers for robot control. Kawasaki incorporated 16-bit microcomputers in the PUMA series in 1982. Other manufacturers which use 16-bit microcomputers include Fujitsu Fanuc Ltd., Hitachi Ltd., Mitsubishi Heavy Industries Ltd., and Tokiko Ltd.

A summary of a few currently available robot control systems is given in Table 8.1.

8.1.2   Special-purpose digital logic

The coming trend in digital electronics for robot control will probably be custom LSI chips [application-specific integrated circuits (ASIC)] designed specifically for robot use. Functions such as the PID algorithms for servomotor control could operate faster in a custom logic form than in a general-purpose microcomputer. The same is true of joint coordinate conversion, calculations that now occupy much computation time.

Two competitive ways of designing special-purpose digital logic for robot control are custom logic arrays (ASIC) and programmed logic arrays (PLAs). The cost and time required for development are greater with ASIC designs than with PLAs. Computer-aided design is almost mandatory for either approach to keep the time and cost within limits. An approach, used by several manufacturers, is to use PLAs for the first production of a new product until field experience confirms the detailed logic required and then to change over to full custom logic for lowest cost and highest performance. Table 8.2 contains a brief listing of ASIC and PLA suppliers.

8.2   Motor Drive Requirements

The digital control logic of an electric robot must be connected to the dc-drive motors by an appropriate power control interface. This interface usually converts the logic-level signals from the controller to pulse-width-

TABLE 8.2   Programmable Logic Arrays and Asic Houses

Name	ASIC	PLA
Advanced Micro Devices		●
Fairchild	●	●
Gould AMI		●
Harris Semiconductor	●	●
Honeywell Solid State		●
Interdesign		●
Intersil	●	
LSI Logic Corporation	●	●
Monolithic Memories, Inc.		●
Motorola	●	●
Signetics	●	●
RCA Solid State	●	
Silicon Systems, Inc.	●	
Texas Instruments	●	●
VLSI Technology	●	
National Semiconductor	●	

modulated (PWM) signals at a level appropriate for the particular motor being driven. PWM techniques are used because of the higher efficiency of the driving circuits.

The semiconductors used in the motor control amplifier can be power MOSFETs, which have nearly ideal switch characteristics, i.e., very low drive power requirements (compatible with logic ICs), fast switching, low off-state leakage, and low on-state resistance. The fast-switching characteristic reduces power dissipation in the device while it is changing from the on state to the off state, or vice versa. MOS devices can also be easily run in parallel to obtain larger current capacities than are available in

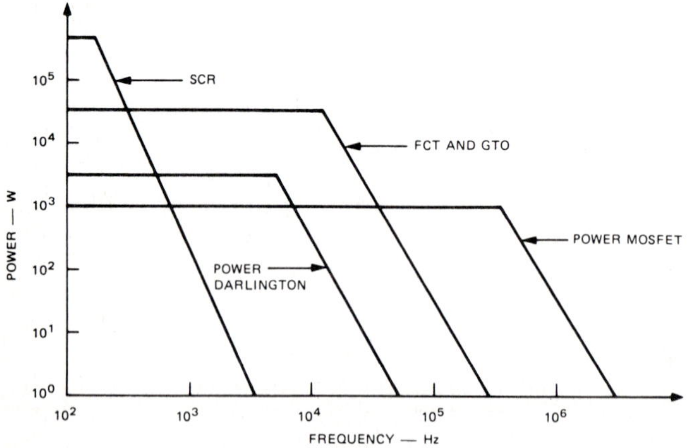

Figure 8.1 Power-switching capabilities of selected semiconductors.

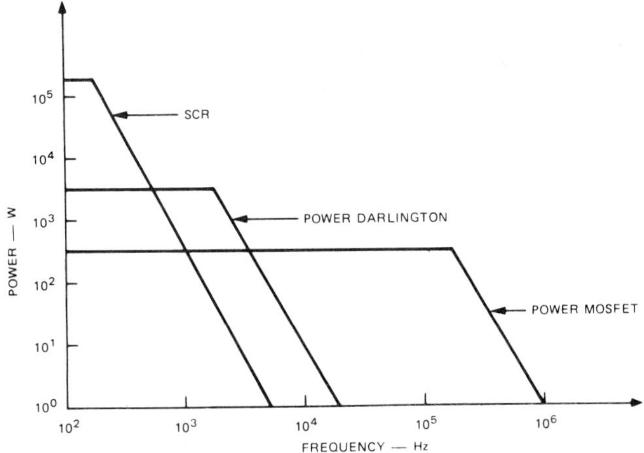

Figure 8.2 Predicted power-switching capabilities of selected semiconductors in 1990.

single devices. Other choices for motor drivers are bipolar power transistors such as power Darlingtons, which have a built-in drive stage so that the drive power required is very low, and silicon-controlled rectifiers (SCRs).

The current-handling capability of power MOSFETs is approximately 50 A (with a voltage rating of about 600 V); power Darlingtons are capable of handling approximately 150 A (with a voltage rating of 800 V). The power MOSFET is expected to have ratings and prices equal to a power Darlington by the year 1990 ($\sim$ 400 A and 900 V by 1990). Figures 8.1, 8.2, 8.3, and 8.4 show existing and future power-handling characteristics of SCRs, power Darlingtons, and power MOSFETs.

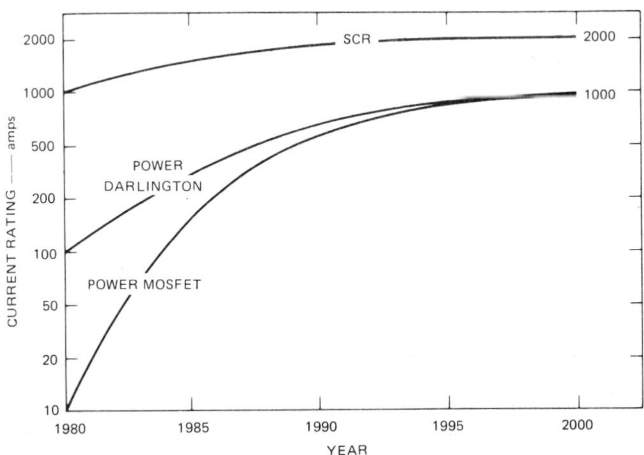

Figure 8.3 Power semiconductor current capability.

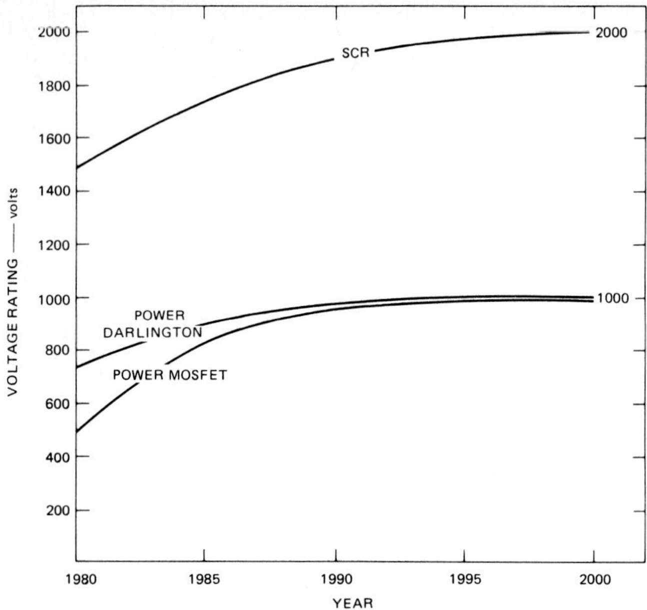

Figure 8.4   Power semiconductor voltage capability.

In small electric robots that are driven by stepping motors, it is possible to arrange the control system in a distributed fashion with a single-chip microprocessor, such as the Intel 8048, assigned to each motor. This allows higher-speed operation than can be obtained with a single microprocessor time-shared among all the motors. The individual motor processors will take care of smoothly changing the stepping rate up from zero when accelerating and down to zero when decelerating. This is essential if any speed is to be attained without risking missed steps, which are intolerable in an open-loop system. The rate of acceleration and deceleration can be adjusted by software to fit the characteristics of the particular motor selected and its associated load frictions and inertia.

8.3   Electromechanical Interfaces

The control system of a robot must interact with a number of electromechanical devices. Above we briefly discussed the electric motor drive circuits. Other types of devices that may have to be interfaced include

- Electrically actuated air valves
- Electrically actuated vacuum valves
- Electrically actuated hydraulic valves
- Limit switches (indicating end of travel of a given coordinate)

- Touch or proximity sensors (indicating presence of a workpiece or other obstacle in path of robot)
- Tachometers (measuring conveyor speeds, for moving-line synchronization)
- Pressure sensors [if used, in gripper(s)]
- Keyboards (for operator control while setting the robot up initially and teaching it a new program)

8.4 Computer Interface

Communication between an external computer and the control unit of an industrial robot is sometimes done via an RS232 interface (bit serial). Many robots today do not make any provision for connection to another computer or programmable controller, but this will be essential in future robot systems. European robots generally use IEC Standard 550 for interfacing to other computers in a control system.

The RS232 serial interface is also used in some robot control systems to connect the robot to a cathode-ray tube (CRT) display and the operator's keyboard for programming purposes. However, in future robot control systems, a faster interface than the RS232 will be very desirable to allow rapid downloading of programs from a higher-level supervisory computer* and to permit rapid updating of status information throughout the system. The RS232C specification covers data rates of 20K bytes/s or less.

An example of a higher-speed serial data bus designed for industrial control systems is the Honeywell LCN. This coaxial cable system uses serial multiplexed communications at a rate of 1.5M baud. The interface circuits at a computer site connected to the LCN convert the 31-bit data format to a parallel bit word structure—24 bits plus parity, for example. Such a bus is needed in a flexible manufacturing system (FMS) using a distributed computing system. Other similar data communication systems are listed in Table 8.3.

Another interface of growing importance is that between the robot control processor and the sensors providing external environment information. This sort of interface is handled in various ways. For example, the Unimation PUMA robots provide eight input lines, one for each of eight sensors. If eight sensors are not sufficient for the job, an external microprocessor can use these eight input channels in a binary fashion, providing $2^8 = 256$ different commands to the PUMA. Because some of the eight channels are dedicated to emergency sensors, fewer than 256 commands are really available. In practice 16 to 32 commands can be used.

---

*Communication between the lowest-level joint servo microprocessors and the first-level processor above them (system supervisory computer) should be done on a byte parallel basis; most microprocessors have suitable input/output buses.

TABLE 8.3 Local Area Networks (LAN)

Network	Company	Mode	Access*	Speed, K bytes/s	Maximum length, km	Number of nodes
CONTROLNET	Siemens-Allis	Base	CSMA/CD	28.8	1.2	32
COPNET	ISSCC	Base	M/S	115	9	256
DECNET	Digital Equipment	Base	CSMA/CD	10,000	2.5	100
Ethernet	Vanous	Base	CSMA/CD	10,000	2.5	100
GENET	General Electric	Broad	CSMA/CD	1,000	5	64,000
LAN 9000	Hewlett-Packard	Base	CSMA/CD	10,000	0.5	100
LCN	Honeywell	Base	Token	5,000	0.3	64
Modway	Gould	Base	Token	1,500	5	250
SY/net	Square D	Base	Token	500	3	200
TIWAY II	Texas Instruments	Broad	Token	5,000	Unlim.	Unlim.
VISTANET	Allen-Bradley	Broad	Token	5,000	7	Unlim.
WDPF	Westinghouse	Base	Token	2,000	6	254
XLAN	Complex Systems	Base	CSMA/CD	1,000	3	192

*CSMA/CD = Carrier sense multiple access/collision detection, std. 802.3.

The PUMA can be used in either of two control modes. The IFSIG instruction in VAL functions as a polling routine and causes the controller to scan the lines regularly and execute the appropriate instruction depending on the state of the eight input lines. The other mode uses the REACT instruction, which causes the controller to stop the ongoing program when an event occurs on the programmed input line. Each line causes a different user-defined VAL subroutine to be executed immediately, as is necessary in an emergency situation. The REACT instruction is somewhat different in that it allows the current program step to be completed before the subroutine is executed (Girod, 1982).

8.5   Internal Multiplexing

As robot grippers and their associated sensors become more capable and complex, the electric wiring required will become quite bulky and difficult to handle. This situation can be alleviated by using data multiplexing on one or a few digital buses that use coaxial cable or fiber-optic techniques to time-share a physical data path. The electronic circuitry necessary to encode the data onto the bus or buses at the control computer and to decode the addresses and data at each sensor or actuator is relatively simple and inexpensive even today. In the future it will be even simpler and more reliable to use such multiplexing techniques. The dc power required by the sensors, and possibly even the actuators, can be carried on the same bus as the data if a coaxial cable is used.

8.6   Fail-Safe Design and Reliability

Because many robots can exert enough force to injure or kill a human operator, it is essential that the robot's control system include a variety of safety features. The electronic circuitry for a robot control system should be designed so that the most likely failure modes do not cause unsafe operation (fail-safe design). This will be difficult to do with a high degree of confidence; therefore, a second level of protection using special sensors should be provided to turn off robot power when an obstacle is detected in an area near the robot. These sensors must be redundant to protect against sensor failure.

Fail-safe design is sufficiently important that it will be prudent to dedicate a special microprocessor to safety functions only. This microprocessor should be able to interpret the safety sensors mentioned above and overrride all other commands to the robot drive motors when necessary. The safety microprocessor can also monitor power supply voltages and currents to detect any out-of-specification performance that might cause a safety problem.

Another technique that can be used is duplicate or triplicate voting redundancy in the control and safety processors. This technique has not yet been used in robot control systems, but is becoming economically feasible as IC costs decrease.

The reliability of a well-designed and thoroughly debugged industrial robot can be on the order of 400 h mean time between failures. The ultimate life of a robot is approximately 40,000 h, with major overhauls performed at 8000- to 15,000-h intervals. Annual maintenance costs are in the area of 11 percent of acquisition costs. Most of the downtime and maintenance costs are attributable to the mechanisms of the robot, not the electronics. It must be recognized that a robot, like any complex mechanism, has wear-out failure modes.

"Graceful degradation," or graceful failure, means that the manipulator performance degrades slowly in response to overloads rather than failing catastrophically. For example, a manipulator will usually be able to operate at a much greater power for a short period than it could on a continuous basis. This is because the thermal mass of the motors smooths out the temperature rise that results from momentary operation at a high power level so that the resulting maximum temperature rise is small. However, if the motors operate at *peak power* continuously, they will ultimately heat up until they fail—usually their insulation breaks down. This is an example of a catastrophic failure mode.

One way to convert this to a *graceful* failure mode is for the control software to monitor motor temperature (say, by means of thermistors) and reduce the speed of operation more and more as the temperature rises. Or, the software might make the manipulator pause between motions to allow its motors to cool a little. This would allow a robot in a manufacturing line to deal with temporary overloads without requiring attention from the operator. It would just "become tired" and work more slowly. The same behavior could also allow the robot to react to certain types of accidental failure. In this example, excessive heating could also result from a shorted turn in the motor winding. The control system would react in the same way as if the arm were being overworked and automatically make it work more slowly. This would protect the motor against catastrophic failure without stopping the production line entirely. It would also be obvious to the operator that something was wrong with that arm if the control software did not report a problem on its own initiative. Repairs could be scheduled before major damage was done.

In general, it should be possible for a robot to protect itself in ways like this by monitoring its internal status. It could react appropriately so that it would not damage itself, and as it inevitably begins to wear out, predict its own failure and request service beforehand. This would help to maintain the productivity of a production line at a high level.

Software Design Methods

9.1 Considerations and Alternatives

The software in a modern computer-controlled, sensor-equipped robot for assembly is extremely complex. It typically consists of tens of thousands of programming statements in a high-level language like C or Pascal. The software must perform reliably in order to be cost-effective, and it must perform correctly in order to avoid damage to equipment and injury to people. The development cost can easily exceed $1 million.

As the user community becomes more sophisticated in their usage of robots and other computer-controlled equipment, they will demand that the capabilities of the software be further improved. The software must be carefully planned in order to meet these demands. The robot manufacturer that commits itself to hastily thought-out, overly simplified control software in order to beat its competitors to market will find itself with an obsolete product in 5 years with no affordable option for improvement to meet market demand.

9.2 Who Works with a Robot?

At least three distinctly different categories of personnel will use the robot control software, each for a different purpose:

- *Equipment operators.*   Line supervisors and blue-collar workers who are responsible for daily operation of the equipment during a production run

- *Task programmers.*   Production engineers who have been trained in the use of a robot programming language and are responsible for programming the robots for a production run

- *System programmers.*   Professional software experts, such as computer scientists, senior programmers, or system analysts, who are responsible for developing software to support task programming and equipment operation

The sections below discuss the role of each employee category in relation to the robot.

9.2.1   Equipment operators

Some of the activities that the equipment operators will carry out with the robots include:

- Setting up the robots and auxiliary equipment at the beginning of a shift. This might include positioning the manipulators, inspecting for broken equipment, seeing that part feeders are loaded, and verifying that all movable pieces of equipment are in their proper positions.

- Starting up the production line and ensuring that everything is running smoothly.

- Stopping the production line whenever they notice a problem.

- Fixing the problem and restarting the equipment.

- Shutting down the equipment at the end of a shift.

Recovery from failures and accidents will be the most demanding part of the equipment operator's job. The operator may sometimes have to seek advice and assistance from the line supervisor. Deciding how to recover from a problem will often require ingenuity and familiarity with the task-specific control software, and sometimes a working knowledge of how the system software operates as well. In advanced manufacturing tasks in which individual workpieces are extremely expensive, it will be important to avoid scrapping a partially completed workpiece if a problem develops. In such extreme cases, the supervisor may decide to call in the task programmer to instruct the robots as to which steps of their program they should try again and which they should skip.

An equipment operator need not, and probably will not, be a programmer. He or she will not have to write programs for robots, but only supply information to programs that have already been written by the task programmer. For example, a robot might have to be told how many of a given product to make. If the robot encounters a problem, it might ask the operator whether it should skip the processing step that failed, try it again, or scrap the workpiece entirely.

Furthermore, in the early stages of factory automation through robots in industry, many equipment operators will be required—perhaps one for every 5 or 10 robots. It would be uneconomical to employ so many programmers. In the later stages of manufacturing automation, programming skills will be less important, because the task-specific programs will be written not by people, but by other programs from CAD data bases and manufacturing knowledge bases.

Equipment operators in FMS will interact with the robotic equipment in much the same way as blue-collar workers now interact with conventional automatic equipment. They will push buttons, adjust knobs, and perhaps type in occasional numbers, codes, or other symbolic information on a simplified keyboard. They will get information from the equipment through annunciators, meters, and character displays.

Some new modes of interaction will also be necessary; some of these will result from the availability of new kinds of controls. Voice input/output equipment, for example, will be used more and more in coming years. As their cost declines, graphic displays will also be increasingly used to replace conventional devices such as meters. They will also be used to convey complex information that is difficult to describe in words or numbers, such as locations in the work space, part shapes, and arm motions. Other modes of interaction will be peculiar to the manipulators themselves, such as methods for remotely controlling them for training, setup, or error recovery. Push buttons are currently giving way to joysticks as remote controls; teleoperator masters may replace joysticks in some applications.

The software needed to support an equipment operator's activities is the task program, which is created by the task programmer.

9.2.2 Task programmers

Some of the activities that the task programmer will carry out include:

- Understanding the design of the product to be assembled and devising a method for manufacturing the product. This might be a set of elementary operations such as component acquisition, fitting, fastening, and inspection.

- Translating the manufacturing plan into a sequence of actions that the robotic equipment can perform.

- Deciding how many robots are required, what type of robot to use, what tooling is required, and how material will flow through the production area.

- Creating one or more programs in a robot programming language to control the various robots and other equipment in the production line.

- Testing the programs.

Initial testing might include computer simulation of how the equipment will carry out the program. A coarse simulation in the style of CSMP or SIMSCRIPT would detect bottlenecks and major errors in the program. A finer-grained simulation could then be run in which geometric models of the parts, manipulators, and tooling would be used to detect collisions, out-of-reach objects, objects in the way of the robot, and similar spatial problems. Final testing would entail operation of the actual equipment with real workpieces.

It is readily apparent that the task programmer's primary skill should be production engineering. Deciding how to manufacture a product presents new problems for each new product. In contrast, the process of translating a given manufacturing procedure into a program in a robot programming language will be much the same from product to product. The language itself will be the same each time. Whenever a programmer develops a software routine to perform a specific manufacturing step, such as installing a C-ring, or a setup step, such as asking the operator to train an arm position, he or she can reuse that routine in other task programs. As the task programmer develops a library of reliable routines, translating each new manufacturing plan into a task program will become progressively easier.

The system software, which is created by the system programmer, is needed to support the task programmer's activities.

9.2.3 System programmers

Some of the activities that the system programmers will carry out include:

- Designing the task programming language and the procedures to be used in task programming.

- Deciding on an overall system architecture (i.e., how software is to be distributed among different computers in the factory and what communication will take place between them).

- Creating software to support task programming.

- Creating software to support testing of task programs.

- Creating software to support execution of task programs. This includes software "drivers" that interface the operator's controls and displays, as well as the manipulators, sensors, and other automatic equipment to higher-level software.

A system programmer will probably work for the company that manufactures the robot as a member of a system programming team, specializing in one or two of the activities listed above.

System software is the most complex software in robot design and contains the most lines of code. It is typically written once and debugged by the robot manufacturer's system programmers and then used for many years in a product line of robots. Such software is modified only occasionally, perhaps every 6 to 12 months, as the robot manufacturer fixes bugs in the software and makes minor improvements in the software's basic capabilities.

The following sections discuss software requirements for robot applications.

9.3 Task Programming

9.3.1 Programming methods

The job of the task programmer is to translate plans for manufacturing procedures into programs to control the automatic equipment. *Symbolic programming* methods (i.e., formal programming methods based on a programming language) have always been used in traditional computation to describe data and algorithms, and they will continue to be important in programming computer-controlled robots. However, a second programming method can also be used with robots: *procedural programming* or simply *training*. In this method, the programmer describes data or algorithms by operating the equipment manually or via remote control. For example, the programmer might move an arm with a joystick to indicate a position in the work space, place a workpiece in view of a video camera so that a vision system can measure its shape parameters for later recognition, or put a tool in the hand to be weighed by a wrist-mounted force sensor.

In such a training procedure, some automatic or semiautomatic process translates the activity of the equipment into a program. The simplest example is training a spray-painting manipulator: The program is just a recording of thousands of successive arm joint positions during the training interval. In material handling and assembly tasks, only a relatively few discrete positions are recorded, and then only on a signal from the operator. Intervening arm motions are ignored. In simpler robots, the order in which those points are trained also implies an algorithm for automatic operation later, i.e., to move through the positions in the order trained. In a computer-controlled robot, however, the control software can assign any useful meaning to a remotely controlled arm motion during training. The most common example is the way modern robot control systems record the cartesian coordinates of the hand relative to some specified reference frame, rather than by joint angles. The reference frame may be in a different location and orientation during execution, or the position might be played back in a different frame. As a more complex example, in Unimation's VAL system, a frame is defined by "training" three points in space—the frame is stored, but the locations of the points themselves are discarded.

Procedural programming or training is often a convenient way for the task programmer (as well as the equipment operator) to supply data to the robots, such as positions, orientations, reference frames, directions of motion, location of tool "tips," trajectory paths, and so forth, to the robot. The numeric values of these data may not be available or may be too complicated to compute from, say, part shape. Procedural programming or training is *not* a convenient way to describe algorithms, except for very trivial record-playback sequences. It rapidly becomes very confusing to try to show the robot alternative and repetitive actions in a moderately complex task.

The task programmer's job will be much simpler if the robot programming language makes it easy to describe activities that occur frequently in the tasks for which the robot is intended. Many different robot programming languages have been developed with this purpose in mind. Good and bad features are found in all of them.

9.3.2   Concepts to describe manufacturing procedures

The level of abstraction of the concepts dealt with in the robot programming language must be adjusted so that it is neither too low nor too high. If it is too low, the task programmer will spend a great deal of time describing manipulator, sensor, and computational activities in unnecessary detail. If it is too high, the programmer will find that he or she does not have sufficient control over the system to perform tasks that are a little out of the ordinary.

Some of the concepts that a task programmer finds useful in describing manufacturing procedures are

- Parallelism

- Reference frames and relative positions of objects

- The action of installing a component

- Attachment relationships between objects

- Visual location and identification of objects

- Tactile-controlled motions

- Procedural data input (training of arm positions or visual prototypes)

- Data input from blueprints (such as CAD databases or geometric models)
- In-process inspection
- Macro operations

Few robot programming languages explicitly support even a few of these concepts today. Instead, the task programmer must translate these natural, or intuitive, manufacturing procedures into lower-level computer concepts such as integer, real, and logical values; loops and conditional branches; subroutine calls; complex arithmetic expressions; and so on. Even when this is done correctly, it is very difficult for another person to tell by reading the program what the equipment is supposed to do. This makes creating and maintaining task programs a complex, time-consuming, expensive, error-prone process.

Therefore, the task programming language should ideally be designed so that activities such as those listed above can be described tersely and in a way that is understandable to someone who reads the program later. Nevertheless, the language should be complete, in the sense that it should also allow the task programmer to describe any algorithm or computation that is needed: In addition to constructs to support the activities mentioned above, the language should thus include more conventional and non-task-specific facilities for computation, control, and input/output. However, there is a limit to how completely the task programmer should have access to the equipment, too. He or she should not, for example, have to (or want to) deal with memory addresses, storage allocation, file structures, or the details of how a hardware device interface operates. Those are the proper concerns of the system programmers.

One of the most troublesome, yet useful, features that a task programming language can have is the capability for parallel operation. As a very simple example, one might describe separate tasks for two adjacent robots to carry out on the same workpiece at the same time. Unless care is exercised in writing and testing such a program, it is very difficult to guarantee that the two robots will never interfere with each other.

For example, if two robots share a tool such as a screwdriver, they might both reach for it at the same time. Or, if they share work space, they might collide with each other. The troublesome aspect of such failures is that they are unpredictable. In the usual implementations, the activities of the two robots would be unsynchronized and would depend upon the timing of such other processes as taking a picture, waiting for a screwdriver to finish tightening a screw, and so on. Thus, through good (or perhaps bad!) fortune, a potential problem may never occur in hours or days of testing. Later, during a production run, some process may take a little more time or a little less time than during testing, instigating the failure, and causing damage or economic loss. Nevertheless, the ability to perform useful activities in parallel can greatly increase the throughput of a robotic work cell, so parallelism is worth considering in a task programming language.

9.4 System Software

The purpose of system software is twofold: to support the activities of the task programmer and to operate equipment during normal production runs. These two types of activity are sufficiently different that it is advisable to consider splitting the system software up into three or more modules. For example, it might be split into modules that consist of the routines common to the following four activities:

1. *Offline* activities, such as editing task programs and simulating their execution

2. *Online nonproduction* activities, such as hardware diagnostics, equipment calibration, data training, program testing, and program revision

3. *Online production* activities, such as reporting production statistics to a management information system in another computer and receiving instructions from the factory control system

4. *Common online* activities, such as operation of the various pieces of hardware (e.g., manipulators, sensors, conveyors, tools, manual controls, and operator displays)

In addition to such automation-specific software, there may also be conventional resident operating systems in one or more of the computers in the system. These would perform general-purpose functions such as file handling, memory management, interrupt servicing, interprocess communication, and (via a network) interprocessor communication.

Actuators

10.1 Considerations

Actuators for robotic applications must satisfy many, often conflicting, requirements. Thus, the selection of actuators for a given robot is much more difficult and more complicated than the selection of actuators for more mundane applications such as machine tools, construction vehicles, or office machines. Such applications generally require excellent actuator performance in one or two areas, with other actuator performance parameters being more or less unimportant. In contrast, robot actuators usually must have excellent performance in all areas. Conventional actuators are often poorly suited to robots, and new actuators tailored specifically to robot requirements are seeing increased use. The performance parameters relevant to robot actuators are each discussed briefly below.

10.1.1 Speed/torque curve

The speed/torque curve is probably the most fundamental performance parameter for electric motors. In many (nonrobotic) applications, the designer need only consult the speed/torque curve to determine if a given motor is suitable for the application. For the robot designer, the speed/torque curve is only one of many parameters that must be considered in the selection of an actuator. The maximum no-load speed will be related to the maximum speed requirements of the robot. The maximum "locked rotor" torque will be related to the maximum acceleration and payload requirements of the robot.

Designers should be especially careful when considering the speed/torque curves for stepper motors. Unusual bumps in these curves are usually indicative of certain resonance speeds. At these resonance speeds, operation of the motor may be erratic. If the motor is driving a robot manipulator (which will have a complex set of resonances of its own), severe instabilities can develop in the robot control system when the stepper operates near one of its resonance speeds. If practical, the best strategy is to confine the steppers to speeds below resonance.

10.1.2 Thermal time constant

The thermal time constant of a motor is equal to the time required for the motor windings to reach 63 percent of the asymptotic temperature when driven with the maximum permissible current. (Note that some motor vendors have used slightly different definitions of this term.) The above definition of thermal time constant is based on the tacit assumption that the thermal response of the motor can be modeled as a simple single-time-constant system. In fact, the thermal behavior of most motors is more complex. The behavior of winding temperature is accurately described only if two or more time constants are included in the thermal model. Because robot actuators are normally operated with a very high ratio of peak-to-average rotor current, the thermal behavior of the motor is a very important consideration in estimating the maximum temperature rise of the actuator. Temperature rise, in turn, is often the dominant constraint on torque and endurance of the actuator. Because of these factors, it appears that robot designers would be wise to investigate the thermal behavior of motors more thoroughly and include the multiple-time-constant behavior of thermal models they use to estimate temperature rise.

10.1.3 Continuous power output

The maximum continuous power output is probably the most often used of motor performance parameters. Ironically, its importance for robot actuators is usually overshadowed by demands for higher peak torque, lower inertia, or lower weight.

Robot designers should be aware that the maximum continuous power output capability of a motor depends on the motor's environment and on cooling provisions for the motor. Robot actuators are often installed inside a robot structure, where they do not receive normal cooling from air convection. On the other hand, clever mounting of the motor can provide cooling by thermal conduction to the robot structure. When this approach is used, the entire robot structure becomes a "heat sink" for the motor, and the continuous output power may be increased above the limit for convection cooling.

Forced cooling has been used on some very high performance servomotors. Air under moderate pressure (10 to 100 cm water) is ducted to the motor to circulate over the rotor. Forced oil cooling has been used in some high-performance spindle drive motors. Forced cooling techniques may be useful in high-performance robots to permit higher continuous power output from actuators.

10.1.4 Rotor inertia

The rotor inertia of the actuator is often the dominant limitation on the achievable acceleration of robot joints. Therefore, low-inertia rotors would generally be preferred for robot applications. Specifically, "ironless rotor" and high-energy permanent-magnet rotor motors would be preferred to wound (iron) rotor and induction motors. Wyllie (1982, p. 36) makes the interesting observation that hydraulic actuators are inherently low-inertia (and, hence, high-acceleration) devices, while electric motors tend to have greater inertia. Only by special design for low inertia can the acceleration achievable with electric motors approach that which is easily obtained with hydraulics.

10.1.5 Efficiency

Efficiency does not have the same dominant importance for robot actuators as for typical constant-speed industrial drive motors. However, as energy costs increase and robots become larger, actuator efficiency will become more important.

Because robot actuators typically operate with low duty cycles, actuator efficiency at partial load and low speeds can be very important to the overall efficiency in this application. In contrast, the efficiency of industrial motors is seldom specified except for full-load full-speed operation. The fact that the efficiency of any actuator at zero speed is zero is often overlooked in the design of robot actuation systems. The resulting system may consume excessive and unnecessary power while the actuator is stopped. For this reason, many modern robots provide brakes for each joint to minimize power consumption when the robot is motionless.

An interesting comparison of the efficiencies of hydraulic and electric actuators has been made for aerospace applications (Wyllie, 1982). The application entails a low-duty-cycle situation, also common in robotic applications. Wyllie shows that the energy consumed by hydraulic actuators is 34 times greater than the energy consumption of an electric actuator. The difference in efficiency is primarily a result of the high energy consumption of the hydraulic system during periods of zero actuator velocity.

10.1.6   Weight

Weight is a very important consideration for robot actuators. Robot actuators are usually mounted near the joint they control. Thus, the weights of the actuators for the more outboard joints become part of the load that must be supported by the more inboard joints. The actuator weights detract from the robot's maximum payload and acceleration capabilities. In most modern electric robots, the weight of outboard actuators is a significant fraction of the total weight of the outboard links. Many newer robots incorporate "tendon" or drive shaft linkages so that actuators for the outboard joints can be mounted closer to the shoulder. Examples of this approach include the Unimation PUMA and the Microbot MiniMover-5 robots.

The weight of electric motors can be decreased in several ways including:

- Improved design, making more efficient use of magnetic materials and conductors.

- Substitution of magnetic materials with improved properties, for example, rare earth permanent-magnet materials.

- Substitution of better conductor materials. For example, aluminum wire is much lighter than copper wire.

Most of these approaches have already been used by motor manufacturers, so obtaining motors significantly lighter than currently available products is very difficult.

The weight of a motor with a given power output capability can generally be decreased by increasing the operating speed of the motor (and simultaneously decreasing the size and torque). Of course, a transmission is then required to decrease the motor speed and increase the torque, but the weight saving in the motor is normally greater than the added weight of the transmission. The weight saving that can be achieved in this way is usually limited only by the maximum practical motor speed. (As motor speed is increased, at some point centrifugal force or some dynamic instability will destroy the rotor.) Conventional motors of less than 1 or 2 hp are usually run at less than 4000 rev/min and often at less than 2000 rev/min. With proper design, rotor speeds up to 10,000 rev/min could be obtained in this power range. Obviously, there is a significant potential here for motor weight reduction.

Most designers agree that the primary advantage of hydraulic motors relative to electric motors is their high torque-to-weight ratios. The robot designer should not forget this fact. When all feasible methods of weight reduction have been applied to an electric actuator and the result is still too heavy for the application, it is time to consider the use of an hydraulic motor.

10.1.7   Brushes

Many important actuator considerations involve brushes. Of course, the most fundamental consideration is whether the actuator uses brushes. Brushless dc and ac induction motors are available; most stepper motors do not use brushes. Brushes are undesirable for several reasons:

- Brushes have a limited lifetime and therefore require periodic inspection and replacement.

- Arcing at brushes can be a problem if the actuator must operate in an explosive environment.

- The cost of a motor with brushes is usually higher than the cost of a comparable motor without brushes. (However, the cost of a brushless motor controller normally exceeds the cost savings associated with the brushless motor.)

- Brushes cause electrical transients that are a source of electromagnetic interference (EMI). (Brushless motor controllers may also generate EMI, however.)

If the designer chooses to use a motor with brushes, parameters such as brush life and replacement procedures should be investigated. Replacement brushes are not usually very expensive, but the cost of labor and robot downtime for brush replacement can be significant. In any case, the designer is advised to avoid motors that do not permit brush replacement or that require elaborate disassembly to change brushes.

10.1.8   Cogging

Cogging refers to the torque pulsations of dc motors associated with commutation. These pulsations occur in both brush and brushless motors. Generally, cogging amplitude is reduced and its frequency is increased if a larger number of poles are used in the motor.

Cogging is of particular concern in robot actuators, which are normally incorporated in a complex servo-control system. Servo stability is always a problem in these systems, and motor cogging only aggravates it. Motor cogging is essentially equivalent to a sinusoidal signal being added to the motor drive current. If the frequency of this excitation is near any resonances in the system or its subsystems, instability may occur. Note that the motor cogging frequency may exceed the response of the servo system, yet the cogging may excite the structure despite electronic filtering.

10.1.9 Hydraulic and pneumatic systems

Hydraulic and pneumatic actuators can also be used for robots. Generally speaking, hydraulic actuators are used in large, high-performance high-cost robots; pneumatic actuators are used more often in small, low-performance low-cost robots. Consideration of the performance capabilities of the two actuator types helps to explain these tendencies.

Hydraulic actuators are generally unsurpassed in the areas of torque-to-weight and horsepower-to-weight ratios. This is especially true if the weight of the hydraulic power supply (pump, motor, fluid reservoir, and filters) is not included in the actuator weight. Because most industrial robots are not mobile, exclusion of the power supply weight when calculating torque-to-weight ratios is appropriate. The hydraulic power supply can usually be located at some convenient location near the robot, and the only weight that must be moved by the robot is the weight of the hydraulic actuator itself.

Hydraulic actuators have a few additional advantages including:

- *Ruggedness.*   The hydraulic actuator is inherently sealed and can be operated in dirty, abrasive, wet, and mildly corrosive environments with few problems. Its tolerance of temperature extremes is also good, but probably inferior to that of pneumatic actuators because of boiling or thickening of the working fluid.

- *Safety.*   Hydraulic actuators use no brushes, so there is no need for concern about brush arcing in explosive environments. Hydraulic actuators are normally controlled by electrically actuated servo valves located near the actuator. Current and voltage requirements of the servo valve are much lower than those of an all-electric actuator. Therefore, it is relatively easy to design a hydraulic actuator system so that no dangerous sparking occurs, even if there is massive damage to or malfunction of the actuator system.

The advantages of hydraulic actuators must be weighed against their disadvantages, which are numerous:

- *Cost.*   Hydraulic actuation systems are generally more expensive than electric or pneumatic alternatives.

- *Maintenance.*   Hydraulic actuation systems are messy. Sooner or later fittings and seals develop leaks, and the robot will develop a coating of dirty, greasy hydraulic fluid. Even in leak-free systems, routine maintenance such as hose, filter, and fluid replacement can lead to hydraulic fluid spills.

- *Noise.*   Hydraulic power supplies are often extremely noisy. This problem can be alleviated by housing the power supply in a soundproofed area, but such soundproofing can limit air circulation and lead to overheating of the power supply.

- *Efficiency.*   Hydraulic actuation systems tend to be inefficient. Some traditional hydraulic systems consume nearly full power even when the actuator is not being used. More modern systems use variable-displacement pumps to achieve good efficiency, but the cost and complexity of these systems is greater.

- *Fluid transport.*   Hydraulic actuation systems normally use elaborate plumbing to bring the working fluid to and from the actuator. This plumbing is expensive (relative to electric wires, for example), and installation of the plumbing in or on the robot usually requires some compromises in the mechanical design of the robot. Hydraulic fluid is usually routed around the robot joints with hoses that are bulky and expensive and require periodic replacement. The hoses may be detrimental to joint performance.

- *Flammability.*   Despite their inherent advantages for operation in explosive environments, hydraulic systems have several safety problems of their own. A ruptured line can quickly douse the robot and everything in its vicinity with flammable hydraulic fluid. If the fluid is ignited, the resulting conflagration can destroy more than just the robot. Small pinhole leaks in the hydraulic system can create intense jets of hydraulic fluid that can injure personnel.

Pneumatic actuators have advantages similar to those of hydraulic actuators in regard to ruggedness and safety in explosive environments. However, their torque-to-weight and horsepower-to-weight ratios are generally rather poor. The primary attraction of pneumatic actuators (aside from safety and ruggedness) is that they offer a simple and low-cost actuation method for linear motions. A good example of this approach is the AutoPlace robot, which is very low cost and primarily uses prismatic (linear motion) joints.

As a result of the compressibility of air, pneumatic actuators are not very stiff and have a relatively slow response (compared to hydraulic or electric actuators). Thus, pneumatic actuators are not well suited to use in high-performance servo-control applications. Consider, for example, the AutoPlace robot. Motions of this robot's axes are not servo-controlled. Instead, the travel of each joint is controlled by limit stops that are adjusted manually at setup time.

10.2   Motors

In keeping with the emphasis of this book, we will discuss the various types of electric motors in some detail and include a briefer description of hydraulic and pneumatic actuators.

Several types of electric motors (dc servo, ironless rotor, brushless dc) are discussed below. Each of these motor types can be configured for linear or rotary motion. Rotary motors are currently far more popular for robotic applications, but more applications of linear motors may occur in the future. In the interest of brevity, only the rotary configuration of each electric motor type is presented below: The reader should realize that linear configurations of each motor type are possible. The performance characteristics of the linear motors will be comparable to those for the rotary motors. Torque of the rotary motor will be analogous to the force of the linear motor, and the revolutions per minute of the former analogous to velocity of the latter.

10.2.1   Dc servomotors

Dc servomotors consist of a stator, rotor, commutator, brushes, bearings, and a housing. The stator creates a constant magnetic field with two, four, six, or more poles. The stator may use either permanent magnets or coils to produce these fields. The rotor of the motor is constructed of iron laminations on which one or more coils are wound. The rotor coils are energized via the commutator and brushes.

Dc servomotors have good torque, speed, and continuous power output performance. They are relatively easy to use in servo-control applications because reversal of the rotor current reverses the motor torque and the motor is well behaved near zero torque or velocity. The primary disadvantage of this type of motor is its large rotor inertia which limits motor acceleration. However, in many applications, the inertia of the load is much greater than that of the rotor. In this case, load inertia dominates and rotor inertia has very little effect on acceleration performance.

Roughly speaking, for a constant stator current, the torque of a dc servomotor is proportional to the rotor current. With no load on the motor, the motor speed (revolutions per minute) is proportional to rotor voltage. These relationships are useful for rough calculations during initial design of the servo-control system.

10.2.2   Ironless rotor dc servomotors

The ironless rotor dc servomotor is similar to the dc servomotor, but it does not use iron in the rotor. Instead, the rotor coils are self-supporting. One popular fabrication method for these coils involves winding the coils on a removable form or bobbin and impregnating this assembly with an epoxy matrix. After the expoy is cured, the bobbin or form is removed. Another approach is to fabricate the rotor using printed circuit techniques.

Performance of ironless rotor dc servomotors is very similar to that of dc servomotors except for two areas:

- Rotor inertia of the ironless rotor motor is much less because there is no iron in the rotor. If the load inertia is not too great, the ironless rotor motor will achieve much better acceleration.

- The continuous output power capability of the ironless rotor motor is inferior to that of the dc servomotor. This is primarily due to the lack of thermal communication between the rotor coils and the iron, which acts as a path for removal of heat. Heat buildup in the rotor coils limits the continuous output power capability. However, *peak* power and *peak* torque are not thermally limited, so the ironless rotor motor performs well in these areas.

10.2.3 Brushless dc motors

The brushless dc motor's name can be misleading, as it suggests a brushless motor that operates on dc power. In fact, both a motor and an electronic motor controller are part of the brushless dc motor system. The motor part of this system is usually constructed with a permanent magnet rotor and a wound stator. A commutator and brushes are not used, but an encoder is an integral part of the motor. The encoder may be either magnetic or optical. The output signal from the encoder is used by the electronic motor controller to switch current to the stator windings in the proper sequence to cause rotation of the rotor.

Brushless dc motors have the expected advantages of a brushless design—reduced maintenance requirements and elimination of brush arcing. Brushless dc motors are more expensive than their brush-equipped counterparts. Rotor inertia is greater than that of ironless rotor motors because of the mass of the permanent magnets; however, continual improvements in permanent-magnet materials have resulted in comparable improvements in rotor inertia.

10.2.4 Stepper motors

Stepper motors are unique because the motion of the rotor is precisely determined by the input signals of the motor. Stepper motor input signals consist of a series of pulses; the rotor advances one step for each pulse. Step sizes from a fraction of a degree up to 15° are common.

Because of the fixed relationship between input signals and rotor motion, stepper motors do not require an encoder and servo-control system in order to achieve the precise positioning required in robotic applications. Elimination of encoders and servo electronics results in a significant cost savings. Low-cost robots such as the Microbot MiniMover-5 use stepper motors for this reason.

Unfortunately, stepper motors have several shortcomings that preclude their use in high-performance robots except in very unique situations. These drawbacks include:

- For a given output torque or horsepower, the weight of a stepper is greater than other types of motors used for robots.

- Rotor inertia is larger than most comparable motors.

- If the motor load exceeds the design limits (even briefly) because of friction, wear of transmission components and bearings, or unanticipated inertia, then the motor may not make every step as commanded. Because encoders are not normally used with these motors, the robot control system has no way of knowing about the missed step(s). Thus, the robot motions will be inaccurate beginning with the motor's first missed step. The error cannot be corrected without stopping the robot and reinitializing counters in the control system.

- Maximum speed of steppers is not normally very high. A maximum speed of about 3000 steps per second is typical for steppers. In a motor with 2000 steps per revolution, this corresponds to a maximum speed of 90 rev/min.

Designers should be aware of some idiosyncrasies of stepper motors. These motors cannot accelerate to full speed instantaneously. The motor control system must be designed so that step commands are not issued faster than the motor can respond. When the motor is started from a standstill, the controller must issue step commands to "ramp" the motor from zero speed to the desired speed at a controlled rate. Depending on the particular stepper used and the inertia characteristics of its load, there may be certain motor speeds where a resonance phenomenon can occur. The motor may miss steps or make extra, undesired steps when operating near this resonance speed.

A stepper motor can be excited in several different modes depending on the stator winding and desired performance (*Machine Design,* 1978):

Two phase.   One entire phase (stator winding) of the motor, end-tap to end-tap, is energized at a given moment in time. Input current and wattage are halved (compared to four-phase excitation). Heat dissipation is decreased. In the two-phase modified mode, both windings (end-tap to end-tap) are energized simultaneously. Energy input in this mode is the same as four-phase, but output performance is increased by about 40 percent. The control for this mode is complex and costly.

Three phase.   Variable reluctance steppers use three-phase windings. Two adjoining phases are excited simultaneously and the rotor is indexed to a minimum reluctance position corresponding to the resultant of the two magnetic fields. Because two windings are excited, twice as much power is required as the standard mode (one phase at a time). Output is not increased, and damping is improved.

Four phase.   Each half winding is regarded as a separate phase, and phases are energized two at a time. The controller is simple but performance is inefficient. Twice the input energy is required compared to single-phase excitation. Torque output is increased by up to 40 percent, and the maximum response rate is increased.

10.2.5   High-torque motors

A relatively recent development in motors is the Motornetics Corporation Megatorque motor. This motor is a type of three-phase stepper motor (although it has also been referred to as a "brushless dc" or "ac reluctance" motor) developed specifically for high-torque low-speed robotic applications (IRI, 1984; Machine Design, 1984; Cushing, 1985). A similar motor is being offered by Yokagawa Electric Corporation. The advantage of the motor is its ability to generate precise motion at very low speeds. When it is used as a direct-drive motor, problems associated with backlash in gear trains are eliminated. The primary disadvantage of the motor is its weight. The high-torque motor is a good choice for a shoulder actuator which is mounted to the robot's base, especially if the axis of the joint is vertical, i.e., the joint is not gravity-loaded. Such motors have been used in the Adept SCARA configuration robots.

10.2.6   Ac induction motors

The ac induction motor uses a stator with (typically) two or three windings. Ac induction motors with two stator windings are used in single-phase applications such as constant-speed drive motors with less than 3 hp. The motors with three stator windings are used in three-phase applications requiring greater than about 3 hp at constant speed. These three-winding motors are also used for variable-speed and servo-control applications and, therefore, are of interest for robotic applications.

The rotor of the ac induction motor is normally fabricated of iron laminations and a simple set of windings. The rotor windings are not connected to a commutator; instead, rotor winding current is induced by the changing magnetic fields created by the stator; hence, the name "induction motor." The most common rotor winding construction, called a "squirrel cage" rotor, consists of cast-aluminum conductors and short-circuiting end rings.

The speed of an ac induction motor can be varied by changing the frequency of the voltage driving the stator windings. Until quite recently, it was impractical to generate the variable-frequency ac required for variable-speed operation of ac induction motors. Recent advances in high-power semiconductor devices now make it practical to design variable-speed controllers and servo-controllers for ac induction motors. Given these advances, it now appears that ac induction motors could be used effectively for robot joint actuation.

Ac induction motors have several advantages relative to dc servo motors for robotic applications, including:

- Absence of brushes and commutators and the attendant maintenance and safety problems.

- Extreme ruggedness.

- Lower cost. (Motor cost is less, but if the costs for both motor and controller are compared, there may not be a cost advantage.)

Ac induction motors began to have an impact on robot design around 1982. The following excerpts from publications of that era give some idea of the excitement that surrounded this technology (*IRI*, 1982*a*).

> Fijitsu [sic.] Fanuc's Ltd. [sic.] alternating current (ac) servo motors for robots (IRI, April 12) will be available to equipment builders in the United States this Fall. They'll be introduced at the international Machine Tool Exhibition in Chicago. . . .
>
> Ac servo motors are said to be superior to dc motors on several counts; higher reliability, greater speed capability, higher load capacity, better power efficiency, and simpler maintenance. The main restraint to their wide use has been lack of equally dependable controllers. F-F clearly believes it and Siemens have licked the controller problems.

Nihon Servo and Toshiba Electric in Japan also announced plans for making the ac servos for robots. In the United States both General Electric and Gould, Inc., introduced both motors and controls for ac induction motor servo systems (*IRI*, 1982*d*).

Two patents relating to ac induction motors for robots may be of interest, as described in *IRI* (1982*c*), p. 8, as follows:

> *US4,321,516. Apparatus for Driving Spindles of Machine Tools with Induction Motors.* Assigned to Toshiba Kikai Kabushiki Kaisha, Tokyo, Japan.

Patent concerns variable-speed ac induction motor with novel torque-control system for driving machine-tool spindles. System controls the speed as well as the angular position of the motor shaft. The motor operates with a short-interval pulse generator coupled to the spindle; a high-speed computation unit calculates stator current of the motor extremely rapidly. Also, a torque instruction applied to the computation unit is calculated from a speed instruction or an angular-position instruction (from a tape-reader or such input) and from data given by a reversible counter (which counts the pulses generated by a pulse encoder coupled to the spindle).

The use of positional and speed feedback systems makes it possible to provide stepless control of the RPMs of a machine-tool spindle, or other device—as well as control of the angular position or orientation of the spindle—solely by electrical means. Consequently, an easily maintained ac induction motor can replace a more costly dc motor.

US4,322,671. Induction Motor Drive Apparatus. Assigned to Fujitsu Fanuc Ltd. Tokyo, Japan.

A new type of drive system for an alternating current (ac) induction motor gives smooth motor rotation in low-torque, low-speed operation by making the actual motor speed coincide with the command speed. This would allow, for example, smooth stopping of a robot arm movement. The new speed control system relies on varying the frequency and effective value of a 3-phase ac voltage applied to the motor. While adequate torque is provided to meet load requirements, controlling the amplitude of a speed-deviation signal below a certain limiting value will limit the torque at low speeds.

Key parts of this Japanese motor-drive apparatus include: 1) inverter to convert dc voltage into 3-phase ac voltage, 2) speed-deviation-signal generator, 3) torque-limit-signal generator, 4) amplitude controller (for the speed-deviation signal based upon the torque-limit signal), 5) slip-signal generator (giving $+$ or $-$ constant voltages depending upon the motor's rotational direction), 6) voltage-signal adder, 7) rectangular-wave-form generator, 8) multiplier (amplitude-controller output $\times$ adder output), and 9) pulse-width modulator.

These developments make it possible to reduce driving torque at low speeds (or in orientation mode) by 50% to 67%, compared to the torque at high speeds or under working load. This design of an ac induction motor also allows for mounting tools on the spindle of a machine without damage to the spindle when it is mechanically stopped in the orientation mode.

The reader is cautioned that the true merits of ac induction motors relative to dc motors have yet to be conclusively determined. For example, it is claimed that "Horsepower for horsepower, ac motors are 30% lighter than their dc counterparts" (*IRI*, 1982c, p. 4). Yet, the same publication also states that brushless dc "motors have higher torque-to inertia ratios, greater power to weight ratios, and better energy efficiency than ac motors" (*IRI*, 1982c, p. 5). Because elements of cost, performance, and ruggedness (and many other factors) are involved in choosing actuators for robots, the final resolution of this conflict will probably be made in the marketplace.

It now appears that ac induction motors will be used extensively where cost and durability are primary considerations. Other types of motors will be chosen when performance is the most important requirement.

10.2.7   Solenoids

The solenoid is a very simple actuator that is seldom used in robots. It is included in this book for the sake of completeness and because it may be quite useful for certain specialized robotic applications.

An electric solenoid consists of a coil and a magnetic circuit passing through the coil. One part of this magnetic circuit (the plunger) is movable in such a manner that its motion changes the reluctance of the magnetic circuit. When the coil is energized, the plunger moves in a direction that minimizes the reluctance.

Solenoids are normally used only for short linear motions. The available stroke is, of course, dependent on the overall size of the solenoid, but stroke lengths of 0.1 mm to 3 cm are typical. The nonlinear force/displacement relationship of solenoids is often a disadvantage, but clever use of "four-bar" linkages and "over center" mechanisms can take advantage of this characteristic.

A fundamental advantage of solenoids is low cost. Thus, they deserve particular attention in robotic applications stressing low cost or requiring many (more than 5 or 10) actuators.

Considering the advantages and limitations of solenoids for robots, we would expect them to be used for fixturing and as grippers, actuators for brakes, and gear shifters in transmissions.

10.2.8   Exotic actuators

Several unusual actuators are described briefly below. It is unlikely that any of these types of actuators would be used in robots, except for a few highly specialized applications; these actuators are generally not suited for robotic applications because of inadequate speed or torque and excessive cost or weight. The actuators are included here mainly for the sake of completeness. If the robot designer is confronted with a very unusual actuator requirement, it would be wise to review this section to determine if any of these exotic actuators might be applicable.

10.2.8.1   Piezoelectric.   Piezoelectric materials such as quartz and barium titanate change shape (expand or contract) when subjected to an electric field. The simplest piezoelectric actuator consists of a single piece of piezoelectric material with metallic electrodes on opposing faces. A voltage applied to these electrodes will cause the material to change shape, and this shape change can be used to move a load. Unfortunately, the stroke of such an actuator is very small—typically less than 0.1 μm. Longer strokes can be obtained by stacking several of these simple actuators. Another configuration, known as a "bimorph," consists of two pieces of piezoelectric material laminated together. The pieces are oriented so that one expands and one contracts under the influence of an applied electric field. This behavior causes the bimorph to bend when energized. The stroke obtained with a bimorph can be much greater than that which can be obtained with the simplest piezoelectric actuator, described above.

10.2.8.2 Inchworm actuator.   This actuator uses piezoelectric devices in a unique configuration to obtain linear motions with long strokes and very good resolution. Figure 10.1 shows the construction of the actuator schematically. The driven rod is the output element of the actuator and moves linearly, as shown by the arrow. Two annular piezoelectric actuators called "grippers" surround the drive rod. When energized, the center hole in a gripper contracts and thereby fixes the gripper to the rod. A third piezoelectric actuator, called the "driver," is also annular, but it expands in the direction of the drive rod axis when energized. Operation of the actuator is accomplished by the following sequence of drive voltages to the three piezoelectric devices:

1. Energize gripper 1

2. Deenergize gripper 2

3. Deenergize driver

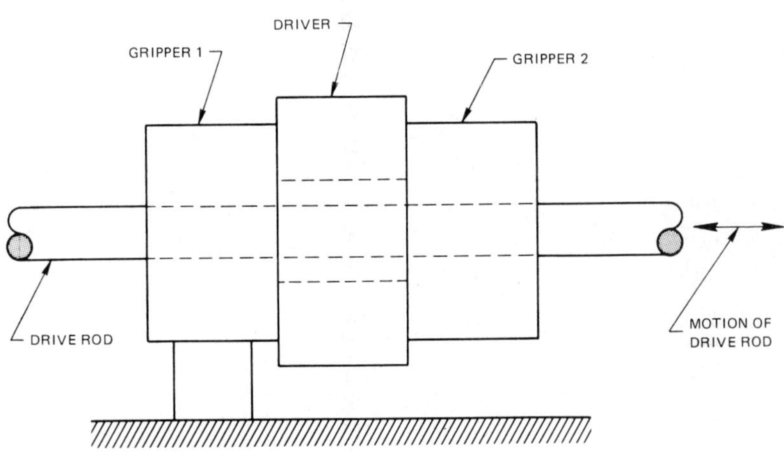

Figure 10.1   Inchworm actuator.

4. Energize gripper 2

5. Deenergize gripper 1

6. Energize driver

7. Go to 1

A similar sequence will move the drive rod in the opposite direction.

Inchworm actuators can have an arbitrarily long stroke, but they are relatively slow and have low output force capability. These actuators are available in the United States from Burleigh Instruments, Fishers, New York.

One application that might be appropriate for Inchworm actuators is in robots used for the assembly of very small devices, such as watches and integrated and hybrid circuits. The extremely good resolution of these actuators (< 0.02 μm) would be ideal for such an application.

10.2.8.3 Thermal. These actuators make use of the fact that most materials expand when heated. Such actuators consist of a rod or block of material with an integral resistance heating element. When the element is energized, the actuator expands as its temperature rises. When deenergized, the actuator contracts. Thermal actuators are simple, rugged, and capable of extremely high output forces. Unfortunately, they are very slow and generally have a very short stroke.

10.2.8.4 Shape-memory alloys. Some of the limitations of conventional thermal actuators are overcome in thermal actuators using shape-memory alloys (Yaeger, 1984). Shape-memory alloys are metals that have the property of returning to a "memorized" shape when heated above a transition temperature. Upon cooling, the metal becomes very soft again and can easily be deformed from the memorized shape. Actuators using shape-memory alloys are characterized by high force for the size of the actuator and have been found to be very reliable. They may find applications in robot grippers or gripper attachment mechanisms for these reasons.

Few organizations have much experience with shape-memory actuators and in dealing with their particular deficiencies and benefits. The engineering procedures for design, including load-fatigue consideration and other special considerations, are being developed. The high price of this type of actuator can be expected to decrease as its use increases.

10.2.8.5 Electrostatic. An electrostatic actuator consists of two conducting plates, separated by a small gap. If a dc voltage is applied to both plates, they will be repelled from each other because of the electrostatic repulsion of like charges. Conversely, if opposite dc voltages are applied to the plates, they will be attracted to each other.

Electrostatic actuators are structurally simple and can have very rapid response time. Unfortunately, they generally have a very short stroke and output force, which would tend to make them unattractive for robotic applications.

10.2.8.6 Magnetostrictive.

The magnetostrictive effect is analogous to the piezoelectric effect, in which an applied electric field causes a material to change shape. In the magnetostrictive effect, an applied magnetic field causes the shape change. Magnetostrictive actuators are limited to very short strokes like piezoelectric actuators. Magnetostrictive actuators are not nearly as popular or as widely used as piezoelectric actuators. Perhaps this is because it is more difficult (i.e., expensive and energy-intensive) to establish a magnetic field in a given volume of sensitive material than to establish an electric field in the same volume. An electric field can be established between two flat conducting plates, whereas a coil and soft magnetic pole pieces are normally used to create a strong magnetic field.

10.3 Brakes

Most types of joint actuators consume a considerable amount of power when the joint is not moving. This is a problem for at least two reasons:

- The power consumed is all converted to heat in the actuator. Because the actuator normally has a low duty cycle, the extra power dissipation while the actuator is at rest effectively decreases the peak and continuous output power capability of the actuator.

- The overall power consumption of the robot is increased. This power consumption is a factor in the total operating cost of the robot and should, therefore, be minimized to make the robot more cost-effective.

The solution to these problems is often the installation of a brake at the joint actuator. While a brake is not, in itself, an actuator, it is clearly a very important part of the actuation system.

Brakes can also be used to advantage when there are certain stability problems in a joint actuator servo-control system. For example, some servo systems evidence slight instabilities as the actuator approaches zero velocity. This problem may be solved by designing the control system so that the brake is applied as soon as the joint velocity is less than a certain threshold.

Several different types of brakes are described below. Most conventional brake devices are configured so that the brake is applied when electric power (or hydraulic or pneumatic pressure) is applied. The brake is released when power is removed. The opposite arrangement (brake applied when power is removed) may be preferable for robotic applications for safety reasons. If the former type of brake is used, the robot may fall under its own weight if power is accidentally interrupted. The fall could damage the robot and the workpiece, and injure human operators who may be near the robot. For these reasons, the brakes used in the PUMA robot are of the latter type. The PUMA, therefore, becomes rigid if power is interrupted.

10.3.1   Electromagnetic friction brake

This brake consists of two or more friction surfaces that are pulled toward each other magnetically when a coil is energized. One of the friction surfaces rotates with the input shaft. The braking torque of this type of clutch is proportional to the friction coefficient μ times the coil current. To the extent that μ can be considered constant, braking torque is proportional to coil current. (The designer is cautioned that constant μ is difficult to achieve, except under laboratory conditions.)

Electromagnetic friction brakes are available with a single moving friction disk and with multiple disks. The single-disk type is more common. Multiple-disk clutches have a greater torque capacity for a given clutch diameter but cannot dissipate heat as well, and, therefore, have very limited duty cycles. Multiple-disk clutches designed for oil immersion are capable of higher duty cycles but are more expensive.

Because electromagnetic friction clutches are actuated with an electrical signal, they are often more convenient for robot applications than the hydraulic or pneumatic types described below. Robot control signals are electrical, and can, therefore, drive electromagnetic brakes directly, without requiring pumps, plumbing, and solenoid valves to interface them to the brake.

A second advantage of electromagnetic friction brakes is their rapid speed of response. Using a technique known as field forcing, response times of 1 or 2 ms can be obtained with the smaller electromagnetic friction brakes.

The friction surfaces of electromagnetic friction brakes experience wear. The wear particles occasionally will present a contamination problem, either to the robot or its environment. Because worn friction surfaces must eventually be replaced, these brakes are a maintenance item.

Electromagnetic friction brakes do not produce sparks under normal operation. However, if the friction surfaces are sufficiently worn, metal-to-metal contact and sparking can occur. For this reason, these brakes should not be used in explosive environments without adequate precautions.

10.3.2 Pneumatic friction brakes

The pneumatic friction brake is similar to the electromagnetic friction brakes described above, but is actuated by air pressure rather than by a magnetic field. Actuators may be pistons or inflatable tubes, bladders, or glands. Pneumatic brakes achieve higher torques than equivalently sized electromagnetic friction brakes, but not as much as hydraulic brakes.

A decided advantage of pneumatic brakes over electromagnetic brakes is the fact that the pneumatic actuator does not generate heat while the brake is engaged. Pneumatic brakes can therefore remain engaged indefinitely without experiencing overheating.

10.3.3 Hydraulic friction brake

Hydraulic friction brakes develop even higher torques than do pneumatic brakes of the same size. Their construction is similar to pneumatic brakes, but the higher pressures used in hydraulic systems (1000 to 5000 lb/in^2 opposed to 50 to 200 lb/in^2 in pneumatic systems) permit higher normal forces at the friction surfaces and, therefore, higher torque.

The pumps, valves, and plumbing required in hydraulic systems are much more expensive and maintenance-intensive than those required for pneumatic systems. This fact is a strong deterrent to the use of hydraulic brakes unless the robot also requires hydraulic power for other functions, such as joint actuators.

10.3.4 Magnetic particle brake

The magnetic particle brake consists of a hollow chamber with a close-fitting disk inside. The remaining space inside the chamber is filled with magnetic particles, such as iron or ferrite dust. A coil is mounted on the chamber so that, when energized, it will set up a strong magnetic field between the disk and the inside surfaces of the chamber. The magnetic particles form dense bridges between the disk and the chamber under the influence of this field and thereby impede rotation of the disk.

The primary advantage of the magnetic particle brake (aside from direct electric actuation) is that its friction surfaces—the magnetic particles—do not wear out. It is, therefore, an essentially maintenance-free device.

10.3.5 Hysteresis brake

The hysteresis brake's moving member is a disk or cup of magnetic material chosen for its large hysteresis loss. An electromagnet and an array of pole pieces generates a series of alternating (north and south) magnetic fields around the periphery of the disk or cup. If the disk rotates, the material is exercised around its hysteresis loop as it passes through the alternating fields set up by the pole pieces. The energy loss associated with the hysteresis in the disk material results in a braking torque exerted on the disk. The hysteresis brake, like the magnetic particle brake, has the advantages of electric actuation and maintenance-free operation for indefinite periods.

10.3.6   Eddy current brake

The eddy current brake is constructed much like a hysteresis brake, but the moving member is constructed of a high-conductivity, nonmagnetic material such as copper. When the moving member rotates, eddy currents are set up in it. The energy loss associated with these eddy currents results in a braking torque being exerted on the moving member.

Eddy current brakes, like hysteresis and magnetic particle brakes, are electrically actuated and require no periodic maintenance. Unlike these other brakes, however, the eddy current brake will not generate a braking torque unless its shaft is rotating. The eddy current brake is therefore of limited use for robots because it cannot hold a joint stationary—the primary function of brakes in robots.

10.4   Selection and Sizing of Actuators

A basic requirement in any practical control mechanism is that adequate force (or torque) is available to control the load in some well-defined manner. To arrive at the specifications for the actuators, we need to examine the nature of the load, the displacements or motions we wish to impart to it, and the geometry of the device that will accomplish this.

If a robot is designed according to the design procedure of Chapter 3, the iteration of the kinematic design, structural design, and structural analysis of the robot will be available to guide actuator selection. These efforts will normally have included the generation of a computer model of the arm, which can be used, together with the robot performance requirements, to determine the performance requirements for each joint actuator. If a computer model of the arm is not available, it is possible to estimate the performance requirements for each joint actuator using rather simple analytical methods.

The various types of loads that must be considered to determine joint actuator performance requirements are discussed below. A highly simplified example of actuator selection is presented (for a specific joint in the Unimation PUMA robot). Because a computer model of the PUMA is not available, the example relies on analytical estimates of joint loads.

10.4.1 Joint actuator loads

It is necessary that a robot designer thoroughly understand the characteristics of the loads seen by each joint actuator and the desired performance of the arm. To this end, it would be desirable to have the following load data referred to each joint in question.

Unbalance or gravity loads.   Unbalance loads will be present at all the joints as a result of the workpiece load having its center of gravity offset from the point of grip by the end effector. Other unbalance loads will be present because of the weight of the individual components of the manipulator arm. All these loads will vary with position as the robot articulates in following any given trajectory. The unloaded versus the loaded condition of the arm presents a large variation of the load to the arm and its joints.

Inertia loads.   The workpiece inertia reflected back to each joint actuator is needed. The workpiece inertia as seen by each joint will vary not only as a function of movement of the arm as it follows a given trajectory, but also as a function of the shape of the load, its orientation, and where it is gripped. The robot structure itself is also an inertial load. Of course, the inertial load "seen" by each joint as a result of the robot structure will also vary as the arm moves. In addition to inertial loads due to linear acceleration or deceleration, centrifugal, and coriolis forces also contribute to the total inertial load seen by each joint.

Friction forces.   Friction forces can at times be bothersome, especially near zero velocity. Friction forces can be of various kinds, e.g., static friction, rolling friction, and viscous friction. Friction forces arise from seals, bearings, and rubbing parts. An additional source of friction, which is often overlooked, is the cables and hoses in the robot that pass through or around each joint. As a joint moves, the cables and hoses near the joint must bend, twist, and slide over each other to allow for the joint movement.

Duty cycle.   The duty cycle for a joint actuator defines a typical and worst-case time history of the loads and speeds experienced by the actuator. Of course, good design practice calls for choosing an actuator based on worst-case loads and duty cycle. However, this is not always done. Design considerations relating to duty cycle are discussed in more detail below.

Note that in the initial design phase some estimates of the torques and inertias must be made, which, as the design emerges, must be later modified or refined. Thus an estimate of actuator weight, torque, and inertia will probably have to be changed from the original estimate to suit the real-world availability of certain stock sizes or shapes of actuators.

10.4.2 Example of actuator selection procedure

This section gives a simplified example of the procedure that would be used for selection and sizing of an actuator for a particular joint of a given robot. We have selected the joint 1 (waist) actuator of the PUMA robot for this example for several reasons:

- The PUMA is a relatively recent (i.e., modern) design from a company with considerable robot experience. (Unimation's previous experience had been with hydraulic, rather than electric, robots. This fact may help explain some of the apparent deficiencies in the particular motor chosen by Unimation for joint 1.)

- Estimation of the loads experienced by the joint 1 actuator is relatively simple because:

  The joint 1 actuator in the PUMA does not experience any gravity loads.

  As a result of the kinematics of the PUMA, inertial loads "seen" by joint 1 as a result of the motion of joints 2, 3, 4, and 5 are zero or small compared to other inertial loads.

For the purposes of this example, we will assume that all elements of the robot design have been fixed except the choice of the actuator for joint 1. Of course, in the real world of robot design, tradeoffs can be made between actuator cost and performance and other parts of the design. To avoid needless digression, we will assume here that no such tradeoffs can be made and will concentrate on selecting an actuator for the robot as it now exists.

Some relevant robot performance parameters of the PUMA are listed in Table 10.1. These parameters are taken from data supplied by Unimation. Determination of which pose or poses of the robot correspond to worst-case loads for the joint actuator can be rather complicated and may require the use of a computer model of the arm to perform the required

TABLE 10.1   Some Puma Performance Specifications

Parameter	Value
Maximum load, including gripper, kg	2.27
Static force at tool, N	58
Maximum tool acceleration, g	1*
Maximum tool velocity, m/s	1
Maximum radius (from shoulder) of gripper mounting surface, m	0.92

* 9.8 m/s^2.

calculations in a reasonable amount of time. Fortunately, we can make good estimates for the loads on the joint 1 actuator because of the convenient geometry.

We will first consider the torque required of the joint 1 actuator to achieve the specified tool acceleration. Note that for a given value of acceleration of joint 1, $d^2\Theta_1/dt^2$, the acceleration of the tool is proportional to r_T where r_T is the perpendicular distance from the joint 1 axis of rotation to the tool. Furthermore, the rotational inertia of a massive tool about the joint 1 axis J_T is proportional to r_T^2. Specifically,

$$\frac{d^2x_T}{dt^2} = r_T \frac{d^2\Theta_1}{t^2} \tag{10.1}$$

and

$$J_T = m_T(r_T)^2 \tag{10.2}$$

where d^2x_T/dt^2 is the tool acceleration and m_T is the tool mass. Let T_1 be the torque at joint 1 required to accelerate the tool

$$T_1 = J_T \frac{d^2\Theta_1}{dt^2} \tag{10.3}$$

Substituting from Equations (10.1) and (10.2) gives

$$T_1 = m_T(r_T)^2 \frac{d^2x_T}{dt^2} \frac{1}{r_T} = m_T r_T \frac{d^2x_T}{dt^2} \tag{10.4}$$

Equation (10.4) implies that the joint 1 torque required to achieve a given tool acceleration is proportional to r_T. Therefore, maximum torque at joint 1 would be required when the pose of the arm is outstretched horizontally. In this pose, $r_T = 0.92$ m (from Table 10.1). This conclusion was reached by ignoring the inertia of the arm structure. Further analysis indicates that the conclusion is still valid if structure inertia is included, except for poses where the elbow is bent and the tool is very close to the shoulder. For the sake of simplicity, we will ignore these bent-elbow poses in our analysis.

Having concluded that the horizontal outstretched pose represents the worst-case load at joint 1 for tool acceleration, we now must determine the rotational inertia for the entire arm in this pose. Estimated inertia of the various parts of the arm in this pose are tabulated in Table 10.2. (Precise values for the inertias would be obtained directly from a computer model of the arm if it were available.) Let J_tot be the total inertia of the arm and tool (or gripper and workpiece if the robot is doing assembly or loading). Equation (10.3) can be rewritten to include arm inertia giving

$$T_1 = J_\text{tot} \frac{d^2\Theta_T}{dt^2} \tag{10.5}$$

TABLE 10.2   Rotational Inertias about the Joint 1 Axis of Rotation

Part of the robot	Inertia, kg/m^2
Rotating part of the tower	0.026
Joint 2 motor and transmission	0.392
Upper arm structure	0.142
Joint 3 motor and transmission	0.528
Joint 4 and 5 motors	0.162
Lower arm structure	0.341
Wrist mechanisms and transmissions	0.634
Gripper and load, 2.27 kg	1.921
Total inertia	4.146

Substituting from Equation (10.1) gives

$$T_1 = J_{\text{tot}} \frac{d^2x}{dt^2} \frac{1}{r_T}$$

$$= (4.146 \text{ kg} \cdot \text{m}^2) \left(9.8 \, \frac{\text{m}}{\text{s}^2} \right) \frac{1}{0.92 \text{ m}} \tag{10.6}$$

$$= 44.2 \, \frac{\text{kg} \cdot \text{m}^2}{\text{s}^2}$$

$$= 44.2 \text{ N} \cdot \text{m}$$

The joint 1 motor drives the joint through a transmission consisting of a pair of idlers driving a large "bull" gear. (The two idlers are used for backlash elimination.) The ratio of this transmission N is about 22:1. Let T_M be the torque exerted by the joint 1 motor. Neglecting losses in the transmission and bearings, we have

$$T_M = \frac{T_1}{N} = \frac{44.2 \text{ N} \cdot \text{m}}{22} = 2.001 \text{ N} \cdot \text{m} \tag{10.7}$$

(In a well-designed robot, transmission and bearing losses should be less than 5 percent of the total torque, so they can be ignored in this approximate analysis.) Thus, the motor must be capable of exerting about 2 N·m of torque to meet the tool acceleration performance specification.

Next, we will consider the "static force at tool" performance specification. The torque at joint 1 required to exert a static force F is given by

$$T_1 = r_T F = (0.92 \text{ m}) (58 \text{ N}) = 53.4 \text{ N} \cdot \text{m} \tag{10.8}$$

and the corresponding torque at the motor is

$$T_M = \frac{T_1}{N} = \frac{53.4 \text{ N} \cdot \text{m}}{22} = 2.43 \text{ N} \cdot \text{m}$$

Next, we will consider the maximum tool velocity specification. The joint 1 motor speed ω_M required for a given tool velocity V_T is given by

$$\omega_M = \frac{V_T}{r_T} N \tag{10.9}$$

Obviously, the joint 1 motor velocity required to achieve the maximum tool velocity is greatest when r_T is small.

Consider a pose of the robot where the tool is above the shoulder and lies on the axis of rotation of joint 1. In this pose, $r_T = 0$, and thus it is impossible to achieve the specified tool velocity in a horizontal direction parallel to the axis of joint 2 (the shoulder). We must conclude that the kinematic design of the PUMA is defective if the maximum tool velocity specification must be met for all poses and all directions of tool motion.

To proceed with this example, we will now assume that the PUMA is not intended to achieve maximum tool velocity for values of r_T less than 0.1 m. (This value of r_T corresponds to the size of the tower of the robot.) Based on this assumption, we can find the maximum motor speed required to meet the performance specification using Equation (10.9):

$$\omega_M = \frac{V_T}{r_T} N = \frac{1.0 \text{ m/s}}{0.1 \text{ m}} 22 = 220 \frac{\text{rad}}{\text{s}} = 2100 \frac{\text{rev}}{\text{min}}$$

At this point, we can summarize the motor performance requirements as follows: To meet the static force and maximum acceleration performance specifications, the motor must be capable of exerting a peak torque of at least 2.43 N·m. To meet the maximum speed performance specification, the motor must have a maximum speed of at least 2100 rev/min.

At this point in the motor selection, the designer would normally consult the literature of several motor vendors to see what type and size of motor might be appropriate. (Alternatively, if the cost can be justified, a motor could be designed specifically for the application.) Table 10.3 lists performance specifications for selected dc motors from various vendors. Based on the speed and torque requirements stated above, several of these motors would be suitable for the application including the Electro-Craft E-542, Clifton Precision DH-2250-B-1 and DH-2250-S-1, and PMI Motors U9M4HT. If the requirements are relaxed slightly, other motors that could be used include the Sierracin SM3 402-12 and SM3 402-13, Electro-Craft E-701, PMI Motors T06LB, Torque Systems MH 3323-001, and Ametek 115869. It is interesting to compare these motors to the one actually used in the PUMA. The PUMA motor is approximately 3 in in diameter and 4.5 inches long. Its estimated weight is 4.8 lb. These parameters are comparable to those of the motors listed above. In particular, the weights of these motors range from 3.0 to 5.5 lb, a range that nicely brackets the estimated weight of the PUMA motor. All the PUMA motors

TABLE 10.3 Performance of Selected Dc Motors

Vendor	Model	Peak torque, N·m	Continuous torque, N·m	Rotor inertia, kg/m^2	Maximum speed, rev/min	Diameter, in	Length, in	Weight, lb
Sierracin	SM3 402-12	2.12*	0.27	0.000205	659	2.8†	3.66	>3.4
	SM3 402-13	2.82	0.30*	0.000268	611	2.8†	4.34	>4.4
	RE5 414-1	2.71	0.21*	0.000918	108	5.3†	1.25†	>2.3
Electro-Craft	E-542	2.51	0.35	0.000044	6000	2.25	5.00	3.0
	E-701	5.30	1.06	0.000706	2000?	4.0	4.81	6.6
	E-702	9.53	2.12	0.001059	2000?	4.0	5.81	8.3
Clifton Precision	DH-2250-B-1	2.54	0.32	0.000053	—‡	2.26	5.68	3.5
	DH-2250-S-1	2.47	0.30	0.000048	—‡	2.26	5.00	3.0
	DH-2250-C-1	2.82	1.73	0.000254	—‡	3.5	4.95	7.5
PMI Motors	TO6LB	2.45	0.98	0.000177	700‡	3.35	4.17	5.5
	U9M4HT	2.75	0.22	0.000056	2700‡	4.37	2.40	—‡
	MC17H	13.56	1.60	0.000777	3900	8.31	4.67	14.3
Indiana General	4310 TENV	17.55*	1.03	0.000706	2200	4.5	6.06	8
Torque Systems	MH 3323-001	2.12	0.53	0.000056	1800‡	3.0	3.90	—‡
	MH 5115-001	4.24	1.38	0.000473	1200‡	5.0	4.75†	—‡
	MH 5130-001	8.47	2.82	0.000847	1200‡	5.0	6.25†	18*
Ametek	115793	3.05	1.24	0.000565	700‡	4.0	6.38	—‡
	115869	5.65	1.24	0.000565	1400‡	4.0	6.38	—‡

* = value estimated from other motor parameters.
† = approximately.
‡ = value in doubt or unknown.

except the joint 1 motor are located in one of the links, and it is therefore desirable to minimize their weight. Because the joint 1 motor is not located in a link, it is not terribly important that its weight be minimized.

Next, we will consider the effect of motor rotor inertia. When a motor drives an inertial load such as a robot through a gear train with ratio N, the effective load inertia seen by the motor J_{eff} is given by

$$J_{\text{eff}} = \frac{1}{N^2} J_{\text{tot}} \tag{10.10}$$

where J_{tot} is the total load inertia about the axis of the joint in question (Tal, 1981). Thus, for the PUMA joint 1 motor,

$$J_{\text{eff}} = \left(\frac{1}{22}\right)^2 \left(4.146 \, \frac{\text{kg}}{\text{m}^2}\right) = 0.00857 \, \frac{\text{kg}}{\text{m}^2}$$

The rotor inertias of the motors mentioned above range from 0.000044 to 0.00071 kg/m^2. We can see that the rotor inertia of any of these motors is much less than the arm inertia seen by the motor. This relationship means that acceleration of the motor and load will be dominated by the load inertia rather than by the rotor inertia, and it will not be important to minimize the rotor inertia for this application. As noted above, these circumstances favor a conventional (iron rotor) dc servomotor over the ironless rotor dc servomotor. In fact, Table 10.3 was compiled with this result in mind—all the motors listed in this table, except for the PMI Motors U9M4HT and MC17H, are conventional dc servomotors. These motors were included because they are "pancake-type" ironless rotor dc servomotors that tend to have high (for ironless rotor motors) rotor inertia.

We can now turn our attention to the issues surrounding the continuous power output rating and duty cycle of the motor. It is noteworthy that Unimation does not specify a duty cycle for the PUMA. This is an unfortunate and—as we shall see—probably deliberate omission. Lacking any performance specifications to the contrary, we can envision two situations that would make worst-case demands on the joint 1 motor's continuous output capability:

- The robot might be commanded to move the tool in a horizontal direction while in the horizontal outstretched pose. If, as a result of an unexpected obstruction in its path, the robot was unable to move as commanded, it would presumably exert the specified maximum force (58 N·m) against the obstruction indefinitely. In this case, the joint 1 motor would have to exert a 2.43-N·m torque indefinitely.

- The robot might be commanded to move the tool back and forth horizontally while in the horizontal outstretched pose. With the proper combination of commands, the tool could be accelerated at 1 g until it reached 1 m/s velocity, then decelerated at 1 g to -1 m/s, then accelerated at 1 g to 1 m/s, and so on. In this case, the joint 1 motor would have to exert a 2.01-N·m torque indefinitely.

The continuous torque rating of a motor is closely related to the continuous power output capability of the motor. Continuous torque ratings are included in Table 10.3. Inspection of the table shows that none of the motors mentioned above would be capable of the continuous torques required by the two worst-case situations described above. The only motor in Table 10.3 capable of these continuous torques is the Torque Systems MH 5130-001; the Electro-Craft E-702, Clifton Precision DH-2250-C-1, and PMI Motors MC17H motors can *almost* exert these continuous torques. Note that these four motors are much larger than the motor actually used in the PUMA—their weights range from 7.5 to 18 lb. While it is conceivable that the motor actually used in the PUMA is a special design capable of these high continuous torques, this is very unlikely in view of its significantly smaller size. It is more likely that the actual PUMA motor is *not* capable of the continuous torque required by the two situations postulated above. If so, then to avoid overheating and damage to the motor, its duty cycle must be limited to something less than about 40 percent. (For example, the continuous torque rating of the PMI Motors T06LB is about 40 percent of its peak torque rating.)

The above observations concerning the PUMA duty cycle are consistent with rumors that the PUMA motors are occasionally destroyed by overheating. Clearly, robot duty cycle is an important performance specification that should not be overlooked in the design of a robot.

This example illustrates an important tradeoff to be considered in the specification and design of any new electric robot. Achieving a duty cycle near 100 percent requires relatively heavy motors. Assuming that at least some of the motors will be mounted in the links, the robot inertia will be increased—and, therefore, its acceleration and frequency response capabilities decreased—if heavier motors are used. Duty cycle must be compromised to obtain optimum acceleration performance, and visa versa. It would appear that innovations that relax this constraint will be important in future generations of robots. One such innovation would be the use of active (air, oil, or water) cooling of the motors.

Seffernick (1982) points out that the weight of an oil-cooled motor can be less than one-tenth the weight of a conventional motor with a comparable continuous horsepower rating.

11

Transmissions

The torque (or force) and velocity available at an actuator differs from that needed to control the joint position and velocity. The transmission is needed to multiply the torque. As will be noted, increasing the torque decreases the speed and increases the effective inertia of the joint, which includes the inertia of the rotating parts of the motor. Furthermore, since it is desirable to reduce the outboard weight (and mass) on a limb, the actuators are often not located at the joints they actuate but closer to the previous joint. The Microbot is an extreme example; all the actuators for subsequent joints are located in the rotating element, including the actuator for the gripper. Means, such as the cables in the Microbot, are needed to transmit the motion from the actuators to the joints.

Figure 11.1 shows a schematic of a joint driven directly by an actuator. Usually gear reduction is needed between the motor and the joint, and a transmission is provided as shown in the schematic of Figure 11.2. In this case, the actuator is further removed from the transmission and is driven through a belt connection. The arrangement sequence of actuator, transfer, transmission, and joint is typical. Locating the transmission at the joint reduces the forces on the transfer components and usually results in a stiffer system with less backlash.

Energy is consumed in transmissions, and this loss increases with the ratio of gearing. A transmission with a high enough loss becomes effectively *non-back-drivable*; i.e., it can be driven from the high-speed side only. The forces required to overcome the starting friction from the low-

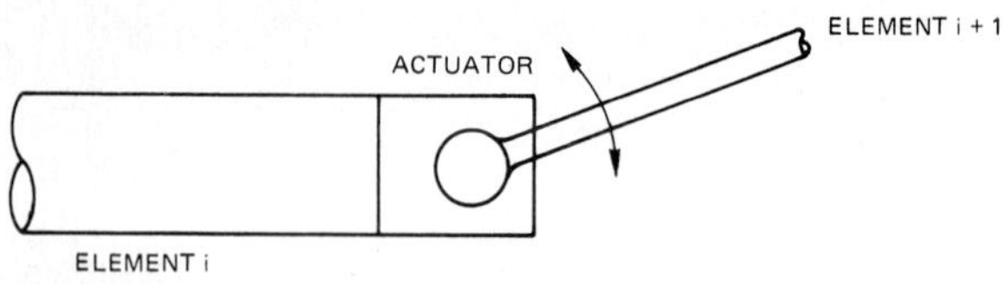

Figure 11.1 Example of a hinged joint.

speed side are destructively large. Transmissions can be made deliberately non-back-drivable. A worm gear can be made self-locking by selecting a gear ratio high enough so that the worm lead angle is smaller than the worm-to-gear coefficient of friction. Hydraulic systems are non-back-drivable as well.

The back-drivable joint is needed if the actuator is to feel and respond to the torque on the joint as required for force control. The self-locking joint, however, eliminates the need for a brake.

The actuator is typically mounted on one element to control the next element, but this need not be the case. The early Unimates had gears that transmitted torque through joints so that the actuator could be located further away than the adjacent element. Several smaller educational manipulators have most of the actuators in the base and use chains or cables as a part of the transmission.

The general advantage of locating the actuators on the element adjacent to the controlled joint is improved precision. The major problems of transmissions—backlash and wear—are minimized. The advantage of positioning on remote elements is that the structure does not have to support the actuators. The structure can be made lighter in weight and faster. These general problems are discussed in the strategy section of Chapter 3, Design Procedure; the focus of this chapter is the transmission elements themselves.

Recent trends have been toward the use of electric actuators. Because the stall torque of electric motors is low, gearing is usually used to multiply the torque. Electric motors have been developed that produce considerably improved stall torque for their weight. These motors are being used for direct drive in the joints of many SCARA robots where the actuators are not gravity-loaded. (Direct drive implies no transmission, since the rotor and stator are parts of adjoining limbs. A gear ratio of 1:1 is also

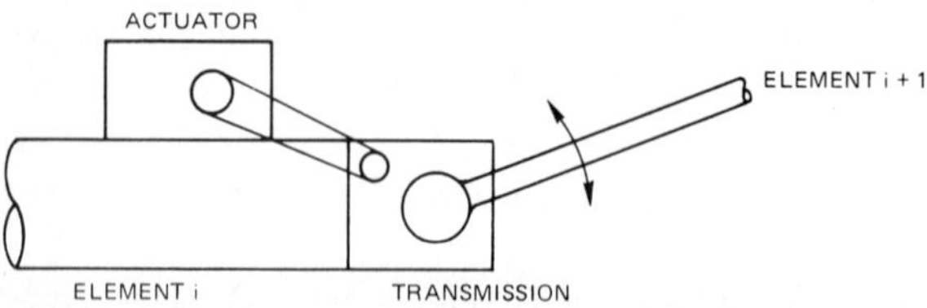

Figure 11.2 If greater force or torque is required, a transmission connects the actuator with the joint to be driven.

referred to as direct drive even though a timing belt may be used.) Usually, however, gear ratios of tens to several hundred are desired. The benefit is that the motor torque is multiplied by that factor so that high torque is delivered for the size of the actuator and transmission. The performance cost is a reduction in speed by the same factor and an effective motor inertia that goes as the square of the gear ratio. Chapter 10, Actuators, discussed the strategy of sizing the motor and selecting the reduction ratio of the transmission. This chapter examines the different types of transmissions and their features.

11.1 Belts, Cables, and Chains

Belts, cables, and chains can be used as tendons to transmit actuator position and force information, or in continuous motion as power transmitting devices. The dentist's drill drive is an example of the power transmission capability of a cable through numerous joints.

11.1.1 Types

Specific configurations are described below.

- *Flat belts.*   Are common for large-diameter crowned pulleys, but not appropriate for power transmission. Not applicable to robots.

- *V belts.*   V-shaped in cross section. The wedging action in the pulley gives greater power-transmitting capability. Belts can slip, which may have the advantage in robotics of providing overload protection.

- *Timing belts.*   Have ridges that run in grooves on the pulleys so that positive, no-slip transmission occurs.

- *Cables.*   Are similar to belts, but because they do not have a preferred bending direction as belts do, they can be used in more than one plane. Some clever designs put studs on cables so that they can be used for timing purposes. Clever arrangements of cables have made it possible to locate all actuators on the base and to control six degrees of freedom (Mini Mover). The Armtron, a toy robot by Tomy, achieves the same result through gear and shaft transmissions. The Armtron is notable in using only a single motor. It should be studied by robot transmission designers!

- *Straps.*   Are a form of flat cable, usually of metal, used as tendons rather than for power transmission, the usual function of belts. Straps can also be used to hold joints together as illustrated in Section 11.1.4.

- *Chains.*   Serve the functions of heavy-duty belts.

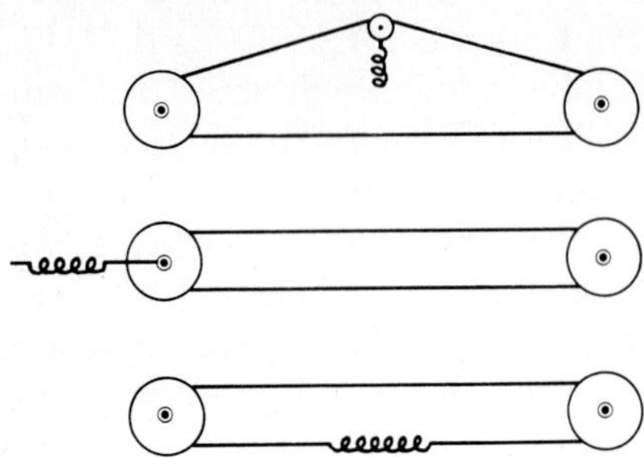

Figure 11.3 Tension adjustment methods.

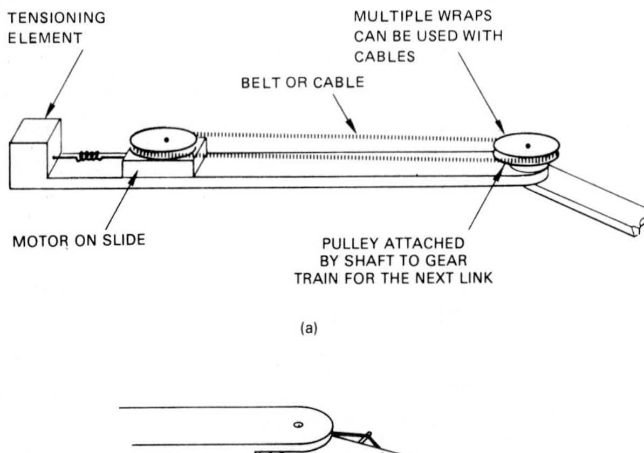

Figure 11.4 Attachment methods. (*a*) Tendon pulley arrangement; (*b*) tendon yoke.

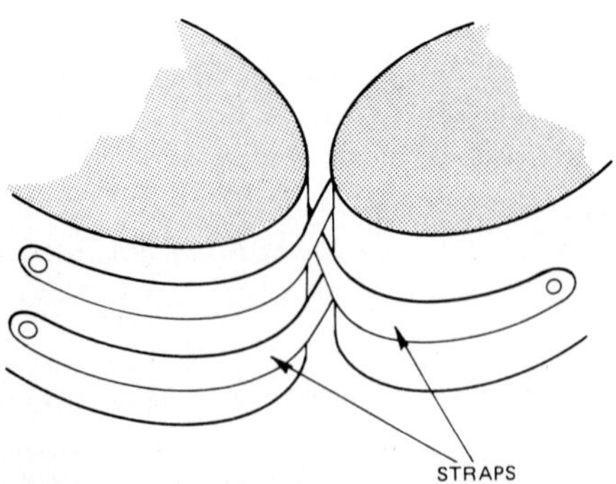

Figure 11.5 Tendon-rolling joint.

11.1.2 Arrangements

Belts, cables, and chains usually have a tensioning arrangement to keep slack out of the system. This is equivalent to backlash in gears. The three common tensioning techniques are shown in Figure 11.3.

11.1.3 Attachments

Figure 11.4 shows two ways to use tendons to control a joint position. The cable can go around a pulley (multiple times if necessary to prevent slip) or can be anchored to the arm being controlled.

11.1.4 Joints

Involute joints (one component rolls on another) can be made using straps as tendons to hold the joint together. Figure 11.5 shows the kind of joint invented by Roth for use in grippers.

11.2 Gear Trains

The most common means of torque or force transmission combines gears into a "gear train," such that the output shaft speed and torque are a linear function of the input shaft speed and torque. The gears themselves transmit rotary motion and torque through the pair of mating gear teeth, the profiles of which are cams acting against one another to produce the desired motion. Several profiles have been developed, of which the involute is the most notable, to transmit uniform rotary motion for all operating positions of the contacting profiles. Contact between gear teeth is primarily rolling, so efficiency is high and back-drivability is possible. For a complete reference on gear tooth profiles, see Buckingham (1949).

11.2.1 Gear types

Gears are generally classified by the general configuration of the gear shafts:

- Parallel shaft
- Intersecting shaft
- Nonparallel, nonintersecting shaft
- Rack gearing

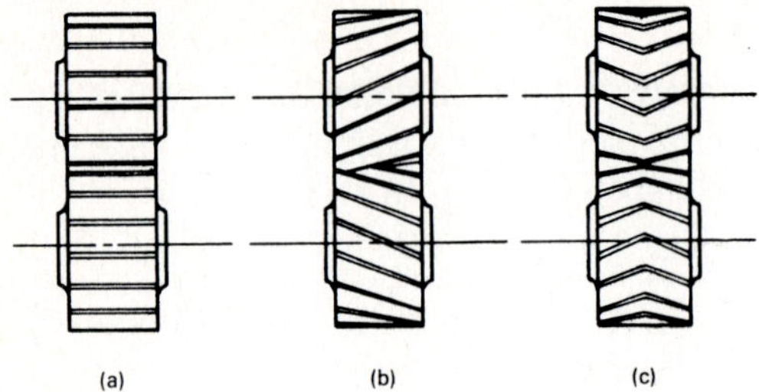

Figure 11.6 Examples of gear shaft configurations.
(*a*) Spur; (*b*) helical; (*c*) herringbone.

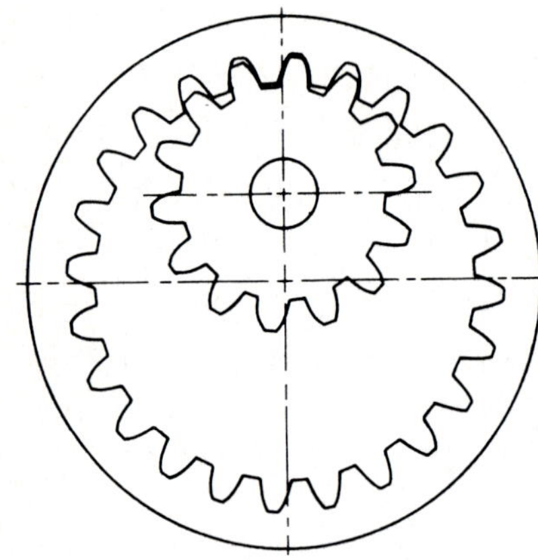

Figure 11.7 Internal gearing.

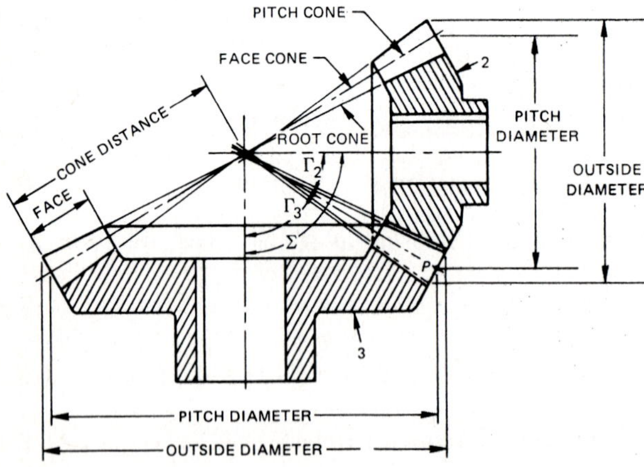

Figure 11.8 Bevel gear geometry.

11.2.1.1 Parallel shaft

11.2.1.1.1 Spur gear. The most common type of gear used in parallel shaft geometry is the spur gear, which is cylindrical in form with teeth parallel to the centerline (see Figure 11.6a). One problem in the operation of the spur gear is that the mating teeth make instantaneous full-width contact at mesh. The sudden contact means that the teeth experience sudden deformation to compensate for the deflection under load of the teeth that were just previously engaged. The resulting robot motion is not perfectly smooth and, therefore, contributes to vibration and shaking.

11.2.1.1.2 Helical gear. The helical gear (Figure 11.6b) provides gradual contact of the engaging teeth, which transmits load more smoothly. The angle of the helix introduces an axial component for force requiring at least one bearing to carry thrust as well as radial load. Helical gears may be arranged in a gear train such that "opposite-hand" (right-handed helix and left-handed helix) gears balance out the axial loads.

11.2.1.1.3 Herringbone gears. The herringbone gear (Figure 11.6c), in which both left- and right-handed helices are cut into a single gear, balances this axial force within the gear itself. The herringbone gear does require very precise axial location relative to the mating gears. The simplest and most common mounting method uses thrust bearings for one shaft only and allows the other shafts to float axially. The herringbone gear tooth forces then automatically position the other gears so that no external thrust forces are present.

11.2.1.1.4 Internal gear. An internal gear has either spur or helical teeth cut on the inside of a gear ring (see Figure 11.7). This configuration is commonly used for reduction gear and planetary gear combinations. The internal gear drive is more compact, and the tooth action is smoother than a conventional spur or helical gear drive. The compact size is a clear advantage in manipulator design.

11.2.1.2 Intersecting shafts

11.2.1.2.1 Bevel gears. The term "bevel" describes gears based upon pitch surfaces that are frustrums of cones. The theoretical extensions of the two cones intersect at the shaft intersection of the rotational axes (see Figure 11.8). When the two shafts are connected with identical gears with pitch angles of 45° (90° between shafts), the gears are called *miter gears*. The action of bevel gears will be the same as that of equivalent spur gears; however, the bevel gears will have a greater contact ratio and will run more smoothly than a pair of spur gears with an equivalent number of teeth.

11.2.1.2.2 Spiral bevel gears. Spiral bevel gears are analogous to bevel gears as helical gears are to spur gears (see Figure 11.9). The spiral geometry improves the tooth action by giving gradual contact and a greater arc of action.

11.2.1.3 Nonparallel, nonintersecting shaft

11.2.1.3.1 Crossed-axis helical gears. Crossed-axis helical gears are used in applications for relatively light loads where it is necessary to transmit torque between shafts that do not intersect and are not parallel (see Figure 11.10). Here the gear pitch cylinders are in contact at only one point, and the subsequent rotation will be accompanied by one gear sliding upon the other, which reduces efficiency and increases wear. Normally, both gears are of the same "hand" (helix angle); however, for small angles, the gears may have opposite hands. As is the case for any pair of mating involute gears, the two gears must have the same pitch and pressure angle in the normal plane of contact.

The angular velocity (ω) ratio between meshing gears is the ratio of gear teeth (N),

$$\frac{\omega_2}{\omega_3} = \frac{N_3}{N_2}$$

It is sometimes desirable to calculate the velocity in terms of the pitch diameters or to calculate the pitch diameters required to yield a particular velocity ratio. For crossed-axis helical gears, the velocity ratio is

$$\frac{\omega_2}{\omega_3} = \frac{D_3 \cos \Psi_3}{D_2 \cos \Psi_2}$$

where D_i is the respective pitch diameter of the gears and ψ_i are the helix angles as shown in Figure 11.10.

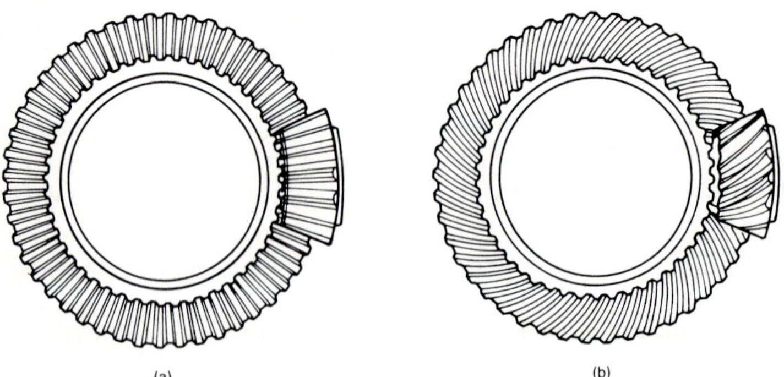

(a)   (b)

Figure 11.9 (*a*) Straight and (*b*) spiral bevel gears.

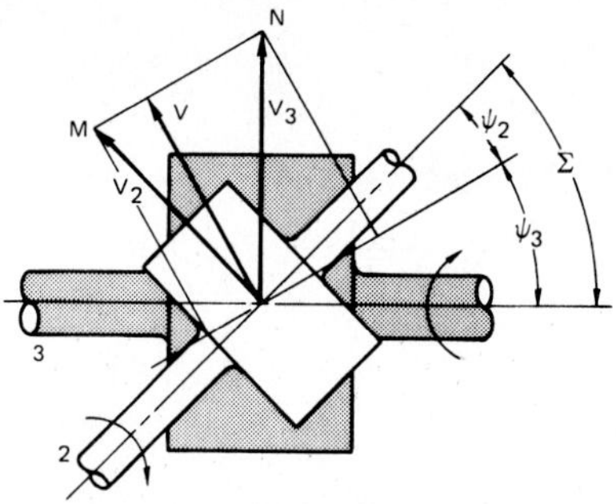

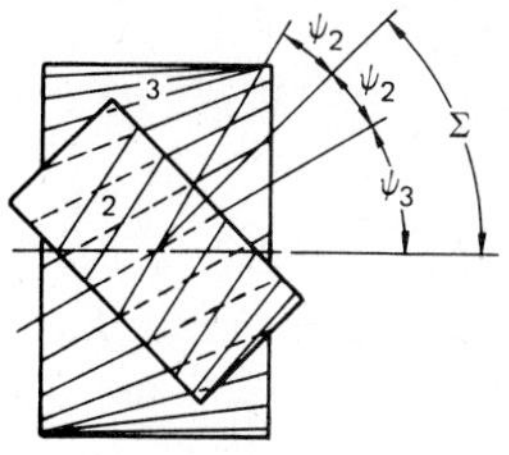

Figure 11.10   Helical gears on crossed axes.

11.2.1.3.2   Worm gears.   In general, worm drives are used to drive nonintersecting shafts which are at right angles to each other. The cylindrical worm (Figure 11.11) is a true involute helical gear with one tooth. The screw-thread appearance results from the combination of a relatively large helix angle and a small gear diameter. The axial pitch corresponds to the pitch of a screw thread; the lead is the distance a point on the screw thread advances during one revolution. The lead equals the pitch for a single-thread (single start) worm and equals twice the pitch for a double-thread (double start) worm. The *lead angle* is the angle at the pitch diameter between a thread and the plane of rotation.

The major advantage of a worm drive is that it is one of the simplest means of achieving very large velocity ratios. When a worm gear is used to drive a spur gear or helical gear, there is only point contact and only light loads can be carried. Use of a worm gear, such as that in Figure 11.12, results in line contact instead of point contact. When the worm is allowed to double-envelope the worm gear, increased line contact results; however, it is critical to align the double-enveloping gears.

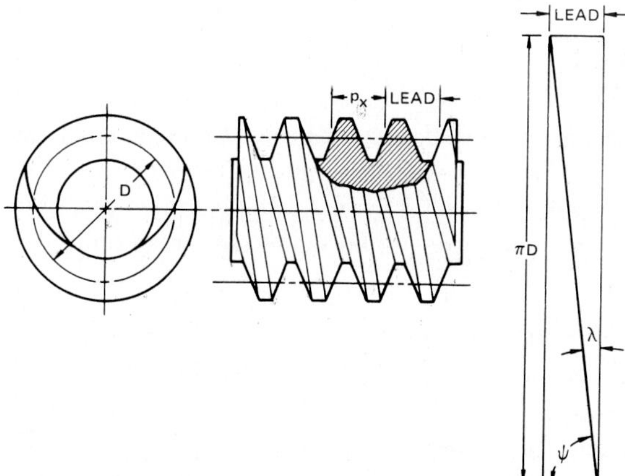

Figure 11.11   Single-thread cylindrical worm.

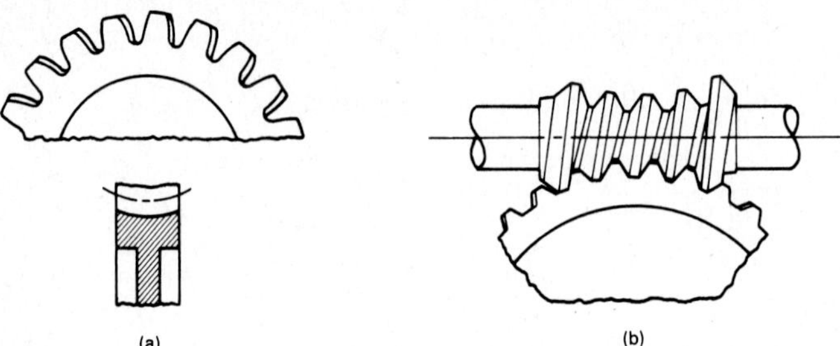

Figure 11.12 Worm gears. (a) Worm gear for use with a cylindrical worm. (b) Double-enveloping worm and worm-gear set.

For shafts at right angles, the efficiency of a worm drive is

$$\eta = \tan \lambda \, \frac{\cos \phi_n - \mu \tan \lambda}{\cos \phi_n \tan \lambda + \mu}$$

where λ = lead angle
ϕ_n = pressure angle
μ = coefficient of friction

Because of the friction losses caused by the sliding and wearing of the teeth, worm drives suffer from high energy losses. Table 11.1 indicates worm gear efficiency as lead angle and speed are varied.

Another feature of a worm drive is that under certain conditions, the drive can be self-locking; in other words, the worm gear cannot drive the worm. The criterion is a function of the lead angle, the pressure angle, and coefficient of friction. As an approximation, the drive will be self-locking if the coefficient of friction is greater than the tangent of the lead angle. Self-locking can be an advantage; joint breaks or holding torques are not necessary.

The low energy efficiency and the associated tendency toward self-locking or lack of back-drivability, however, mean that motor torque cannot be translated into joint torque. This may be an advantage in simple robots designed merely to move from one location to another, but it is a serious drawback if the aim is to apply a force with the manipulator or to implement more sophisticated systems where forces need to be controlled. Furthermore, the worm gear can sometimes shake when the arrangement is working to unload torque.

11.2.1.3.3 Hypoid gears.

Hypoid gears are similar to spiral bevel gears; however, the shafts are nonintersecting. This gear form has advantages in automobile differential design because it lowers the drive shaft and thus the floor tunnel of the automobile. The hypoid pinion (see Figure 11.13) is considerably larger and stronger than the spiral bevel counterpart; thus, these gears have increased utilization in heavy loading drives. The pinion

TABLE 11.1 Relation of Lead Angle, Speed, and Efficiency and Worm-Gear Drives

Rubbing velocity, ft/min*	Efficiency at designated lead angle of worm					
	5°	10°	15°	25°	35°	45°
5	0.37	0.53	0.62	0.71	0.74	0.75
10	0.42	0.59	0.68	0.76	0.79	0.79
20	0.49	0.65	0.73	0.80	0.83	0.83
40	0.55	0.71	0.78	0.84	0.86	0.87
80	0.61	0.76	0.82	0.87	0.89	0.90
100	0.63	0.77	0.83	0.88	0.90	0.90
150	0.66	0.79	0.85	0.89	0.91	0.91
200	0.67	0.80	0.85	0.90	0.91	0.92
300	0.68	0.81	0.86	0.90	0.92	0.92
400	0.71	0.82	0.87	0.91	0.93	0.93
500	0.72	0.84	0.88	0.92	0.93	0.94
750	0.71	0.82	0.87	0.91	0.93	0.93
1000	0.68	0.80	0.86	0.90	0.92	0.92
1250	0.65	0.79	0.84	0.89	0.91	0.91
1500	0.63	0.77	0.83	0.88	0.90	0.90
1750	0.61	0.76	0.82	0.87	0.89	0.90
2000	0.60	0.74	0.81	0.86	0.88	0.89
2500	0.57	0.72	0.79	0.85	0.87	0.88

*ft/min $\times$ 0.3048 = m/min.

SOURCE: *Kent's Mechanical Engineer's Handbook,* Design and Production Volume, 12th ed., Wiley, New York, pp. 14–44.

offset is usually less than 20 percent of the cone distance of an equivalent bevel gear combination. The offset may be up to 40 percent for normal power drives.

The hypoid teeth are subject to sliding. Consequently, the efficiency of a hypoid drive is on the order of 90 percent, and the efficiency of a corresponding bevel drive is on the order of 99 percent. Under heavy loading conditions, an extreme-pressure lubricant may be required to prevent "welding" of the gear surfaces. Wear and back-drivability are reduced.

Figure 11.13 Hypoid gear.

Figure 11.14   Gear-and-rack gripper.

11.2.1.4   Rack gearing.   Teeth on a gear may mate with teeth on a plane or on a rack. This provides one way to convert from rotary to linear motion. Other methods include screws, cams, cranks, and other linkages. The gear and rack is sometimes used in simple grippers, as shown in Figure 11.14.

11.2.1.5   Ball screw.   Use of ball bearings has made a useful linear transmission out of the screw jack, a device related to the worm gear. The ball screw is best used in a linear joint but has been used in some manipulators as would a hydraulic cylinder to activate a hinged joint; see Figure 11.15. Use of a linear actuator in a rotating joint gives an effective transmission ratio that varies as the position of the joint. This may slightly increase the computational effort in some cases.

11.2.2   Simple gear trains

A simple gear train, as shown in Figure 11.16, transmits motion through a series of individual contacting gears. Intermediate gears are called "idler gears." The velocity ratio depends only on the ratio of the number of teeth

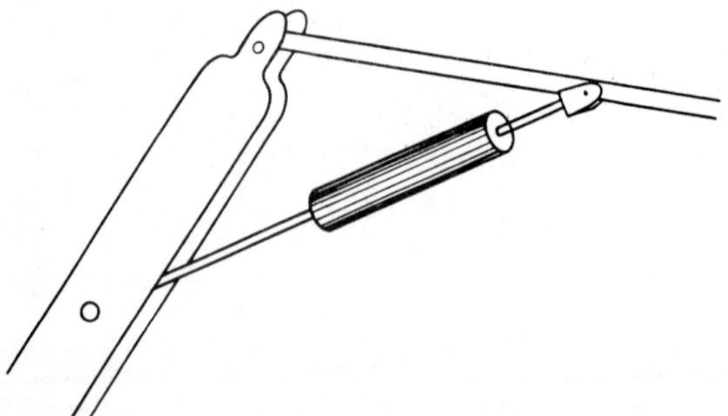

Figure 11.15   Hydraulic cylinder or ball screw with a hinged joint.

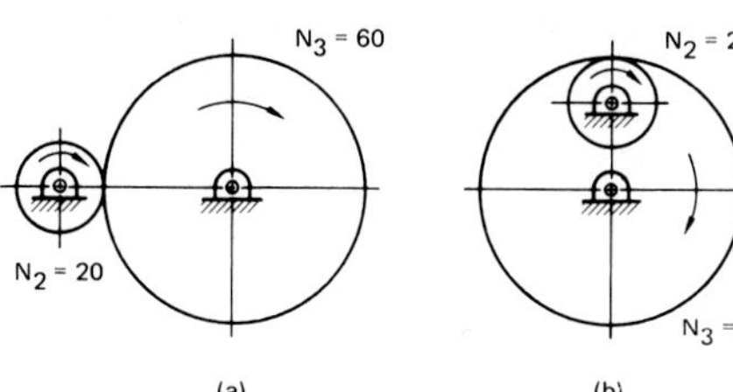

Figure **11.16** Simple gear trains. (*a*) Reversed rotation direction; (*b*) preserved rotation direction.

in the first and last gears. However, an odd number of idler gears will preserve shaft rotation direction; an even number of idler gears will reverse shaft rotation direction. The velocity ratios for the examples in Figures 11.16*a* and 11.17*a* are identical. From Figure 11.16*a*,

$$\frac{\omega_3}{\omega_2} = \frac{N_2}{N_3} = \frac{20}{60} = \frac{1}{3}$$

From Figure 11.17*a*,

$$\frac{\omega_4}{\omega_2} = \frac{N_2}{N_3} \cdot \frac{N_3}{N_4} = \frac{N_2}{N_4} = \frac{20}{60} = \frac{1}{3}$$

Similarly, the velocity ratios for the examples in Figures 11.16*b* and 11.17*b* are identical. From Figure 11.16*b*,

$$\frac{\omega_3}{\omega_2} = \frac{N_2}{N_3} = \frac{20}{60} = \frac{1}{3}$$

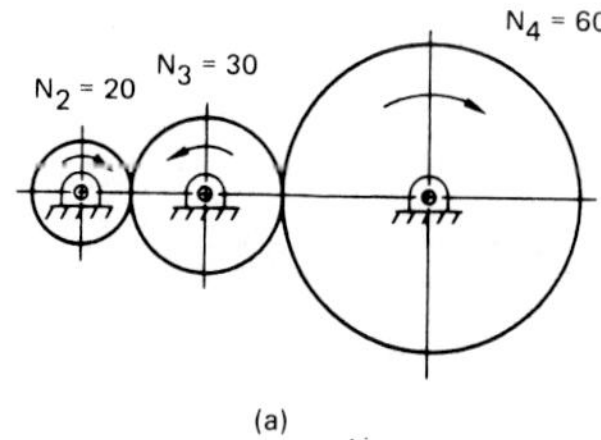

(a)

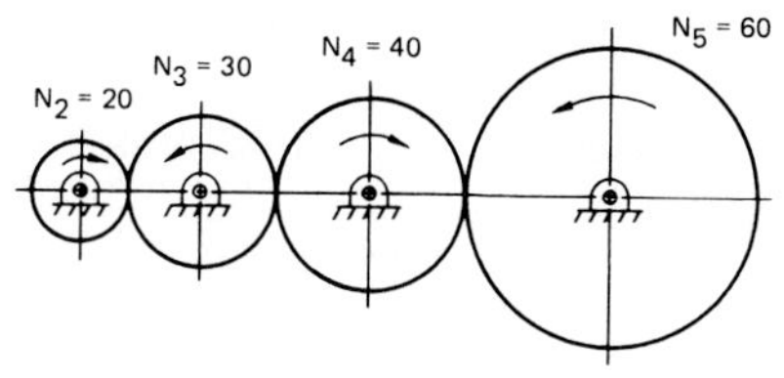

(b)

Figure **11.17** Idler gear trains. (*a*) An odd number of intermediate gears preserves rotation direction. (*b*) An even number of intermediate gears reverses rotation direction.

From Figure 11.17b,

$$\frac{\omega_5}{\omega_2} = \frac{N_2}{N_3} \cdot \frac{N_3}{N_4} \cdot \frac{N_4}{N_5} = \frac{20}{60} = \frac{1}{3}$$

The obvious disadvantage of a simple gear train is that as the velocity ratio increases, the size of the last gear—and thus the overall physical size of the train—also increases.

11.2.3 Compound gear train

To use large velocity ratios with economy, it is often desirable to use a compound gear train, as shown in Figure 11.18. In this configuration, several gears are rigidly mounted to a common shaft. Note that intermediate gears are now considered in the velocity ratio calculation:

$$\frac{\omega_5}{\omega_2} = \frac{N_2}{N_3} \cdot \frac{N_4}{N_5} = \frac{20}{60} \cdot \frac{20}{60} = \frac{1}{9}$$

Figure 11.18b shows a gear train in which the center distance for gears 2 and 3 equals the center distance for gears 4 and 5. Such a configuration may have the input and output shafts located on the same centerline. This arrangement is known as a "reverted gear train." Reverted gear trains are particularly useful for securing the maximum drive capacity in a minimum space.

11.2.4 Epicyclic gear train

An epicyclic gear train is essentially a gear train in which meshing gears have a motion of rotation about their own axes and a rotation about the axis of a fixed gear. The name is derived from the fact that points on the revolving gears trace an epicyclic curve. The gear trains are more commonly referred to as "planetary gear trains," with the fixed gear referred to as the "sun gear" and the revolving gears referred to as "planet gears."

Referring to Figure 11.19, calculation of the velocity ratios for planetary gear trains is complicated by the double rotation of the planet gears. We will present two methods of determining the angular velocity ratio of output to input using as examples the two configurations shown in Figure 11.19a and 11.19b.

The first method uses the equations for relative angular velocities. Let $\omega_{a/b}$ be the angular velocity of body a relative to body b. We want $\omega_{2/1}$ for case a. We may write

$$\omega_{1/4} = \omega_{1/2} + \omega_{2/4}$$

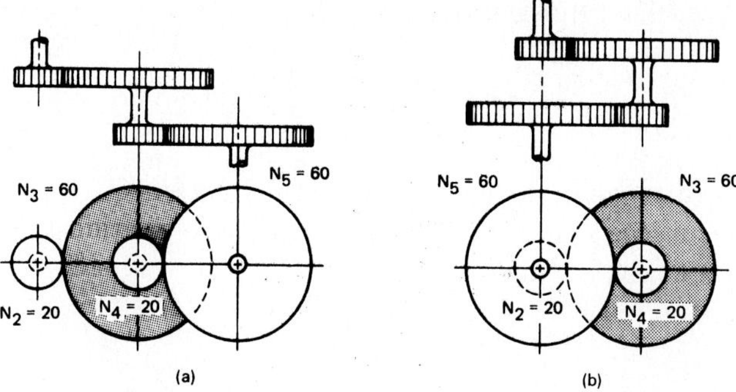

Figure 11.18 Compound gear train. (*a*) Typical configuration; (*b*) reverted gear train.

and divide through by $\omega_{2/4}$,

$$\frac{\omega_{1/4}}{\omega_{2/4}} = \frac{\omega_{1/2}}{\omega_{2/4}} + 1$$

If we consider the planet gear carrier 2 to be a fixed frame, then the other gears become a simple idler train. We may use this fact, first by substituting $\omega_{4/2} = -\omega_{2/4}$, so that

$$\frac{\omega_{1/4}}{\omega_{2/4}} = 1 - \frac{\omega_{1/2}}{\omega_{4/2}}$$

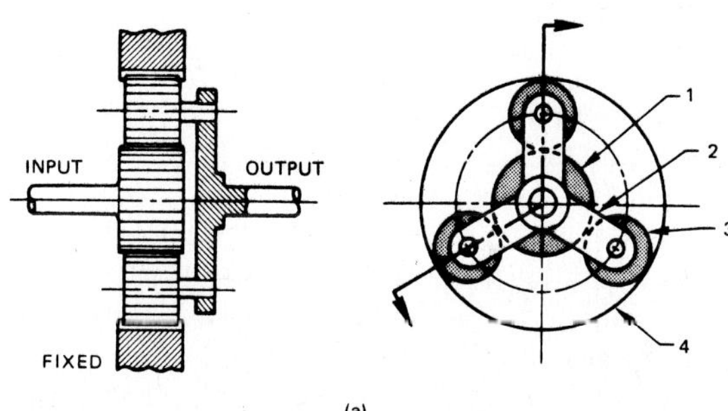

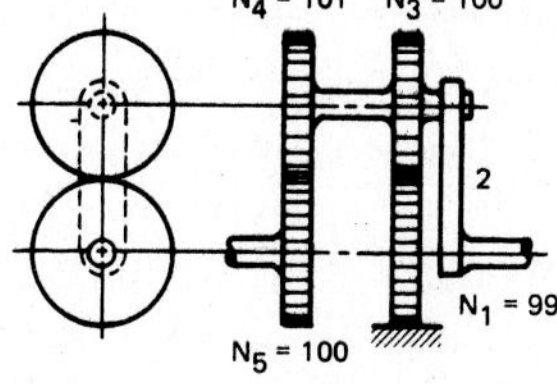

Figure 11.19 Epicyclic, or planetary, gear trains. (*a*) Simple gear train; (*b*) compound gear train.

From the simple gear train referenced to 2,

$$\frac{\omega_{1/2}}{\omega_{4/2}} = -\frac{N_4}{N_1}$$

where the negative sign indicates a change of direction. Finally

$$\frac{\omega_{1/4}}{\omega_{2/4}} = \frac{\omega_1}{\omega_2} = 1 + \frac{N_4}{N_1}$$

We want the inverse of this,

$$\frac{\omega_2}{\omega_1} = \frac{N_1}{N_1 + N_4}$$

To perform the same calculation using the second method, we tabulate the rotations of the gears relative to the carrier:

1. First consider the train "locked" and rotated with the carrier.
2. Then consider the carrier "fixed" and rotate the gear, which has a known absolute motion, the number of revolutions so that the sum of steps 1 and 2 nets the known absolute motion.

If, for instance, one gear were fixed, then the known motion is that the net motion is 0. Note that the entry for carrier 2, relative to the carrier, is 0. For the example, see Table 11.2, which shows

$$\omega_{1/2} = 1 + \frac{N_4}{N_1}$$

and therefore

$$\omega_{2/1} = \frac{N_1}{N_1 + N_4}$$

The next example uses a compound epicyclic gear train. Very large gear reductions are possible within a minimum volume. Using the analytical method to determine the output gear 5 relative to the input shaft (connected to carrier 2):

$$\omega_{5/1} = \omega_{5/2} + \omega_{2/1}$$

$$\frac{\omega_{5/1}}{\omega_{2/1}} = \frac{\omega_{5/2}}{\omega_{2/1}} + 1$$

TABLE 11.2 Velocity Ratio Using the Tabular Method

Rotate	Number of revolutions of indicated member			
	1	2	3	4
With carrier	$+1$	$+1$	$+1$	$+1$
Relative to carrier	$+\dfrac{N_4}{N_1}$	0	$-\dfrac{N_4}{N_3}$	-1
Net	$1 + \dfrac{N_4}{N_1}$	1	$1 - \dfrac{N_4}{N_3}$	0

Because $\omega_{2/1} = -\omega_{1/2}$,

$$\frac{\omega_{5/1}}{\omega_{2/1}} = 1 - \frac{\omega_{5/2}}{\omega_{1/2}}$$

The compound gear train yields

$$\frac{\omega_{5/2}}{\omega_{1/2}} = \frac{N_1}{N_3} \cdot \frac{N_4}{N_5}$$

Therefore,

$$\frac{\omega_{5/1}}{\omega_{2/1}} = \frac{\omega_5}{\omega_2} = 1 - \frac{N_1}{N_3} \cdot \frac{N_4}{N_5}$$

Using the tooth numbers of the example,

$$\frac{\omega_5}{\omega_2} = 1 - \frac{99}{100} \cdot \frac{101}{100} = \frac{1}{10,000}$$

Table 11.3 shows this method in tabular form. Note that gear 1 is fixed;

TABLE 11.3 Velocity Ratio Using Tabular Method

Rotate	Number of rotations of indicated member				
	1	2	3	4	5
With carrier	$+1$	$+1$	$+1$	$+1$	$+1$
Relative to carrier	-1	0	$+\dfrac{N_1}{N_3}$	$+\dfrac{N_1}{N_3}$	$-\dfrac{N_1}{N_3} \cdot \dfrac{N_4}{N_5}$
Net	0	$+1$	$1 + \dfrac{N_1}{N_3}$	$1 + \dfrac{N_1}{N_3}$	$1 - \dfrac{N_1}{N_3} \cdot \dfrac{N_4}{N_5}$

therefore, we want the resultant motion to be 0. The table lists ω_5/ω_2 directly as

$$\frac{\omega_5}{\omega_2} = 1 - \frac{N_1}{N_3} \cdot \frac{N_4}{N_5}$$

Planetary and reverted gear trains are usually designed so that two or more pairs of gear teeth mesh at all times. Though this is advantageous for load capacity, there are limitations to prevent gear tooth interference. The basic rule is that

$$\frac{\text{Number of}}{\text{sun-gear teeth}} + \frac{\text{number of}}{\text{ring-gear teeth}} \div \frac{\text{number of}}{\text{planets}} = \text{integer}$$

The velocity ratio of the sun gear to the carrier has an upper limit because of possible interference between teeth of adjacent planet pinions. This maximum ratio is given as

$$\left. \frac{\omega_{\text{sun gear}}}{\omega_{\text{carrier}}} \right|_{\text{max}} = \frac{2N_s - 4}{N_s[1 - \sin(180/n_p)]}$$

where N_s = number of gear teeth on sun gear
n_p = number of planet gears
$180/n_p$ = angle in degrees

Table 11.4 gives the speed ratio for various combinations of sun and planet gears, together with the integer number of equally spaced planet pinions that can be used with each combination.

Epicyclic gear trains have the advantages for robotics of being compact (for the reduction ratio), efficient (hence back-drivable), durable, and smooth. The disadvantage, common to most gear trains, is that the structure is relatively heavy.

11.2.5 Power losses in gear trains

In power-transmitting gears, losses are associated with friction between the teeth, friction in the bearings, turbulence in the lubricant, and windage.

Tooth friction. The loss may be estimated as $P_{\text{loss}} = \omega_T \cdot v \cdot L$

where ω_T = tangential tooth load
v = speed of teeth of driven gear relative to a common perpendicular to the axes, ft/min
L = loss factor

TABLE 11.4 Velocity Ratio and Allowable Number of Planets*

Number of teeth in one planet gear	Number of teeth in sun gear								
	12	13	14	15	16	17	18	19	20
12	4.000	3.846	3.714	3.600	3.500	3.411	3.333	3.263	3.200
	2,3,4	2	2,4	2,3	2,4	2	2,3,4,5	2	2,4
13	4.167	4.000	3.857	3.733	3.625	3.529	3.444	3.368	3.300
	2	2,4	2,3	2,4	2	2,3,4,5	2	2,4	2,3
14	4.333	4.154	4.000	3.867	3.750	3.647	3.556	3.474	3.400
	2,4	2,3	2,4	2	2,3,4,5	2	2,4	2,3	2,4
15	4.500	4.308	4.143	4.000	3.875	3.765	3.667	3.579	3.500
	2,3	2,4	2	2,3,4,5	2	2,4	2,3	2,4	2,5
16	4.667	4.462	4.286	4.133	4.000	3.882	3.778	3.684	3.600
	2,4	2	2,3,4	2	2,4	2,3	2,4	2,5	2,3,4
17	4.833	4.615	4.429	4.267	4.125	4.000	3.889	3.789	3.700
	2	2,3,4	2	2,4	2,3	2,4	2,5	2,3,4	2
18	5.000	4.769	4.571	4.400	4.250	4.118	4.000	3.895	3.800
	2,3,4	2	2,4	2,3	2,4	2,5	2,3,4	2	2,4
19	5.167	4.923	4.714	4.533	4.375	4.235	4.111	4.000	3.900
	2	2,4	2,3	2,4	2	2,3,4	2	2,4	2,3
20	5.333	5.077	4.857	4.667	4.500	4.353	4.222	4.105	4.000
	2,4	2,3	2,4	2	2,3,4	2	2,4	2,3	2,4,5
21	5.500	5.231	5.000	4.800	4.625	4.471	4.333	4.211	4.100
	2,3	2,4	2	2,3,4	2	2,4	2,3	2,4,5	2
22	5.667	5.385	5.143	4.933	4.750	4.588	4.444	4.316	4.200
	2,4	2	2,3,4	2	2,4	2,3	2,4	2	2,3,4
23	5.833	5.538	5.286	5.067	4.875	4.706	4.556	4.421	4.300
	2	2,3,4	2	2,4	2,3	2,4	2	2,3,4	2
24	6.000	5.692	5.429	5.200	5.000	4.823	4.667	4.526	4.400
	2,3,4	2	2,4	2,3	2,4	2	2,3,4	2	2,4
25	6.167	5.846	5.571	5.333	5.125	4.941	4.778	4.632	4.500
	2	2	2,3	2,4	2	2,3,4	2	2,4	2,3

*The upper figure of each pair is the velocity ratio; the lower figures indicate the number of equally spaced planets which may be used.
SOURCE: *Machine Design.*

For spur and helical gears with parallel shafts,

$$L_a = \left(\frac{1}{N_p} + \frac{1}{N_g} \right) \Big/ 5 \sec \psi$$

where N_p = number of teeth of pinion
N_g = number of teeth of gear
ψ = helix angle of spiral at reference circle

If the configuration involves an internal gear, N_g is negative.
For bevel gears,

$$L_b = \left(\frac{\cos \gamma_p}{N_p} + \frac{\cos \gamma_g}{N_g} \right) \Big/ 5 \sec \psi$$

where γ_p is the cone angle of the pinion and γ_g is the cone angle of the gear.

For crossed-axes gears,

$$L_c = \frac{\mu \nu_s}{\nu}$$

where ν_s is the sliding speed in feet per minute and μ is the coefficient of friction of materials at speed ν_s.

For speeds of ν_s up to 3000 m/min:

- Steel on bronze: $\mu_1 \approx 0.1 - 0.4(\log \nu_s)^{1/2}$
- Cast iron on bronze: $\mu \approx 1.2\mu_1$
- Cast iron on cast iron: $\mu \approx 1.35\mu_1$
- Steel on steel: $\mu = 2\mu_1$

Bearing friction. The frictional torque on a sleeve bearing is approximately

$$T_f \approx \frac{W_J D_J}{400}$$

where W_J is the journal load and D_J is the journal diameter.

Oil turbulence. The power loss by oil turbulence is usually assumed to be of the same order as the power loss in the bearings.

Windage. The windage losses are usually assumed negligible at ordinary speeds.

11.2.6 Load-carrying capacity of gears

Many factors affect the load-carrying capacity of a gear train. The compiled data are generally empirical because of the interactions of such factors as:

- Gear materials
- Working stresses
- Contact stresses
- Sliding contact
- Lubrication
- Dynamic load
- Critical speed

Two excellent references for sizing gears include:

- *Kent's Mechanical Engineer's Handbook,* pp. 14-24 through 14-48
- *Mechanical Design and Systems Handbook,* pp. 32-37 through 32-44

11.2.7 Gear backlash

Backlash is the amount by which the width of a tooth space is wider than the thickness of the engaging tooth as referenced to the pitch circles. Gear backlash is a problem in achieving accuracy in robot servo system design. Backlash introduces a complicating nonlinear phenomenon. Consider a robot control system commanding a motor, through a gear train, to move a robot arm. The control system applies voltage to the motor and expects a change of angle and angular velocity at the joint; however, because of gear backlash, the driven gear, attached to the position sensor, has not moved. Consequently, the control system accelerates the motor to effect the desired charge. When the gear teeth finally mesh, extreme stresses could occur in the gear teeth or the inertia of the motor and transmission could cause position errors when feedback finally is achieved. For consistent performance and to compensate for wear, backlash should be eliminated.

Tables 11.5 and 11.6 give the standard tolerances and backlash allowances for coarse- and fine-pitched AGMA gears. Note that fine-pitched gears have backlash designations, increasing letters associated with less backlash allowance.

Several methods for eliminating all or part of the backlash component of gear train errors are described below.

Antibacklash gears. Each of these gears is a spring-loaded gear pair suitable for light-duty control gear trains (see Figure 11.20*a*, *b*, and *c*). These gears eliminate all backlash and allow liberal tolerances for gear manufacture and gear assembly.

Adjustable gear centers. Adjustable gear centers (Figure 11.21) provide for no backlash only at the condition of tightest mesh. These gear trains can be used to transmit substantial torque. Because of the skill required for adjusting the gear centers, assembly and field replacement are likely to be troublesome.

Gear preload. Figure 11.22 shows a technique that uses an auxiliary torque motor to preload a gear train so that gears tend to favor one side of the gear tooth. If the loading on the gear train can be designed to always have one directional sense, then backlash will not be a problem.

TABLE 11.5   Backlash Tolerances for Coarse-Pitch Gearing
(Inches Measured in Normal Plane)

	Normal diametral pitches				
Center distance	0.5 to 1.99	2 to 3.49	3.5 to 5.99	6 to 9.99	10 to 19.99
Up to 5	—	—	—	—	0.005–0.015
Over 5 to 10	—	—	—	0.010–0.020	0.010–0.020
Over 10 to 20	—	—	0.020–0.030	0.015–0.025	0.010–0.020
Over 20 to 30	—	0.030–0.040	0.025–0.030	0.020–0.030	—
Over 30 to 40	0.040–0.060	0.035–0.045	0.030–0.040	0.025–0.035	—
Over 40 to 50	0.050–0.070	0.040–0.055	0.035–0.050	0.030–0.040	—
Over 50 to 80	0.060–0.080	0.045–0.065	0.040–0.060	—	—
Over 80 to 100	0.070–0.095	0.050–0.080	—	—	—
Over 100 to 120	0.080–0.110	—	—	—	—

Tapered gear mesh.   Figure 11.23 indicates a method of tapering the mesh of a pair of spur gears so that the teeth tend to engage in loaded contact. This type of backlash correction is expensive and requires careful assembly and adjustment.

Variable-tooth-thickness worm.   Figure 11.24 indicates how a variable-thickness worm design tends to preload the worm gear, thus eliminating backlash. Again, this is an expensive technique and should be used only in special circumstances.

Dual idler.   A variation on the technique of Figure 11.20 is shown in Figure 11.25. An input pinion drives pinions on two idler shafts. A second pinion on each idler shaft meshes with the output gear. The two idler

TABLE 11.6   Backlash Tolerances for Fine-Pitch Gearing
(Inches Measured in Normal Plane)

Backlash designation	Normal diametral-pitch range	Tooth thinning to obtain backlash, per gear	Resulting approximate backlash, per pair
A	20–45	0.0020	0.0040–0.0080
	46–70	0.0015	0.0030–0.0070
	71–90	0.0010	0.0020–0.0055
	91–200	0.0008	0.0015–0.0030
B	20–60	0.0010	0.0020–0.0040
	61–120	0.0008	0.0015–0.0030
	121–200	0.0005	0.0010–0.0020
C	20–60	0.0005	0.0010–0.0020
	61–120	0.0004	0.0008–0.0015
	121–200	0.0003	0.0006–0.0010
D	20–60	0.0003	0.0006–0.0010
	61–120	0.0002	0.0004–0.0007
	121–200	0.0001	0.0002–0.0004
E	All pitches	Do not provide tooth thinning	0.0000–0.0002

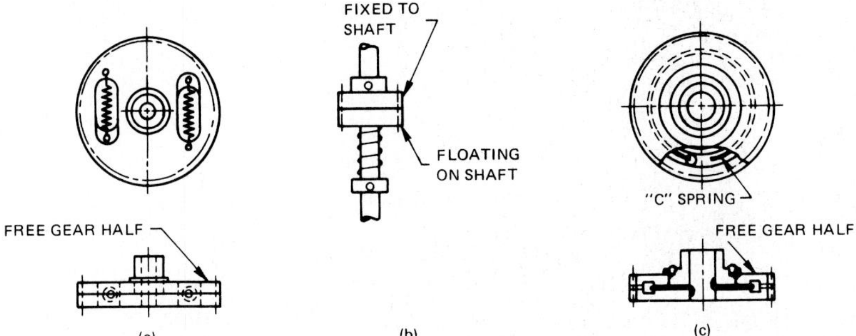

Figure 11.20 Spring-loaded antibacklash gears. (*a*) Extension- or compression-spring type; (*b*) torsion-spring type; (*c*) C-type spring.

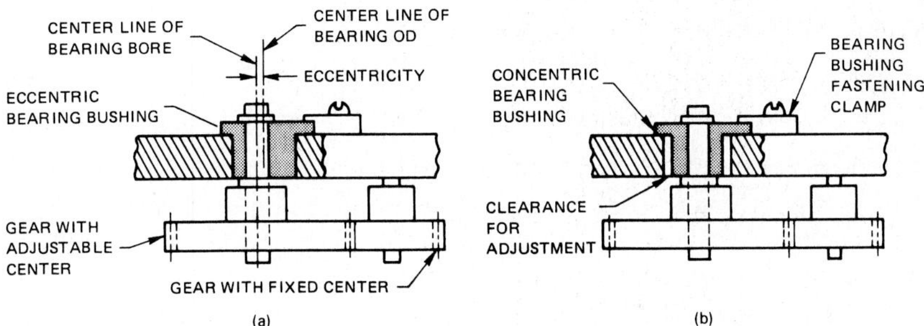

Figure 11.21 Adjustable gear centers. (*a*) Eccentric bearing; (*b*) floating bearing.

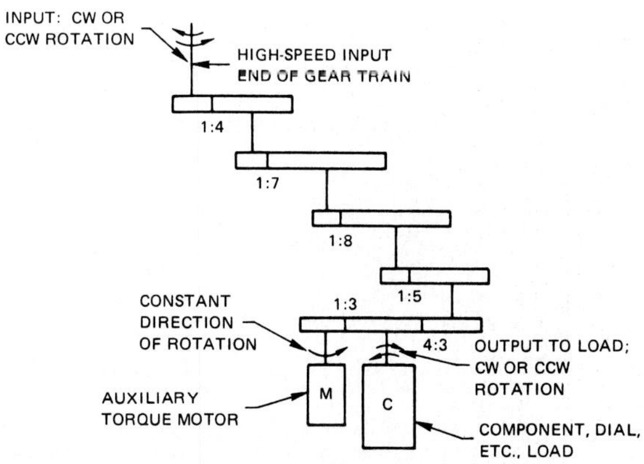

Figure 11.22 Gear preload using an auxiliary torque motor.

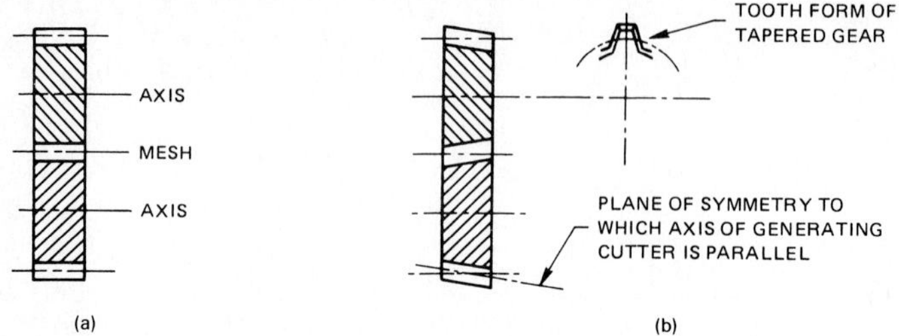

Figure 11.23   Tapered gears. (*a*) Normal gear; (*b*) tapered gear.

shafts and their pinions are identical except that one idler shaft has a
reduced diameter, as shown in the figure. The angular relationship
between the input pinion and output gear is determined by the thick idler
and its two pinions. The pinions on the thinner idler shaft are mounted
so that the shaft will be twisted slightly when installed. In other words,
the thinner idler shaft acts as a torsion spring to eliminate any backlash
in the gear train. A gear train similar to that of Figure 11.25 is used in
joint 1 of the PUMA robot for backlash elimination.

11.2.8   Multiple gear arrangement

As was suggested in Figure 3.7, more than one actuator may be needed to
control a joint. Figure 11.26 shows a common technique where two shafts
are used to control two joints. Each joint depends on the position of both
shafts, and only by coordinating the motion of both shafts together can
the joints be made to move independently.

11.3   Harmonic Drive

Harmonic drives (which are patented devices) utilize toothed components
in a unique manner to provide large velocity ratios and high torque capac-

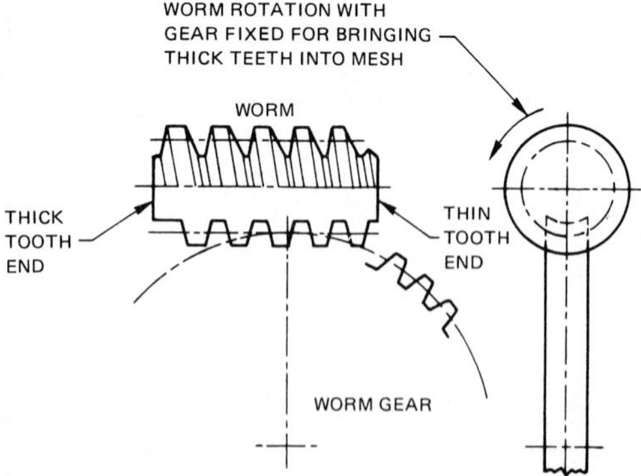

Figure 11.24   Variable-tooth-thickness worm.

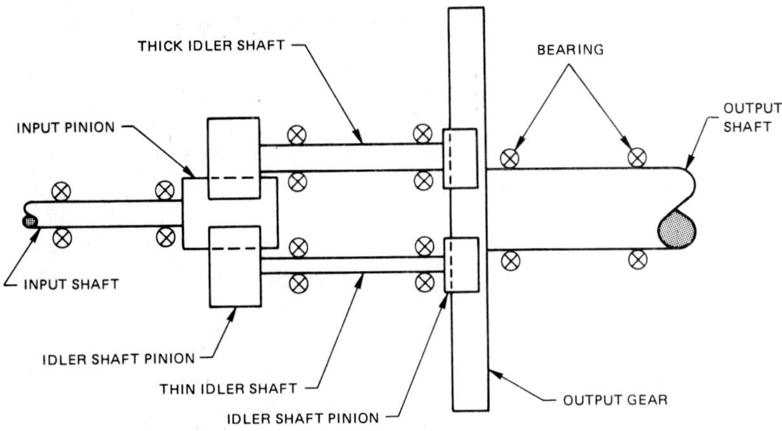

Figure 11.25 Dual idler used for backlash elimination.

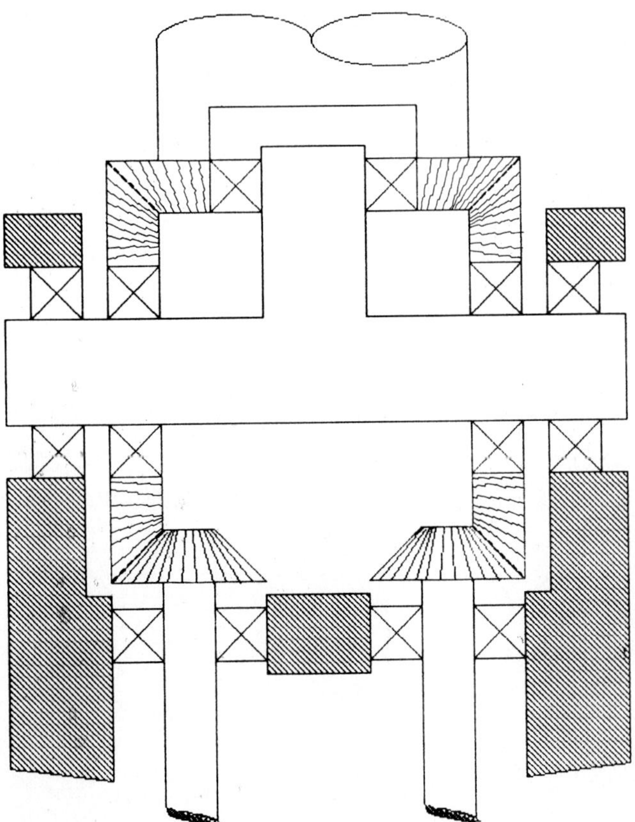

Figure 11.26 Dual shaft pitch-and-roll control

ities in a very compact space. The three basic components shown in Figure 11.27a include a rigid circular spline, a flexible circular spline, and an elliptical wave generator. The flexible circular spline, called a "flexspline," normally has two fewer teeth than the rigid circular spline. With the unit assembled as in Figure 11.27b the wave generator deforms the flexspline, thus engaging teeth at diametrically opposite points concident with the major axis of the elliptical wave generator and disengaging points at the minor axis. If the rigid circular spline were fixed, then rotation of the wave generator would produce a reverse motion of the flexspline; if the rigid spline has 202 teeth and the flexspline has 200 teeth, then the single revolution of the wave generator will precess the flexspline backward two teeth. In this case, the velocity ratio would be 100:1. When the flexspline is fixed and the circular spline is the output, then the input and output shafts rotate in the same direction. The general equation for the reduction ratio is

$$\frac{\omega_{\text{in}}}{\omega_{\text{out}}} = \frac{N_o}{N_c - N_f}$$

where N_o = number of teeth on output member of flexspline or circular spline

N_c = number of teeth on circular spline

N_f = number of teeth on flexspline

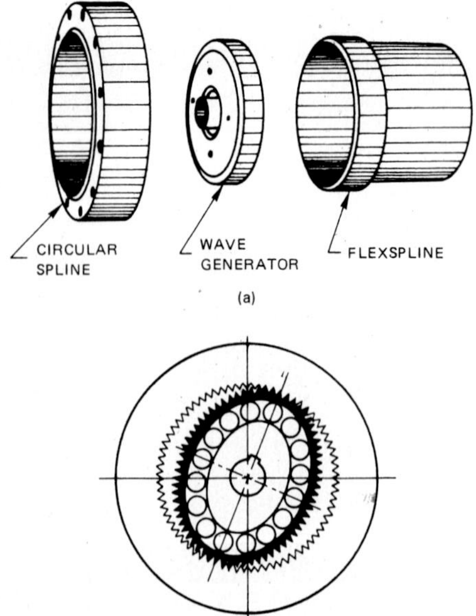

Figure 11.27   Harmonic drive. (a) Components; (b) assembled.

Standard units are available with velocity ratios from 64:1 to 320:1 and with torque output capacities of up to 2.4×10^4 lbf/in.

The harmonic drive has been very popular in robot applications because of its high reduction ratios and because it is lightweight compared with standard gear systems. The efficiency is moderately good; applied torques can be expected to transfer at the reduction ratio. However, because of the high ratios, back-drivability is limited.

The harmonic drive has two major drawbacks that become increasingly evident in more sophisticated systems. The motion is not perfectly smooth but has a small ripple corresponding to the drive frequency. The ripple, as with spur gears, is due to deformations under load. These ripples

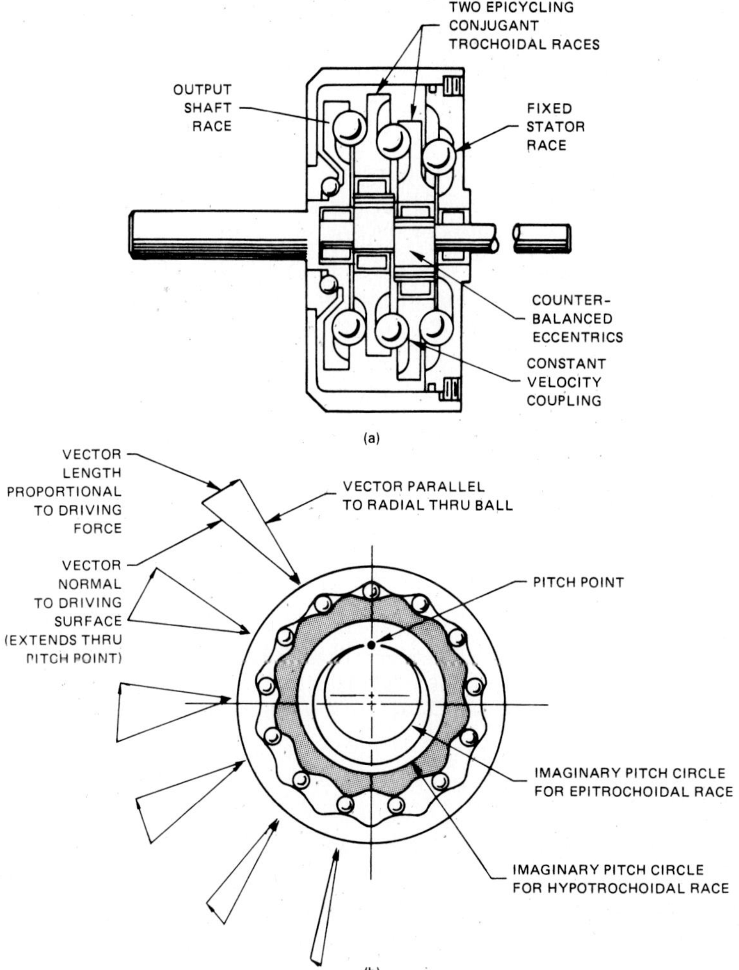

Figure 11.28 Configuration of the anti-friction drive. (*a*) Cross section; (*b*) third-element configuration.

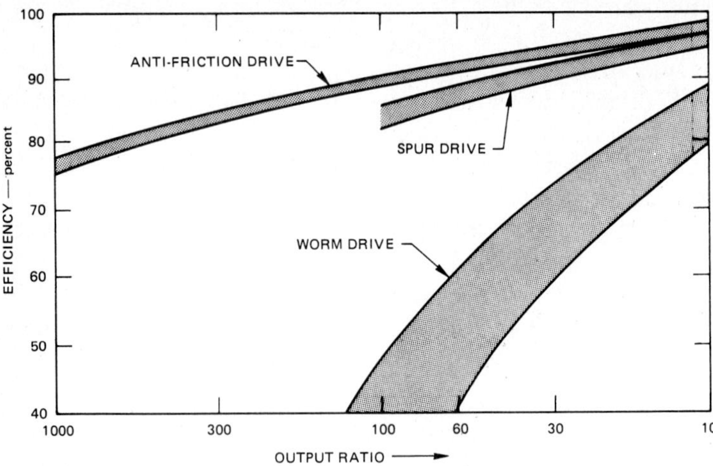

Figure 11.29   Efficiency of the antifriction drive as a function of ratio.

are larger than with spur gears, as might be expected, since the harmonic drive function is via deformation and cannot be rigid. The amount of the ripple increases with the applied load. This may lead to oscillations in the manipulator system. The second drawback is that the spring constant of the harmonic drive is not linear but decreases with increasing load. Variable stiffness with load is a complicating feature in the analysis of performance and design of control systems. The harmonic drive is more flexible than corresponding gear systems (has a lower spring constant). Flexibility tends to complicate position- and high-speed-control considerations. Thus flexibility is now a drawback. However, the drawback may become an advantage in later generations of manipulators, particularly as computational, and hence control, capabilities improve.

It is of interest to speculate whether harmonic drives can be used in pairs to reduce ripple or whether such a strategy will introduce new problems. Such an approach is not advocated by the manufacturer, although it is not specifically discouraged. Perhaps the manufacturer is attempting to eliminate the ripple—the major drawback of an otherwise excellent transmission for robots.

11.4   Cyclo Drives

A concept in competition with the harmonic drive has been called the "cyclo drive." This planetary concept is in many ways similar to the harmonic drive, except that there are no deformations. Contact is made between carefully shaped rows by means of balls. The friction introduced is therefore very low, as in ball bearings. The stiffness is high.

Sumitomo Machinery Corporation has a product of this type called the

SM-Cyclo Drive. They advertise single-stage reductions from 6:1 to 87:1 in the horsepower range of 1/16 to 135.

Another transmission of this type is under development as the Anti-Friction Drive, by Advanced Energy Technology, Inc., of Boulder, Colorado. Figures 11.28 and 11.29 show the concept. The company is aiming its development effort at robot-sized transmissions.

Sensors

Sensors are used to determine the internal state of the robot (proprioception) and the robot's environment (perception).

Proprioception is the sensing of the state of the joints and gripper where the state may include position, velocity, force, and torque. Some knowledge of the state is essential in the most elementary robot, if only to know whether the end of a commanded position has been reached. For example, it is often useful to know whether the gripper is open or closed. Increased proprioception will be required as the sophistication of the control system increases and as the robot is required to do more sophisticated tasks, tasks that require dexterous motion or perhaps force control. Even when the end position can be determined directly—and perhaps more accurately—by an external transducer such as a camera, it is necessary to know the joint states in order to determine what adjustments are necessary to change the robot's motion.

Perception is the robot analog of the senses. Although taste and smell might not be important to most robots, other nonhuman senses (e.g., measuring radiation) may be important. Perception is all-important to a survey or sentry robot and is essential to any robot that needs to respond to its environment. Perception sensors are on the frontier of robotics. This frontier is as much the processing and use of perception information as it is the gathering of it. Issues include processing speed and how to deal with redundant and sometimes conflicting information.

12.1 Performance

Sensors used in robots are specified with the same performance parameters as are used for sensors in other applications:

- Accuracy

- Resolution

- Hysteresis

- Noise

- Frequency response

These parameters should be familiar to instrumentation engineers.

There is a tendency to overdesign robot sensors because of naive notions of how the robot will be used. An appreciation of the variety of robot applications can therefore give important guidance to the robot sensor designer. For example, an assembly robot may be used for inserting a shaft into a bearing. A typical bearing clearance is 0.02 mm; one might then conclude that the robot joint-position sensors should have sufficient accuracy to permit the robot to move the shaft to within about 0.01 mm of the desired location (the bearing journal). In fact, few, if any, modern robots have 0.01-mm positioning accuracy. There are several reasons for this:

- Positioning accuracy of ± 0.01 mm is virtually impossible to achieve in practical robotic manipulators larger than about one-third human size.

- In most practical applications, the location of a fixtured workpiece (e.g., a bearing journal) relative to the robot will not be consistent from one workpiece to the next. A variation in the location of a bearing journal of ± 0.5 mm is not unusual.

- In practice, it would be difficult to devise a gripper capable of holding a shaft so that the location of the shaft relative to the robot is known to an accuracy of ± 0.01 mm.

In many practical applications of robots doing insertions, the control strategy requires that the shaft be brought close enough to the hole to allow a chamfer on the shaft to partly enter the hole. The entry is detected by force or torque sensors, and these sensors are also used to guide the shaft into the hole. Using this strategy, robot positioning accuracy of about 1 mm may be quite adequate.

Because of such considerations, a robot designer must consider all aspects of the performance of robotic sensors.

Absolute versus relative measurement.   For encoders, this is the difference between "absolute" and "incremental" devices. Absolute sensors are generally preferable to relative ones for robotic applications. If relative sensors are used in an industrial environment, the problem of "missed counts" can be difficult to solve. Therefore, a robot using relative sensors must periodically check for missed counts by returning to a "home" position. This requirement can be accommodated, but the increased software and control system complexity required may not outweigh the cost advantage of relative sensors. Furthermore, a robot using relative sensors will probably damage itself or the workpieces when missed counts occur.

Accuracy.   The example of the insertion task described above should illustrate the futility of striving for high accuracy in robot sensors. Resolution and hysteresis (see below) are often much more important than accuracy. The accuracy claims of robot vendors should be viewed with suspicion. Often, the claimed accuracy is in fact the repeatability, and the actual accuracy is much worse.

Bandwidth.   Most engineers will recognize that the bandwidth of sensors should be at least as great as that of the joint servos. In fact, sensor bandwidth greater than 20 times the servo bandwidth is often required because of numerous factors. Fortunately, the required sensor bandwidth can usually be obtained; robot servo bandwidths are normally less than 100 Hz.

Drift.   Drift is actually a form of noise (see below), but drift is seldom included in the sensor noise value that is specified. Drift can be a special problem in force and torque sensors. These variables must often be measured accurately in the region near zero. In this case, a small drift in the sensor output can result in unacceptable errors.

Hysteresis.   Excessive hysteresis in robot sensors can be disastrous. Hysteresis is often larger than the sensor resolution, in which case the useful resolution is essentially equal to the hysteresis. Hysteresis in a servo system can make an otherwise stable system become unstable. Often, the only acceptable solution to these instabilities lies in eliminating the hysteresis.

Monotonicity.   Sensors with digital outputs (and occasionally analog sensors) can have nonmonotonic input/output relationships (i.e., the slope of the input/output curve changes sign). Such sensors cannot be recommended for use in robots because they almost invariably cause servo instability.

Noise.   Noise can be an unexpected problem in robot sensors, especially analog ones. Like hysteresis, excessive noise can decrease a sensor's useful resolution. In the advertised characteristics of sensors, noise is sometimes specified for an unrealistically small bandwidth to make the sensor appear better than it is. The above comments about sensor bandwidth also apply to measurement of sensor noise. (Noise measurement bandwidth should be greater than 20 times the servo bandwidth.)

Repeatability.   The robot designer should not be misled by the notion that repeatability is nearly equivalent to accuracy. In principle, elaborate calibration tables can be stored in the robot in order to effectively increase sensor accuracy to the level of sensor repeatability. The robot designer should be cautious of the practical consequences of this approach: requirements for additional circuitry in the robot's electrical subsystem and compromised interchangeability of parts. There is one robot application where repeatability is important. It is feasible to "teach" small robots by manually leading the robot through the desired movements. If the same physical robot is used for both teaching and working, then sensor repeatability will be more significant than accuracy. On the other hand, if teaching is done with one robot and others do the work, then sensor accuracy is the important parameter.

Resolution.   This is often the most important performance parameter for robot sensors. Robot position sensors are normally used "inside" a servo loop. In this case, the position resolution of the robot is limited by the resolution of the position sensors. Requirements can be surprisingly severe. For example, consider a robot with a 1-m cube working volume. To have a positioning resolution of 0.01 mm (not unreasonable if insertion tasks must be performed), the resolution of sensors such as the shoulder rotation sensor must be better than 1 m/0.01 mm $= 10^5$. This corresponds to about 17 bits of resolution or a signal-to-noise ratio of 100 dB!

12.2 Proprioception

Proprioception is the sensing of the internal affairs of the manipulator. The variables needed for control, the joint positions and velocities, are the ones emphasized in this section. Other information such as internal forces and torques, temperatures, and wear measurements may be incorporated in more advance machines. This information will be useful for sensing problems and indicating the need for maintenance.

12.2.1 Position

All but the most rudimentary robots include position sensors at each joint and at the gripper. These sensors, even if they are as simple as a limit switch, serve the obvious purpose of letting the robot know where it is. Joint position is key information in the servo systems that permit the robot to move in a controlled fashion.

By far the most commonly used position sensors in robots are encoders and resolvers. *Encoders* use marks to indicate the position. Their resolution is limited to the step size of the marks. *Resolvers* are a class of position (or angle) sensors with resolution much finer than the step size.

12.2.1.1 Encoders. Absolute encoders contain all the information necessary to obtain the position. The typical encoder would have a track for each binary digit of information. The arrangement of markings on four tracks for a 4-bit 16-position rotary absolute encoder is shown in Figure 12.1.

Incremental encoders have two readers on a single track. The readers (a light-emitting diode and a corresponding sensor in the case of an optical encoder) are displaced about a quarter of the binary wavelength as shown in Figure 12.2. The offset allows a finer resolution than would be provided by a single reader, but more important, comparison of the two reader signals provides information on the direction of motion. The position can then be updated in an external circuit.

Both rotary and linear encoders and resolvers are available, but linear devices are seldom used. Linear joints are usually actuated with a rotary motor, and the encoder is mounted on the motor.

A discussion of several types of encoders with their advantages and disadvantages follows.

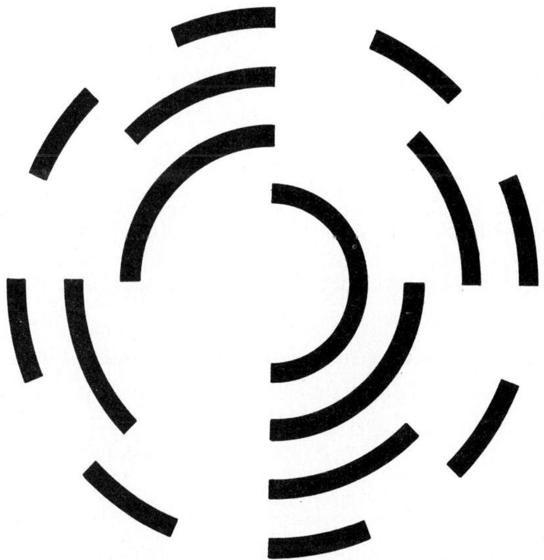

Figure 12.1 Absolute rotary encoder.

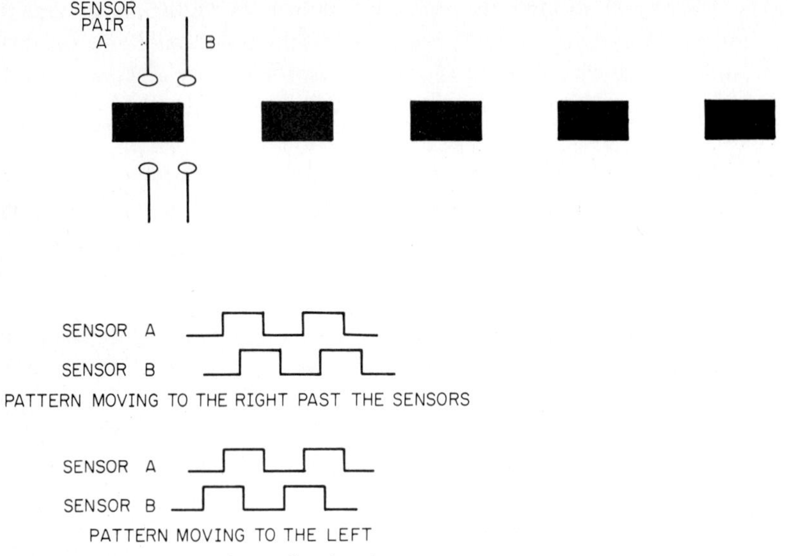

Figure 12.2   Incremental encoder signals.

12.2.1.1.1 Brush encoders. Brush encoders consist of an insulating substrate with conductive stripes or bands to define the encoder steps. Electric contacts, called brushes, are used to sense the passage of the conductive stripes as the encoder moves. Brushes may be metallic, carbon, or conductive polymers. Brush encoders are relatively simple and inexpensive, but they have limited life: Endurance of 100,000 to 1 million cycles is typical. Brush encoders are seldom used in robots because of their limited lifetime and poor reliability in adverse environments: They are rather sensitive to dirt or corrosion.

12.2.1.1.2 Optical encoders. Optical encoders consist of a transparent substrate with opaque stripes or bands to define the encoder steps. One or more light sources and light detectors are located on opposite sides of the transparent substrate. The motion of the encoder substrate is determined from the interruption of the light beams. Because there are no brushes in optical encoders, their lifetime is nearly infinite. The fundamental limitation to optical encoder lifetime is normally degradation of the light source; incandescent lamps eventually fail, and the output of light-emitting diodes (LEDs) slowly decreases with age. Despite these problems, readily available optical encoders have lifetimes exceeding 10 years.

In dirty environments, the transparent substrate of these encoders can become sufficiently soiled to impair operation. Thus, optical encoders used in dirty environments usually have shaft seals to keep dirt away from the substrate.

12.2.1.1.3 Laser optical encoders. A new type of optical encoder uses a laser diode for illumination and exploits the principles of diffraction (Nishimura and Ishizuka, 1986*a,b*) in order to achieve angular resolution that is orders of magnitude greater than would be possible with a conventional optical encoder of comparable size. The laser encoder may be the ideal for applications where the space available for encoder mounting is limited.

12.2.1.1.4 Magnetic encoders. There are two types of magnetic encoders. The first uses a ferromagnetic moving part with teeth that pass through a detector area. The presence of a tooth in the detector decreases the magnetic reluctance of a flux path, resulting in an increased magnetic field. The second type of encoder uses a moving part that is permanently magnetized with several alternating north and south poles on its surface. The magnetic field in both types of magnetic encoders is usually detected with a Hall effect sensor. A coil is occasionally used as the sensor in magnetic encoders, but this approach is not desirable for robots because the output signal is velocity-dependent—at low joint velocities, the encoder may miss one or more counts.

Magnetic encoders are relatively immune to dirty environments but are certainly affected by magnetic particles (e.g., iron dust and chips from machining operations). Shaft seals can often be used to exclude these particles from the working parts of the encoder. Compared to optical encoders, the resolution of magnetic encoders is poor; resolutions of less than 1° are not commonly available. This resolution limitation may have been overcome (Miyahara and Uchida, 1982) in a patented design. The magnetic encoder uses a nonmagnetic encoder disk upon which a magnetic film is deposited. The magnetic film is magnetized by a permanent magnet at the center of the disk. A toothlike pattern formed in the magnetic film is used to achieve the desired field variations around the disk. Apparently, very high resolution magnetic encoders can be produced this way.

More recently, improvements in injection-molded ferrite magnets have made possible magnetic encoders with magnetic pole spacings as small as 100 μm (Campbell and Wasson, 1986). This is similar to the resolution of many optical encoders.

The resolution obtained with encoders is usually equal to the size of the steps on the moving substrate of the encoder. *Synchros* and *resolvers* are a class of position (or angle) sensors with resolution much finer than the step size.

12.2.1.1.5 Synchros. Synchros are similar in construction to three-phase wound-rotor motors. The rotor has a single winding and the stator has three windings, usually connected in a star configuration. In operation, the rotor of the synchro is energized with an ac voltage, usually between 50 Hz and 10 kHz (50, 60, and 400 Hz are the most common). The voltages induced on the three stator windings are precisely related to the angle of the synchro rotor. An electronic circuit known as a "synchro converter" is used to convert the synchro output voltage to an analog or digital representation of the rotor angle. Accuracy of commercially available synchros is typically 10 to 50 minutes of arc.

12.2.1.2 Resolvers.

Resolvers are similar in construction to synchros. The rotor has a single winding and the stator has two windings. The two stator windings in a resolver are offset 90° from each other, whereas the three windings in a synchro are offset by 120°. The rotor of a resolver is energized with an ac voltage, and an electronic circuit known as a "resolver converter" is used to convert the resolver stator voltages into an analog or digital representation of rotor angle.

Most resolvers generate unique output signals for 360° of rotor revolution. These are known as "single-speed" resolvers. There are also two-speed, four-speed, ten-speed, and other multispeed resolvers that generate unique output signals for 180, 90, or 360°, respectively, of rotor revolution. Thus, the multispeed resolvers are equivalent to a single-speed resolver driven through an appropriate gear train. The advantage of the multispeed resolvers is that they have no backlash, whereas the accuracy of a gear-driven single-speed resolver would be compromised as a result of backlash in the gears. Accuracy of commercially available resolvers is typically 2 to 20 minutes of arc.

Several types of resolvers are discussed below. The names for various types of resolvers are not well standardized. Two companies may use different names to refer to similar devices.

12.2.1.2.1 Reactasyn resolver.

Recently a new rotation sensor has been developed (Cushing, 1985), primarily for robotic applications. The sensor is similar to both synchros and multispeed resolvers, but uses a three-phase wound stator with no windings on the rotor. The reactasyn was developed specifically for use with a low-speed high-torque motor intended for use in the shoulder and waist joints of industrial robots. Cushing claims that the accuracy of the reactasyn is 30 seconds of arc, and could be improved if a look-up table were used for linearization.

12.2.1.2.2 Inductosyn (inductive).

Both the rotor and stator of an Inductosyn consist of a serpentine conductor pattern on a nonconducting substrate. The operating principle of the Inductosyn is described by Farrand (1981a). The electrical output signals from an Inductosyn are similar to those from a multispeed resolver but their amplitude is much smaller. Inductosyns typically are 20- to 1000-speed devices. Accuracy of commercially available Inductosyns is typically 1 to 10 seconds of arc.

The inductive Inductosyn is rather sensitive to ambient magnetic fields.

This effect should be considered, especially if the Inductosyn is to be mounted near an electric motor or other source of magnetic fields.

12.2.1.2.3 Inductosyn (capacitive). It is unfortunate that the capacitive Inductosyns are often called by the same name as the inductive type, because this usage can cause confusion. The operating principle of the capacitive Inductosyn is described by Farrand (1981*b*). A similar device is described by Meyer (1974).

The capacitive Inductosyn is a relatively new development, but it has several advantages (relative to the inductive type) that suggest that it will become very popular. The cost of both the sensor capacitive device and the necessary signal-processing electronic circuits are much less. The capacitive device is also insensitive to ambient magnetic fields. Accuracy of commercially available capacitive Inductosyns is typically 1 to 10 seconds of arc.

12.2.1.3 Others. Many other types of sensors can be used for proprioception in robots. Some of these are described below.

12.2.1.3.1 Potentiometer. The potentiometer is a well-known position or angle sensor. The low-precision potentiometers are especially inexpensive. The primary disadvantages of potentiometers are their low precision (relative to encoders and resolvers) and limited lifetime. Hysteresis is another common flaw in precision potentiometers. Accuracy of commercially available potentiometers is typically 0.05 to 5 percent of full scale, depending on the price.

12.2.1.3.2 LVDT. The linear variable differential transformer (LVDT) consists of a differential transformer with a movable, magnetically permeable core. Movement of the core causes corresponding changes in the transformer secondary voltages. A suitable electronic circuit is used to convert the secondary voltages to analog or digital form. LVDTs are generally more expensive than potentiometers, but they have essentially infinite life and zero hysteresis. Accuracy of commercially available LVDTs is typically 0.1 to 1 percent of full scale.

12.2.1.3.3 RVDT. The rotary variable differential transformer (RVDT) is similar to the LVDT, but the movable core rotates instead of moving linearly.

12.2.1.3.4 Differential inductor. The differential inductor sensor consists of two ferrite cores wound as inductors and connected in series. A permanent magnet is attached to the input shaft. The field from this magnet is sufficient to saturate part of the ferrite cores and thereby decrease the inductance of the two inductors. As the input shaft moves, the inductance of one inductor increases while the other one decreases. The sensor is excited with an ac voltage; the output is an ac voltage with amplitude proportional to the input shaft displacement. This type of sensor can be configured for linear or rotary motion and is described in Sidor (1976), Bernin (1976), Bowman et al. (1978), and Genz (1978).

The differential inductor has the advantage, relative to potentiometers, of nearly infinite life and zero hysteresis. The cost of this sensor is comparable to that of LVDT or RVDT sensors, but requires less expensive electronic circuitry to convert its output to a useful analog or digital form. Accuracy of commercially available differential inductor sensors is typically 1 to 5 percent of full scale.

12.2.1.3.5 Capacitance.   The capacitance sensor consists of two or more conducting plates that move relative to each other in response to motion of the input shaft. The capacitance between the plates is measured with suitable electronic circuitry, and the measurement is then converted to an analog or digital representation of input shaft position. Capacitance sensors have been designed for detection of both linear and rotary motion. One of the primary advantages of capacitance sensors is that the moving part of the sensor can have very low inertia. It is therefore especially suited to use where the motion of a very light structure must be measured at high speed. Capacitance sensors also have excellent resolution and zero hysteresis. Accuracy of commercially available capacitance sensors is typically 0.1 to 2 percent of full scale.

12.2.1.3.6 Eddy current.   The eddy current sensor, in its simplest form, consists of a coil of wire. More elaborate sensors include a second sense coil and pole pieces. The coil is energized with an ac current. The moving part of the sensor consists of a conductive plate. As the plate moves closer to the coil, the ac magnetic field produced by the coil induces eddy currents in the conductive plate. The magnitude of the energy dissipated by the eddy currents is determined by suitable electronic circuits connected to the coil. The advantages of eddy current sensors are similar to those of capacitance sensors, but eddy current sensors are usually slightly less accurate. Accuracy of commercially available eddy current sensors is typically 0.2 percent to 2 percent of full scale. Eddy current sensors are normally used only when the motion to be measured is small—typically less than 1 cm. Rotary eddy current sensors are possible, but not widely used.

12.2.1.3.7 Optical displacement sensors.   These sensors could, in principle, be used for proprioception in robots. However, because these sensors are very expensive, it is unlikely that they will be used for proprioception in the near future.

- *Amplitude-modulated laser radar.*   An amplitude-modulated laser beam is reflected by a target. The reflected beam is detected by a receiver (usually) located near the transmitter. The distance to the target can be determined from the phase difference between the transmitted and received signals. This type of sensor is described by Nitzan et al. (1977) and Smith (1980).

- *Laser interferometer.* The optical interferometer is a well-known device used for extremely accurate distance measurements. It measures distance using interference between a sample of the transmitted beam and a beam reflected from the moving target. Interferometers can achieve measurement accuracy of up to 10^{-8} m.

12.2.2 Velocity

It is often desirable to measure the velocity of one or more joints in a robot manipulator. Conventional servo design requires that the servo controller include a "velocity term" in its transfer function if the servo loop is to be stable. Without the velocity term, a servo system will usually exhibit an undamped, resonant behavior and will be highly unstable. Including the velocity term will produce the desired damped response.

In principle, the signal from a joint position (or angle) sensor can be electronically differentiated to obtain joint velocity. This approach is sometimes practical, but often is not. If the joint position sensor has a noisy output, or if it has a digital output (such as encoders have), differentiating the position sensor signal can effectively magnify the noise sufficiently to make the servo system unstable or unreliable. In these cases, a separate velocity sensor should be used.

Several types of sensors can be used for measurement of robot joint velocity. These are described below.

12.2.2.1 Tachometer generator.
The tachometer generator is simply a small dc generator. These devices are usually constructed with a permanent-magnet stator and a multipole wound armature. The armature is fitted with commutators similar to those used in dc motors and generators. The tachometer generator functions as a velocity sensor because its output voltage (when driving a high-impedance load) is proportional to the rotation speed of the armature. The designer should carefully consider the number of poles used in a tachometer and the output ripple, because excessive ripple at undesirable frequencies can upset the stability or performance of a servo system.

A minor disadvantage of tachometer generators is that they usually use brushes for commutation. Brush life in tachometers is quite long, but not infinite. Arcing at tachometer brushes is minimal because of the very small current, but is an important consideration if the robot will be used in explosive atmospheres.

12.2.2.2 Tachsyn.   The tachsyn has been developed for use with brushless dc motors (Luneau, 1986). The device acts like a three-phase permanent-magnet alternator, the alternator output voltage being proportional to the rotation speed.

12.2.2.3 Linear velocity transducer (LVT).   The linear velocity transducer consists of one or more coils, which are usually stationary. The moving part of the LVT is usually a permanent magnet that slides inside the coils. The LVT works by the well-known principle of electromagnetic induction: The voltage generated by the coils is proportional to the velocity of the moving permanent magnet.

In some applications, it is possible to derive joint velocity from an accelerometer (acceleration sensor). One example of this approach is described by Maiorca and MacNeil (1979). The acceleration signal is electronically integrated to derive a velocity signal. Fortunately, the process of integration tends to remove noise from the acceleration signal, so this approach does not suffer from the problems which arise when differentiating a position signal (see above).

In traditional robots, each joint is controlled by an independent servo loop. In this case, the velocity of each joint (i.e., relative velocity of the two attached links) is needed by the servo control systems. Unfortunately, this information can be readily derived from acceleration only for the first (shoulder) joint in most robots, because an accelerometer measures acceleration of a link relative to inertial space, rather than relative to another link.

One can envision an advanced robot control system that does not treat the joints independently. In such a system, a three-axis accelerometer mounted near the gripper might be used to supply velocity feedback to the control system, and thereby help stabilize the response of all six joints in the robot. A few types of accelerometers are described briefly below.

- *Piezoelectric.*   This type of accelerometer consists of a small mass attached to a piezoelectric generator. The piezoelectric accelerometer generates a charge proportional to acceleration, but is not sensitive to dc (continuous) accelerations or gravity. The low-frequency response of these accelerometers typically extends to about 0.1 to 10 Hz. Piezoelectric accelerometers are simple, rugged, and relatively inexpensive.

- *Force-balance.*   Force-balance accelerometers use an electromagnet to hold a small mass in a fixed location relative to the housing of the accelerometer. A servo loop is used to vary the coil current to maintain the position of the mass in the presence of accelerations. The coil current is then proportional to the acceleration. The frequency response of force-balance accelerometers typically extends from dc to an upper limit of about 50 to 500 Hz. Force-balance accelerometers are relatively complex, fragile, and expensive. However, they have excellent accuracy and resolution.

- *Strain gauge.*   Strain gauge accelerometers usually have a small mass mounted on the end of a cantilever beam. Strain gauges, mounted on the beam near its base, detect bending of the beam that is proportional to acceleration. The frequency response of these accelerometers typically extends from dc to about 200 to 10,000 Hz. Because recent developments in semiconductor fabrication now permit the fabrication of strain gauge accelerometers from monolithic silicon using integrated-circuit fabrication technology, these accelerometers could eventually be the least expensive of all the types discussed here.

- *Potentiometer.*   These accelerometers use a small mass supported by a spring. The mass is attached to a potentiometer wiper, and wiper displacement is proportional to acceleration. This type of accelerometer generally has poor accuracy and resolution, but the output is a high-level signal that does not require electronic amplification.

- *Capacitance.*   The capacitance-type accelerometers use a small mass restrained by a spring. The mass also acts as one plate of a capacitor, and acceleration is inferred from changes in capacitance. The frequency response of these accelerometers typically extends from dc to 10 to 1000 Hz.

12.2.3   Force and torque

Many robots currently in production do not have a means to sense force or torque. In many practical applications, force and torque sensors must be added to the robot (such as the wrist sensor, described below) if the robot is to perform the desired task. Incorporating force sensing in the robot itself will probably be a trend for the future.

Techniques that have been devised for torque and force measurement include piezoelectric, potentiometric, and capacitive sensors. However, only two techniques are likely to be used for "built-in" force and torque sensing in electric robots. These are described below.

12.2.3.1   Strain gauges.   Strain gauges can be mounted at appropriate points on the robot structure to determine the strain at these points. If the mounting locations have been properly chosen, the forces or torques exerted by the various joints (or by the gripper) can be determined from the strain gauge signals. To obtain improved accuracy or resolution, the designer is advised to modify the robot structure slightly to create sites with particularly large or uniaxial strains. Strain gauges are then mounted in these preferred locations.

The robot designer should remember several characteristics of strain gauges that will influence where and how they should be mounted. Strain gauges are very sensitive to temperature changes. They should not be mounted near heat sources, such as motors. The effective temperature sensitivity can be decreased if strain gauges are mounted in pairs. The two-arm and four-arm bridge configurations are often used with opposite bridge arms mounted in locations of equal and opposite strain. Strain gauges are very fragile and subject to attack by numerous chemicals and solvents. Therefore, the strain gauge mounting locations should be protected from mechanical and chemical abuse. Ideally, all strain gauges should be mounted within hermetically sealed chambers on the robot.

12.2.3.2 Motor current measurement. Many robots are powered by dc servomotors. The output torque of such motors is linearly related to the armature current in the motor. Thus, it is possible to measure the torque at each robot joint (indirectly) by inserting a suitable resistor in one lead of each servomotor. The voltage across this "sense" resistor is proportional to the motor current and is therefore related to motor torque. This technique has been used on the Stanford Arm (Scheinman, 1969).

Motor current measurement is simple and inexpensive but is not without drawbacks. The measurement accuracy is affected by any friction in the motor bearings, associated gears, and joint bearings. These frictions can change from relatively large values to nearly zero as the lubricants are heated during operation of the robot. For these reasons, accuracy of torque measurements by this method is seldom better than 3 to 10 percent over the normal operating temperature extremes.

12.3 Perception

12.3.1 Vision

Vision is perhaps the most useful of the human senses and is clearly a perception means applicable to robots. Vision equipment and methods provide a large topic and can only be touched upon here.

12.3.1.1 Equipment. The development of photodiode and charge-coupled-device (CCD) array cameras has made vision increasingly attractive since these cameras are much smaller, more rugged, less expensive, and use less power than those based on the vidicon tube. In the vidicon tube, the image is focused on a photoconductor on the face of the tube. An electron beam within the tube scans the photoconductor: Electrons passing through the conductivity image in the photoconductor and collected on a conductive coating constitute the image-bearing signal. The vidicon method requires a vacuum tube many inches long, a power source for the electron-beam cathode, and the means to deflect the beam.

The diode array is an array of photodiodes on a chip. The image is focused onto the array, and the signal is read as the photocurrent in the diodes. In addition to the cost, power, and size advantages mentioned above, the electronic image is geometrically stable; it does not depend on focus or raster adjustments as does the vidicon tube. This feature permits position calibration of the image. Additional advantages have to do with operational factors such as blooming (saturation of the photoconductor) and afterimage (slow photoconductor response relative to image motion). Diode arrays simply have better performance characteristics than the vidicon tube. Diode arrays are so small that the lens is often the largest part of the camera (see Figure 12.3). Diode arrays as large as 512 × 512 pixels are commercially available.

Scanning is an alternative method for obtaining the same scene information obtained by a camera. Scanners move a beam of light over the scene using galvanometer mounted mirrors; the intensity of the reflection is detected by a photodiode. Scanners are commonly used to read bar codes, where the form of the information is well defined and designed for information processing.

A scanner and a diode array can be used in combination to obtain three-dimensional information. Other techniques, such as the use of two cameras (for stereo vision), can also be used to obtain three-dimensional information but is usually more difficult because comparative image processing is necessary to extract the information.

12.3.1.2 Processing. Vision involves large amounts of information. A 512 × 512 array has 262,144 pixels, each of which gives light-level information. Eight bits of light level (256 levels) multiplies the information by 8, and color further multiplies the value by 3. A pair of cameras, for three-dimensional information, would then double the data. The prodigious

Figure 12.3 Reticon Model MC520 100 × 100 matrix camera (body 3.58 × 3.8 × 3.75 in).

amount of information in a single image means that the extraction of information from a scene is time-consuming. Resolution can be sacrificed for speed. Smaller arrays and the use of binary light level (black and white only) can speed processing. Even elementary vision processing is so slow that it is not yet feasible to use it for position servoing of robot joints.

Deriving useful information usually begins by processing the image for primitive features, e.g., the location of edges and corners. The images are enhanced by filtering in order to highlight the desired features and to suppress spurious information. A simple filter for an edge would look for dark pixels next to light ones. A typical filter considers one small array at a time, the pixel in question and its nearby neighbors. Higher level features are obtained from the combination of features derived from filtering. Identification of objects involves pattern matching, analysis of objects by geometric characteristics such as area and perimeter, and recognition of features in combinations. The process is complicated by overlapping parts and ordinary lighting.

When a central processer is used, all the filtering on the image data must be done sequentially, one pixel being considered after another. One method being explored in research labs to speed up vision processing is *parallel processing*. In this approach, each pixel would be filtered for information by its own processor. The filtering algorithms could be hardwired into the equipment. The parallel-processed information would then be transmitted to a subsequent sequential-processing algorithm where high-level pattern-recognition decisions are made.

12.3.2 Proximity

Many problems in robotics call for some form of proximity sensor. One example is the need for short-range (approximately 10 to 100 mm) sensors to prevent collision of a robot arm with fixtures, the workpiece, or other robots. A second example is the need for long-range (approximately 100 mm to 100 m) sensors for guidance, navigation, and collision prevention for robot vehicles.

Discussion of long-range sensors is beyond the scope of this book, but several types of sensors useful for short-range proximity detection are discussed below.

12.3.2.1 "Whiskers." The common domestic cat is equipped with very effective proximity sensors on its face called whiskers. Analogous sensors, which can be used in robots, consist of long, thin, flexible rods attached to the robot with force or displacement sensors of some kind. In the past, simple switches have usually been used at the base of whisker sensors. Despite all precautions, whisker sensors can be bent or broken. Therefore, the sensor should be designed so that damaged whiskers can be replaced easily and inexpensively.

12.3.2.2 Eddy current. Eddy current sensors can be used to detect the proximity of electrically conductive materials. The principle of operation of these sensors is described in Section 12.2.1.1.6. Eddy current sensors are most effective for very short ranges, such as 1 to 20 mm.

Bahr (1986) has built an array of eddy current sensors for robotic applications. An unusual feature of the array is the arrangement of actuating coils in pairs. Current can be fed to paired coils to amplify the field produced or to cancel it. In the canceling mode the response becomes sensitive to edges near the coil pair. Thus, the sensor can be switched from proximity mode to edge-detection mode.

12.3.2.3 Capacitance. Capacitance sensors are also able to detect only the proximity of conductive materials. Operation of this type of sensor is described in Section 12.2.1.1. Capacitance proximity sensors are effective only at very short range, typically less than 5 mm.

12.3.2.4 Ultrasonic sonar. This sensor consists of an ultrasonic transmitter and receiver and associated electronic circuitry. (In fact, in many sonar sensors, the same transducer is used as both transmitter and receiver.) The sensor transmits a burst of ultrasonic energy (typically between 20 and 60 kHz) that will be reflected back toward the receiver by any objects it strikes. The distance to an object can be determined directly from the time interval between transmission and receipt of the burst. The ultrasonic reflectivity of nearly all objects is sufficient to allow proximity detection by this method. For example, SRI has developed an ultrasonic proximity detection system capable of detecting a person wearing a fur coat at a distance of about 1 m. (Fur is one of the poorest ultrasonic reflectors.) Ultrasonic proximity detectors are most effective at intermediate ranges from about 10 mm to 10 m. An ultrasonic ranging device developed for use in a camera has recently been made available by Polaroid Corporation for other applications on an OEM basis. This device has received considerable attention because of its (relatively) low cost and ease of use.

12.3.2.5 Optical reflection. Proximity sensors using optical reflection can be grouped into two categories as shown in Figure 12.4. These sensors consist of a light source (usually infrared) and a light detector. When an object with sufficient reflectivity is present in the area of intersection of the transmit and receive beams, a signal is generated in the receiver indicating the presence of the object. The coaxial type of sensor has a large area of beam intersection, which enables a single sensor to cover a relatively large area. Unfortunately, the range of the coaxial sensor is strongly dependent on the reflectivity and size of the object being detected.

The triangulation type of optical sensor has a relatively small intersection area and, therefore, is relatively insensitive to variations in object size or reflectivity.

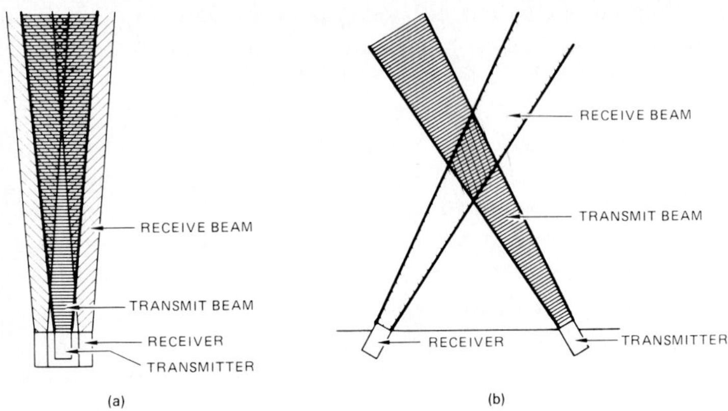

Figure 12.4   Optical reflection proximity sensors.

Optical proximity sensors cannot detect nonreflective objects, but most objects (including many that appear black to the eye) have sufficient reflectivity to be detected optically. Specular surfaces (such as mirrors or highly polished metals) are detectable with optical sensors only if they are properly oriented.

12.3.2.6   Optical transmission.   This sensor consists of an optical transmitter that directly illuminates the receiver when no object is present. An opaque object in the beam will block the light entering the receiver and, thereby, will be detected.

12.3.3   Touch

Touch sensors may be used in robot grippers for various purposes:

- Ensuring that the workpiece is properly grasped (i.e., properly positioned relative to the gripper)

- Ensuring that the gripper exerts the right force on the workpiece so as not to crush it or drop it

- Inspecting the workpiece for gross defects

Tactile sensor technology has aimed first at duplicating aspects of human tactile performance. In order to do this it is necessary to have a spatial resolution of 1 mm, a pixel sensitivity of 1 g in the range up to 1000 g, and a bandwidth up to as much as 1000 Hz (Pennywitt, 1986). The robotic hand also must have thermal-sensing capability. Shear stress may also be a useful attribute (Hackwood et al., 1983).

Most touch sensors and applications developed in research laboratories have remained there. The reasons include high cost, poor spatial resolution, low reliability, and lack of repeatability of the sensors. The sensors are not so well developed as vision sensors. In addition the data generated by a touch sensor are complicated and not easily used in controlling the manipulator. Use of tactile information requires knowledge of the environment as well as models that are not available except in simplified cases. However, the obvious improvements in performance that could be achieved using tactile sensing suggest that manipulators eventually will use touch.

The last few years have seen an explosion of tactile sensor development in the research laboratories. Allan (1985) documents 14 research efforts, each pursuing a different approach. Hopefully these efforts will lead to low-cost high-resolution high-reliability easy-to-use touch sensors for industrial needs.

Of course, any of the force sensors described in the previous section can be used for touch sensors. Several other sensors developed or suggested specifically for use as robot touch sensors are described below.

12.3.3.1 Switches. An array of electric switches can be mounted on a robot gripper to detect contact with the workpiece. A significant disadvantage of switch touch sensors is their binary output. It is impossible to determine the exact grasping force. Some designs have attempted to alleviate this problem by providing one set of switches that closes at low force and a second set that closes at a higher force level, and perhaps additional sets for still higher forces. For example, investigators at Carnegie-Mellon University have fabricated an array of switch touch sensors each capable of resolving 16 levels of force (Allan, 1985). The substrate for this tactile array was fabricated photolithographically in silicon, and the switch contacts include conductive elastomers.

12.3.3.2 Air ports. A surface, such as the grasping surface or outer surface of a gripper, can be provided with one or several small ports that are fed with air at relatively low pressure (e.g., 5 lb/in^2) and flow rate. If the surface touches a part so that the port is blocked, the air pressure behind the port will increase. A suitable pressure sensor or pressure switch located behind the port can then be used to indicate when the gripper is touching a part. The balancing circuit shown in Figure 12.5 can be used to make more sensitive measurements and even measure distances from the surface with the nozzle to another surface in terms of a reference distance. Despite its disadvantages (including the need for pressurized air and a relatively expensive pressure sensor), an air port is probably the most rugged of the available touch-proximity sensors.

12.3.3.3 Discrete strain gauges. An arrangement similar to the wrist sensors described later (discrete strain gauges mounted on support beams) has been used to determine the total force and center of pressure

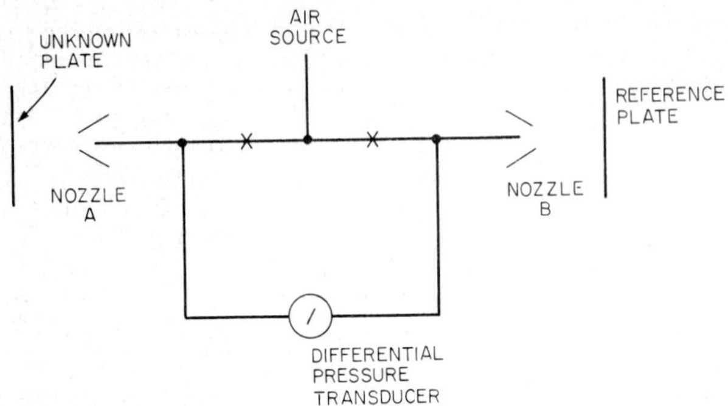

Figure 12.5   Air jet sensing.

exerted on a gripper or inspection surface. This approach has been used by Eaton Corporation in a 2 × 2 in sensing pad for example. A severe limitation of this approach is its inability to determine the outline or shape of a part being gripped.

12.3.3.4 Integrated-circuit (IC) strain gauges.   Integrated-circuit fabrication techniques can now be used to fabricate mechanical structures with embedded strain gauges in monolithic silicon. Arrays of pressure sensors have been fabricated in this way (Eckerle and Newgard, 1977) with spatial resolution of about 0.5 mm and accuracy of about 1 percent of full scale.

Transensory Devices, Inc., has fabricated 3 × 3 arrays of force sensors from silicon using IC fabrication techniques (IRI, 1984a). Spatial resolution of these devices is 2 mm. Guckel and Burns (1984) have been fabricating pressure-sensor arrays in a polysilicon layer grown on a silicon substrate.

12.3.3.5 Conductive elastomers.   Several researchers have suggested using conductive elastomers for robotic touch sensors. These elastomers have an electric resistance that decreases with increasing applied pressure. This property should make a proportional force sensor possible. One problem has been poor accuracy and long-term drift. Barry Wright Corporation offers a 16 × 16 tactile array sensor with 2.5-mm spatial resolution that they claim has overcome the problems of accuracy and long-term drift (IRD, 1983a).

12.3.3.6 Photoelectric.   Several tactile sensor designs have emerged employing interruption or deflection of a light beam as the transduction modality. Lord Corporation produces a 10 × 16 tactile sensor with a spatial resolution of 1.8 mm. In this sensor, the light beam is interrupted by small bumps in an elastomeric gripper surface. A similar sensor, using reflection rather than transmission of the light beam, has been described by McCammon (1984).

Tactile Robotic Systems is producing a 16 × 16 photoelectric tactile sensor with spatial resolutions from 1.3 to 25 mm (IRI, 1985a). Deflection of a small metal beam modulates the light passing into an optic fiber.

A group at MIT has developed a 34 × 35 tactile sensor array of optical fibers and a two-layer elastomeric "skin" (Allan, 1985). The upper skin layer is white (i.e., reflective), and the lower layer is transparent. Transmitting and receiving optical fibers are arranged beneath the lower layer. When the top layer is depressed, the intensity of light at the nearby receiving fibers increases. A TV camera in the data-processing method receives light from the fiber bundle.

A similar, more elegant tactile sensor has been developed by British Robotic Systems (Sensor Technology, 1985). In this sensor, light travels transversely through a clear acrylic plate. Forces on the sensor surface cause localized changes in the acrylic index of refraction, and some light is thereby scattered out the bottom of the plate. A camera captures the scattered light for visual processing.

12.3.3.7 Capacitive. Groups at Stanford University, Bell Laboratories, and MIT have each investigated capacitive tactile sensors. These sensors employ two layers of orthogonal conductive strips separated by a compliant layer. The capacitance between any two orthogonal strips increases when a force is applied near their intersection. Spatial resolutions of 2.5 to 3.0 mm have been achieved.

12.3.3.8 Thermal sensor. A thermistor, a device in which the resistance is a strong negative function of temperature, can be mounted in a gripper so that it will be in contact with any parts being held. If the thermistor is heated above the ambient temperature, heat will be transferred at different rates to the parts being held depending on their thermal properties. Characteristics of that heat transfer can be used to detect part presence, determine the material, and detect slip. The thermistor may be operated at constant power so that the temperature and temperature transient yield the desired information. An alternative is constant temperature, where the power needed is used as a measure of the heat transfer. The constant power technique is preferred because it reduces the effect of the thermal time constants of the sensor itself.

Figure 12.6 shows the SRI-developed constant-temperature thermal sensor (Shrader, Kornbluh, and Andeen, 1986) in a gripper. When the gripper closes on an object, the thermistor begins to heat a spot on the object's surface. The transient behavior depends on the thermal diffusivity of the object. High power is needed until a hot spot is achieved. The steady-state power depends on the thermal conductivity. When the object is moved relative to the sensor, colder material is brought into contact with the thermistor and more power is required to keep the thermistor at a constant temperature. This change in power indicates slip.

Figure 12.6 Thermal sensor in a gripper face.

The thermal sensor is simple, inexpensive, and rugged. Yet it gives considerable information that is useful for verifying part identity (by thermal property), location, and movement in the gripper.

12.3.3.9 Miscellaneous. An unusual tactile sensor has been studied by Bonneville Scientific (IRI, 1983*b*). In this sensor the local thickness of an elastomeric gripper pad is measured by ultrasonic sonar. The ultrasonic transducer is a polyvinylidine membrane.

Cochlea Corporation has developed a unique robotic sensor with some attributes of both tactile and vision sensors (Buckley, 1985). The sensor "illuminates" the sensing area with ultrasound in the 20 to 40 kHz range. An array of ultrasonic detectors is used to determine the acoustic signature of interference patterns caused by any object in the sensing area. This system has been used for parts sorting.

Herga Electric Ltd. has developed a touch sensor fabricated from fiber-optic cable components (Prahl and Tracey, 1986). The sensor includes a spiral fiber wrapped around a central light-transmitting fiber. A contact force applied to any point along the sensor will cause significant attenuation of the transmitted light. This sensor is useful for applications such as pressure-sensing safety mats and bumpers for automatic guided vehicles (AGVs).

12.3.4 Wrist force-torque

Several robot applications (including assembly, glue application, riveting, and grinding) are most easily accomplished if the robot is equipped with a device known as a wrist sensor. The wrist sensor normally mounts between the robot arm and the gripper and measures three (vector) components of force and three torque components. Most wrist sensors have used strain gauge technology but have been designed in different geometries.

One of the early wrist sensors (developed at SRI) is shown in Figure 12.7a. All pivot points in this sensor are implemented with flexures rather than conventional bearings. This approach practically eliminates hysteresis in the sensor.

The sensor shown in Figure 12.7a is generally in the shape of a hollow cylinder, which permits nesting the actuator components of the gripper inside the sensor to obtain a compact sensor/gripper assembly.

An important performance parameter for wrist sensors is the degree of cross talk between axes. In principle, all cross talk can be eliminated in software by suitable matrix manipulation of the sensor output signals, but this approach may not be practical in industrial use. (It requires elaborate calibration of each sensor.)

The wrist sensor designer should also give careful consideration to the overload capability of the sensor; the robot will eventually damage the sensor if it is physically possible for it to do so.

Some alternative designs for wrist sensors for the Stanford Arm are discussed in Scheinman (1969). In later work, Scheinman developed a completely different wrist sensor for this arm. The later design was in the shape of a cross with strain gauges at the vertex as shown schematically in Figure 12.7b.

Work at SRI that led to the wrist sensor shown in Figure 12.7a concluded that the optimum shape for a wrist sensor designed for mass production might be a very thin walled (e.g., 0.1 to 1 mm) tube. (The sensor of Figure 12.7b was never intended for mass production.)

The number of sensors must be equal to or greater than the number of measurements to be made. Three sensors would be required to measure only the force vector or only the torque vector. Six sensors would be required to measure the force and torque vectors together. The wrist sensor device in Figure 12.7a has eight pairs of strain gauges, or essentially eight sensors for the measurement of six values.

In theory, a large number of sensors could be applied to any structure and the desired information extracted by multiplying by a calibration matrix. For example, four measurements (m_1, m_2, m_3, and m_4) could be multiplied by a suitable matrix $\mathbf{A}$ to yield the force vector (F_x, F_y, and F_z).

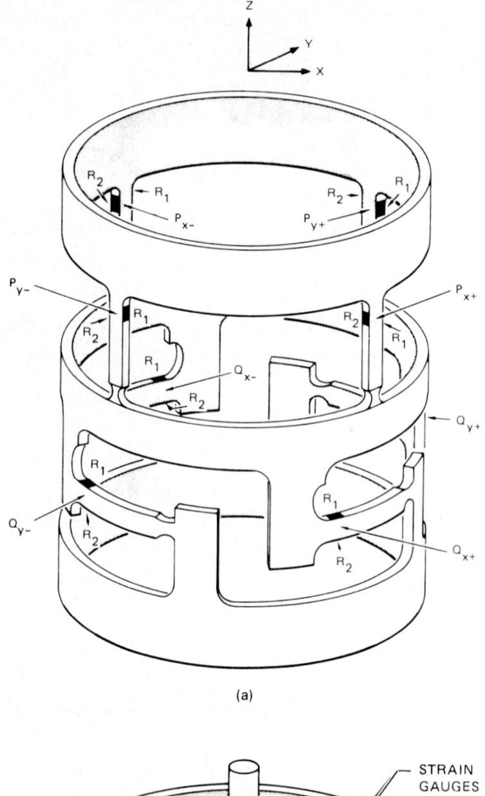

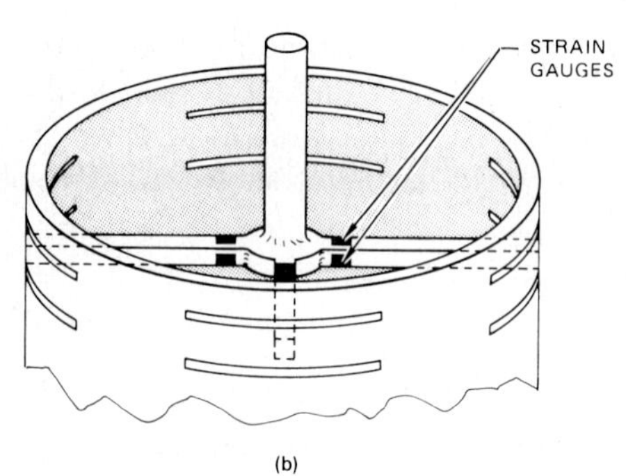

Figure 12.7 Wrist sensors. (a) SRI-developed gauge wrist sensor; (b) schematic of Scheinman-developed wrist sensor.

$$\begin{bmatrix} a_{11} & a_{12} & a_{13} & a_{14} \\ a_{21} & a_{22} & a_{23} & a_{24} \\ a_{31} & a_{32} & a_{33} & a_{34} \end{bmatrix} \begin{bmatrix} m_1 \\ m_2 \\ m_3 \\ m_4 \end{bmatrix} = \begin{bmatrix} F_x \\ F_y \\ F_z \end{bmatrix}$$

The flaw in the general approach is that no effort has been made to make the most effective use of the sensors or to minimize the processing time. In particular, the sensors may be operating in a region of low signal-to-noise ratio, or perhaps in a nonlinear region. A simpler matrix would not require computer calculation but could be implemented in hardware.

The wrist sensor of Figure 12.7a has been designed to have a matrix with a large number of zero values:

$$
\begin{bmatrix}
0 & 0 & a_{13} & a_{14} & 0 & 0 & 0 & 0 \\
a_{21} & a_{22} & 0 & 0 & 0 & 0 & 0 & 0 \\
0 & 0 & 0 & 0 & a_{35} & a_{36} & a_{37} & a_{38} \\
a_{41} & a_{42} & 0 & 0 & 0 & 0 & a_{47} & a_{48} \\
0 & 0 & a_{53} & a_{54} & a_{55} & a_{56} & 0 & 0 \\
a_{61} & a_{62} & a_{63} & a_{64} & 0 & 0 & 0 & 0
\end{bmatrix}
\begin{bmatrix}
P_x+ \\
P_x- \\
P_y+ \\
P_y- \\
Q_x+ \\
Q_x- \\
Q_y+ \\
Q_y-
\end{bmatrix}
\begin{bmatrix}
F_x \\
F_y \\
F_z \\
T_x \\
T_y \\
T_z
\end{bmatrix}
$$

where P and Q refer to the sensor locations shown in Figure 12.7a. In practice, the zero terms may not be zero, but should be small.

The design strategy for the SRI wrist sensor shown was to make a structure in which an applied force would result in a large strain that could be measured by a pair of gauges. The clamped-free beam is a structure that maximizes measurable strain at the base for an end loading across the beam. Figure 12.8 shows the strain gauge relationship to the beam and the electric circuit to give the output of a pair of gauges. The beam is not clamped-free but is clamped partially free. The material at the end of the beam has been "necked down" to provide flexibility without introducing hysteresis that would be associated with a rubbing contact.

The beams were arranged so that they would be deflected according to particular forces or torques. In this case, the beams were located in an annular region for nesting. The size of the beams was selected according to the material yield strength, a safety factor, and the anticipated load. Because the strain gauges respond to strain, the highest reasonable values of stress at expected loads should be used to maximize sensor output. Foil gauges were used in the SRI design. Semiconductor gauges would have increased the sensitivity with only a small amount of nonlinearity.

The possibility of overload and damage to the wrist sensor can be protected against by shear pin mounting, stops, or a combination of the two techniques. The implementation first used at SRI is shown in Figure 12.9.

A clever technique for stop protection for the ring wrist sensor is shown in Figure 12.10. In this design, the P beams are nested within the Q beams. Thus the deflection of the P and Q beams is adjacent to the original mount. Stop pins at this point, therefore, prevent excessive deflection and overload of the beams.

The nesting concept was an important design consideration in the SRI wrist sensor. Other wrist sensors based on crossed beams have been designed; they are excellent provided that the center portion can be

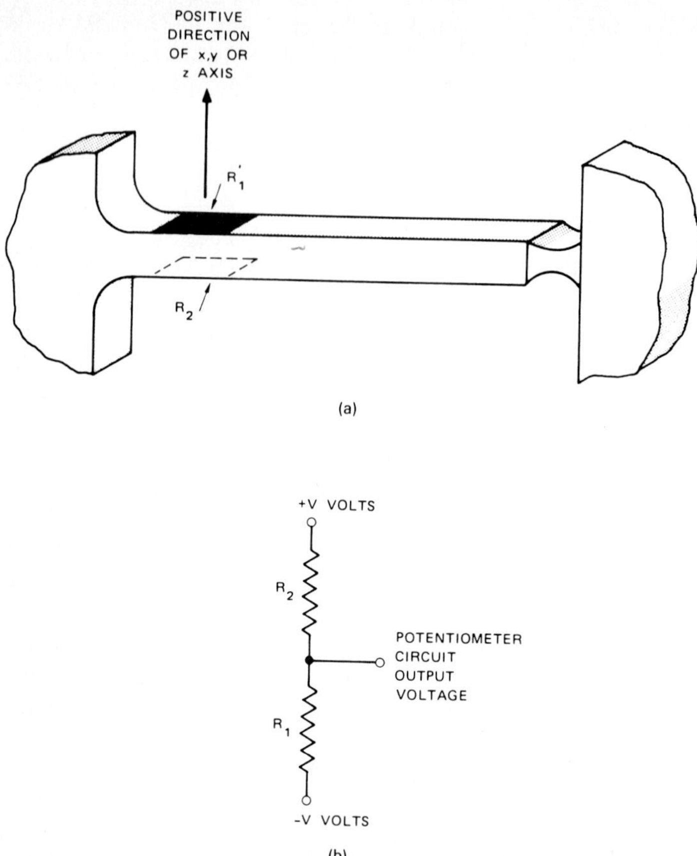

Figure 12.8 Two strain gauges for each beam. (*a*) Mechanical mounting; (*b*) electrical connection.

blocked. Once again we see that the basic design philosophy plays a significant role in implementing details.

12.3.5   Accurate position sensing

Most measurement techniques give accuracies proportional to the size of measurement. This is analogous to describing the resolution of a camera by an angle rather than by a length (which may be microscopically close or at a great distance). To make accurate measurements, one restricts the space in which the measurement is made. (Some measurement techniques such as laser ranging, for example, apparently escape this general rule.)

Making an accurate determination of the position of the robot end effector within the entire work envelope may be difficult. Usually, how-

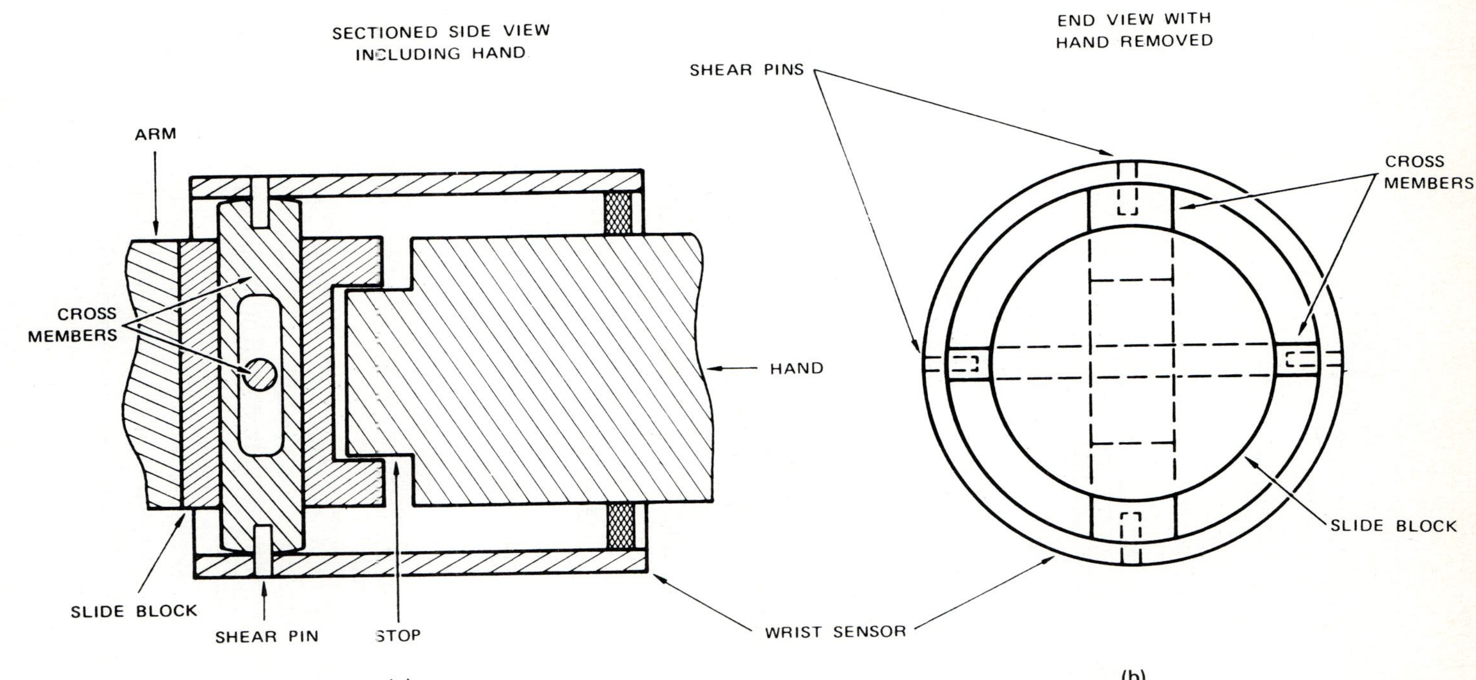

Figure 12.9 Overload protection mechanism. (*a*) Sectioned side view including hand; (*b*) end view with hand removed.

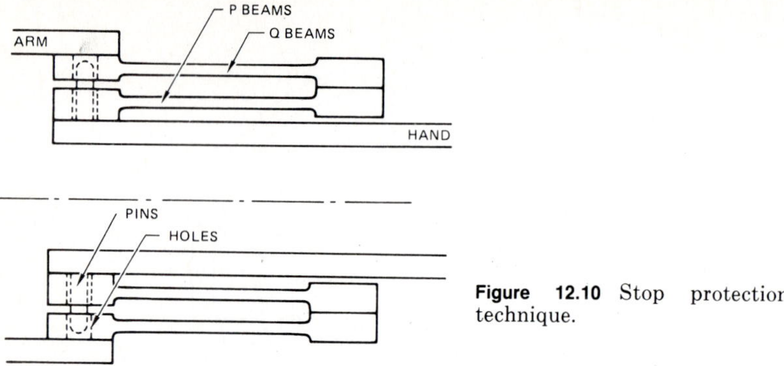

Figure 12.10 Stop protection technique.

ever, accurate measurements are necessary only in a smaller volume, such as in the region of an insertion. As with an elevator, where the exact position is immaterial except in the neighborhood of a floor, the control of the manipulator should be transferred to local position transducers that can provide better information than can be inferred from joint positions. An accurate measurement station can also be used to provide an occasional check on the accuracy of the working robot. Several stations can be used during the development of a robot so that accuracy in widely separated positions in the work envelope can be determined.

Final position adjustment should be made by the movement of most local joints. Special end effectors may be provided that are capable of fine positioning over short distances. An alternative for very accurate positioning is to have the work positioned on an xy table, capable of small accurate movement. The robot then positions as accurately as it can, and the table does the final high-precision maneuvering of the work to the robot.

Some of the kinds of devices that could be used for accurate measurement are:

- Physical contactors of the kind used in part metrology. These devices can note when an object touches them, or they can be advanced to strike a stationary body.

- Diode array cameras, using optics. Can be one-dimensional or two-dimensional. Arrays are available at up to 1000 per in, which should increase as technology advances. The resolution of the measurement depends on the optics.

- Interrupted or scanning beams. The beams have a position that can be well known. Scattering of light by an object indicates position.

- Interferometry is possible for fine measurements.

Chapter

13

Grippers

13.1 General Considerations

The guiding philosophy of robotics has been that general-purpose equipment can do specific tasks through changes in software. Robot manufacturers can supply the same piece of equipment for many tasks. Furthermore, the robot can change tasks during its life simply by being reprogrammed. Nowhere has this philosophy proven to be more of a challenge than in the area of grippers. Although the same manipulator may be used in a bakery for handling doughnuts and in a light-bulb plant for handling hot glass parts, different end effectors are most often required. The end-effector design sometimes provides the cost-effectiveness, for example, by providing multiple part holding as in the case of low-value items (the doughnuts) or by providing two tools on the same effector to save the time of going back to a tool changer. The lack of a universal gripper means that there are many companies supplying grippers according to the task and many providing custom design services. These gripper suppliers are often smaller companies that do not make the manipulators. The lack of a universal gripper also increases the need for quick-change interfaces.

Gripper prices are generally 3 to 10 percent of a robot price, but are sometimes as low as 0.1 percent or as high as 20 percent for smarter or more specialized grippers.

13.2   Kinds of Grippers

Grippers or end-effectors can roughly be classified into five categories.

1. *Mechanical.*   Two or more fingers or jaws for external/internal grasping
2. *Vacuum.*   One or more vacuum cups for handling flat or nearly flat workpieces that are reasonably smooth and clean
3. *Magnetic.*   One or more electromagnets (can be permanent magnets), for ferrous materials only
4. *Expandable cuff.*   Inflatable bag or cuff to handle odd shapes and fragile materials, also inflatable prehensile fingers
5. *Adhesive.*   For materials that are lightweight and not suitable for vacuum techniques (like cloth)

A chuck to hold a variety of specially designed tools might be considered as a sixth category. The first three types, mechanical, vacuum, and magnetic, are discussed further.

13.2.1   Mechanical

Motive power for robot grippers is often pneumatic; a four-way valve controls air at pressures of 50 to 100 lb/in^2 (350 to 700 kPa) to drive the jaws open or closed as illustrated in the designs shown in Figures 13.1 and 13.2. Pneumatic actuation provides fast motion and allows the jaws to close on an object with nearly the same force, which is independent of the size of the object. Similar designs can be actuated by hydraulics. Since hydraulic pressures are generally an order of magnitude higher, the actuators can be miniaturized and made lighter. The disadvantages are a loss of simple force adjustment and the complications of leak-free hydraulic disconnects.

Figure 13.3 shows the gripper mechanism used in the Microbot educa-

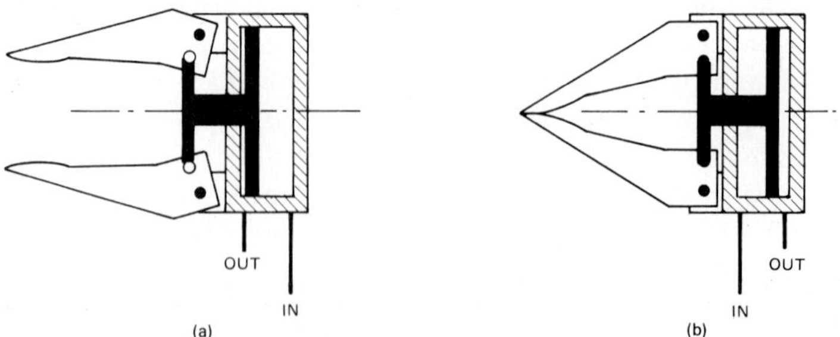

Figure 13.1 Grippers operate on pressure from four-way air valves. (*a*) Pressure to open; (*b*) pressure to close.

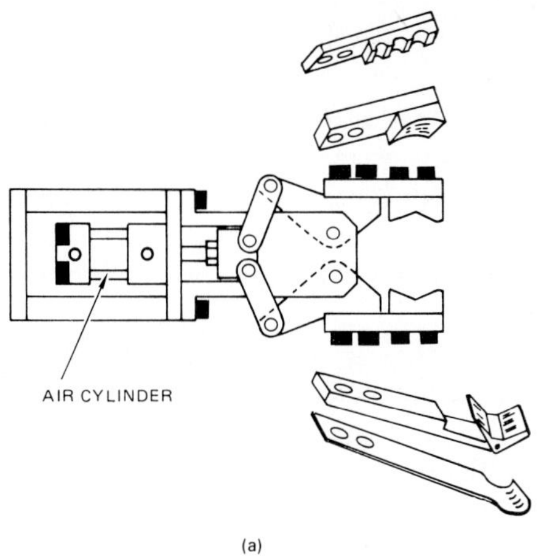

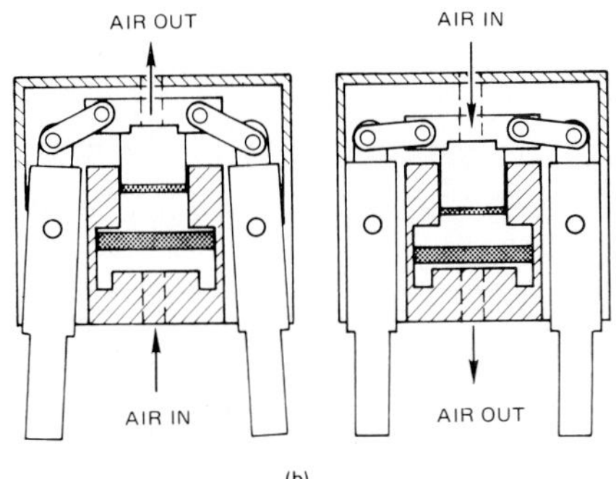

Figure 13.2 Pneumatic grippers. (*a*) Gripper; (*b*) operation with mechanical levers.

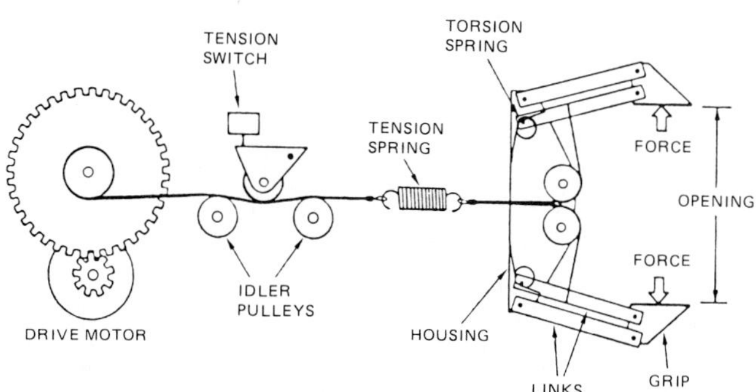

Figure 13.3 A simple method for measuring force exerted by a parallel-jaw gripper. (*Source: Microbot Corp.*)

tional robot. It uses a stepping motor to drive a cable or tendon. The spring in the tendon allows for force adjustment. A dc motor would allow force control. The actuator in this design only closes the jaws; return torsion springs open them. The idler pulley tension measurement mechanism is used to stop closing. The pads of the gripper of Figure 13.3 always remain parallel, which is often desirable when grasping materials of different thicknesses. The jaws of this gripper do not move directly toward one another, but move in arcs.

Tendons, cables, or strips running in sheaths provide the ability to remotely locate the actuators. This is advantageous if the gripper has several degrees of freedom (actuators). Tendon designs are especially popular among researchers attempting to duplicate the dexterity of the human hand (Jacobsen et al., 1986). It is likely that compact actuators, perhaps from shape-memory alloys, will be useful for some of the slower or less often used joints of dexterous end effectors.

Some of the early pioneering work in end effectors was for prosthetic devices; the intention was to duplicate the form as well as the function of the hand. More recent work has focused on how best to grip objects given their mass, inertia, center of gravity, shape, and coefficient of friction; and how to manipulate them (such as turn the object over in one's hand or transfer it between fingers).

Development of a unique multiple prehensile manipulator system

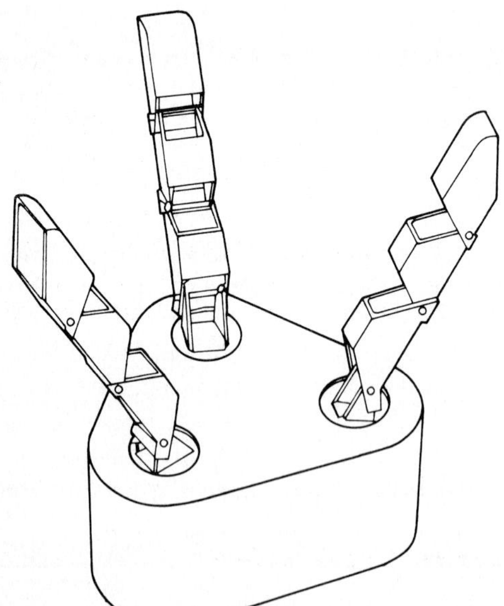

Figure 13.4   A sketch of the Skinner hand built for NASA.

(MPMS) is described in Skinner (1975). His design, referred to as the "Skinner hand," using three multiple-articulated fingers (three links constitute one finger) is shown in Figure 13.4. A human hand has 22 degrees of freedom and 48 actuators. The Skinner hand is simpler but still can perform most of the six prehensile patterns of the human hand shown in Figure 13.5. Figure 13.6 shows the modes available to the three-fingered hand. A simpler version of Skinner's three-fingered hand is shown in Figure 13.7.

Salisbury (1984) has advanced the Skinner-type hand by adding sensing that allows the hand to do dexterous manipulation. He has used a miniaturized force/torque sensor of the type normally used in wrists (see Section 12.3.4) as fingertip sensors. He is able to find the forces, including shear, on each finger.

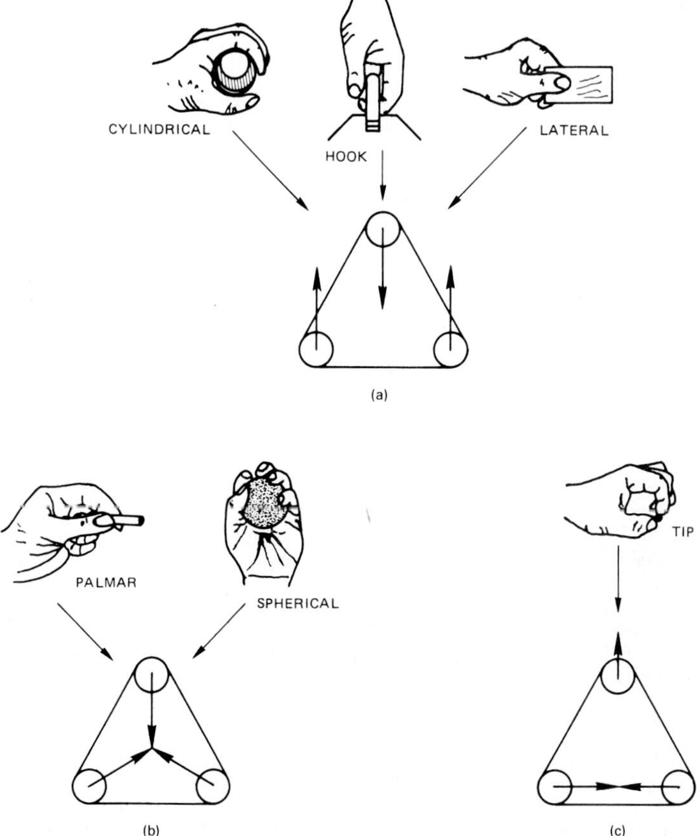

Figure 13.5 Six prehensile patterns and their mechanical equivalents. (*a*) Wrap (top view); (*b*) three-jaw (top view); (*c*) tip (top view).

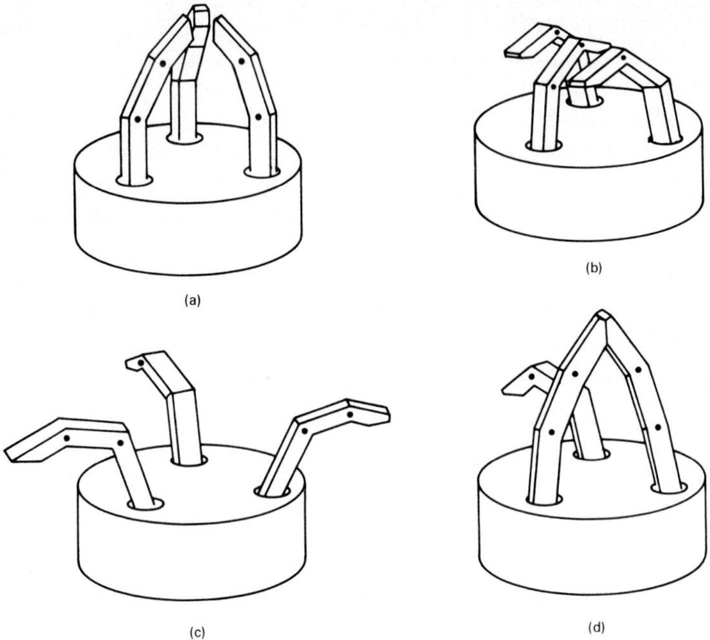

Figure 13.6 Prehensile modes available with the Skinner hand. (*a*) Three-jaw position; (*b*) wrap position; (*c*) spread position; (*d*) tip position.

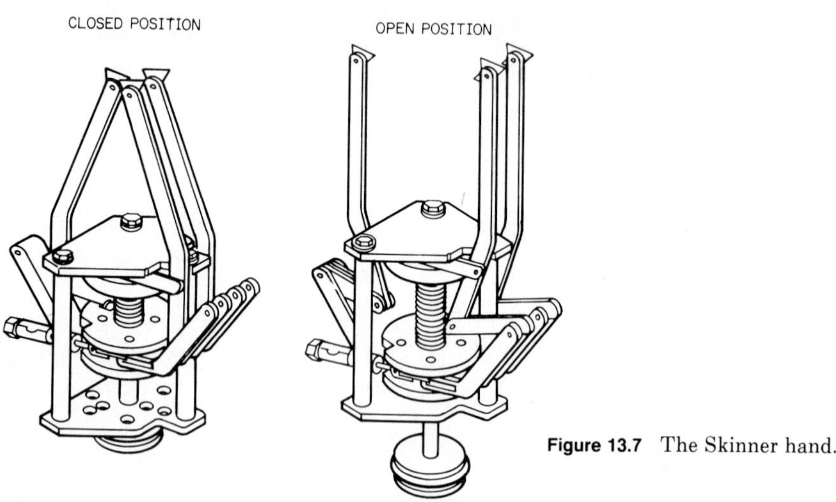

Figure 13.7 The Skinner hand.

13.2.2 Vacuum lifters

Vacuum lifters generally use cups made of rubber (synthetic or natural) or polyurethane material. The force obtainable is the area of the cup multiplied by the pressure differential between the inside of the cup and the outside atmosphere: $F = A \times \Delta P$. Small vacuum lifters are used to handle IC wafers and dice during processing and in assembly into systems.

There is a tradeoff between the cup size and the differential pressure used—larger cups and lower vacuum often give greater speed. Release speed is increased by applying a positive pressure inside the cup immediately after the vacuum is turned off.

The cups in vacuum lifters are often spring-mounted so that contact is made with the part before the robot arm is in the final pickup position. This allows the cups to stick to the part while the arm is decelerating.

The vacuum for use in lifters can be generated by either an electromechanical pump or a venturi operating from shop air. The motor-driven vacuum pump provides higher vacuum than a venturi and is quieter in operation, but the venturi generally costs less than the motor-driven pump.

13.2.3 Magnetic lifters

Magnetic lifters are useful for handling ferrous metal parts. The force of attraction for a magnetic lifter is proportional to the mass of the part.

Electromagnets are suited to remote control and can provide moderate pickup and release speeds, but they need a source of dc power. The control system generally reverses the polarity of the power supply for a short interval to speed release.

Permanent magnets are sometimes used instead of electromagnets. Permanent-magnet lifters need a device to remove the part from the magnet. It is possible to design the permanent magnet to penetrate only to a small depth in the part (as low as 0.031 in).

Magnetic lifters impose some extra requirements for suitable workpieces. The part to be handled magnetically should have a large flat area for the lifter to work on. If the part is rounded, rather than flat, a stronger magnetic lifter will be required. Temperature is also a problem: The part to be handled must not be too hot (300°F is generally safe).

13.3 Common Gripper Configurations

Figure 13.8 shows examples of a variety of manipulator/gripper configurations. These are further described in Table 13.1.

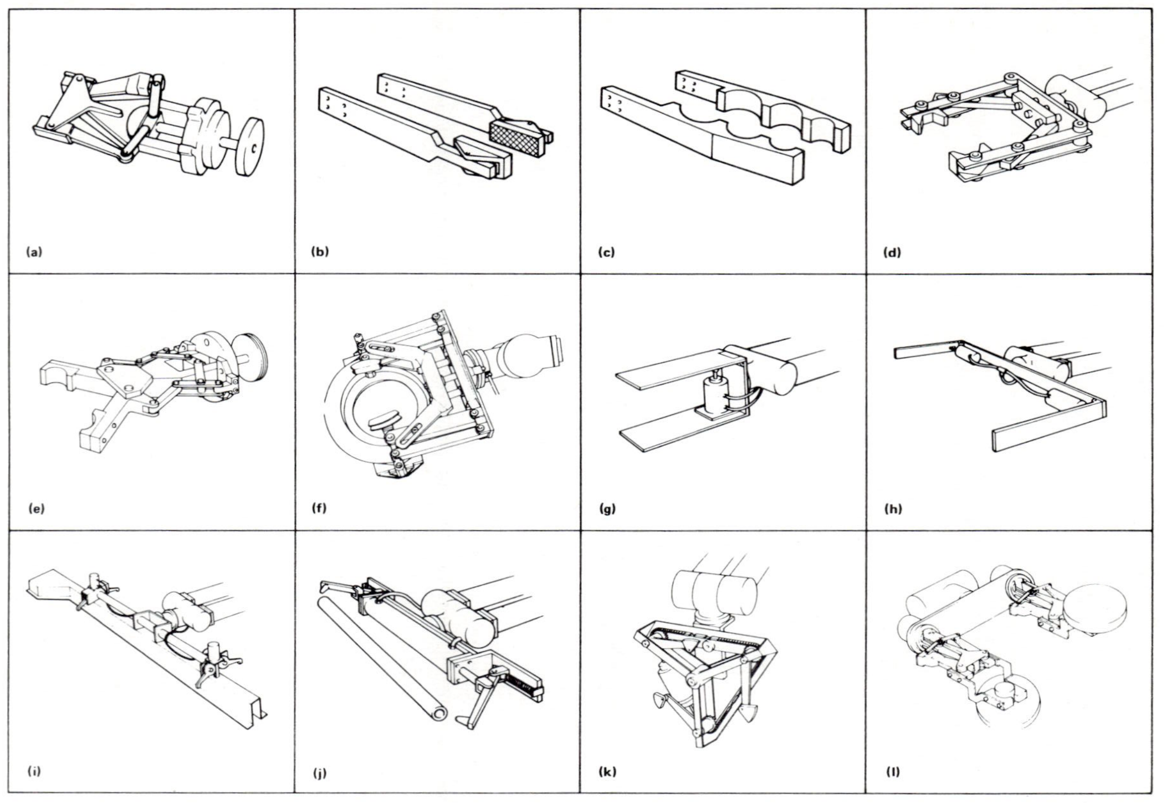

13.8

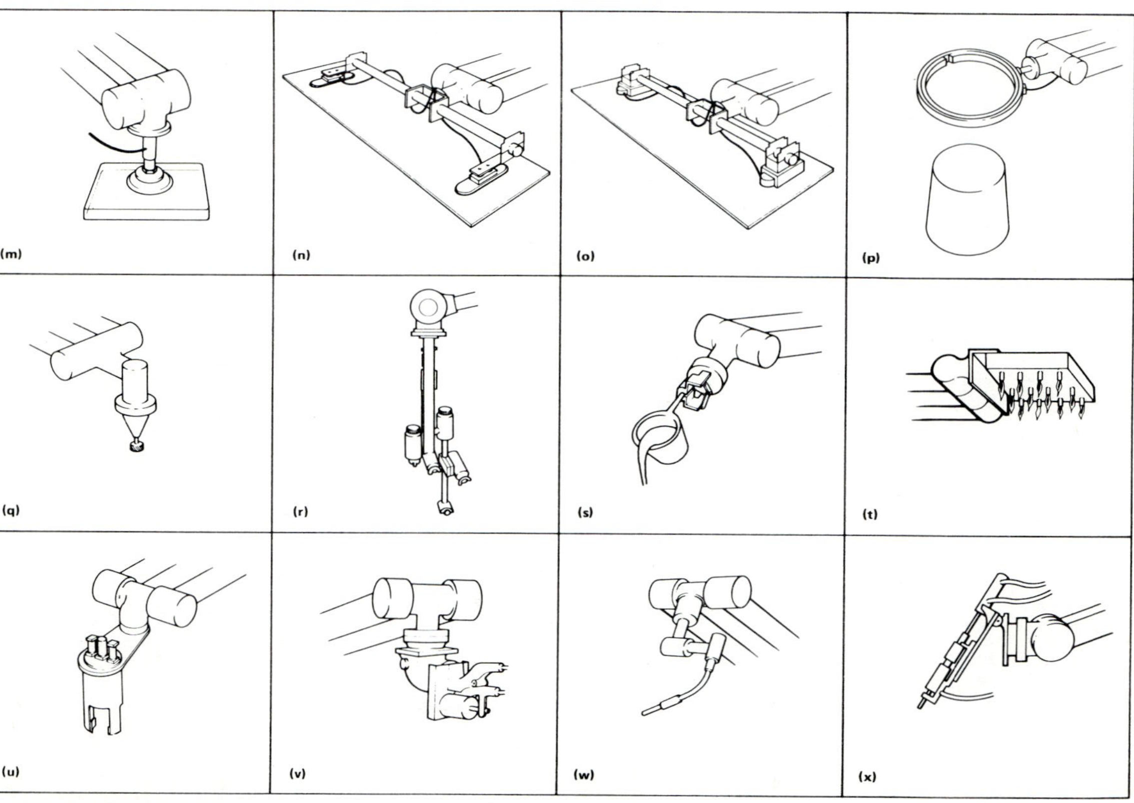

Figure 13.8 Some typical grippers.

Figure key	Gripper	Comments
a	Standard hand	Inexpensive and all-purpose; accepts a variety of custom fingers tailored to parts to be manipulated.
b	Self-aligning fingers	Self-aligning pads ensure secure grip on a flat-sided part.
c	Fingers for grasping different size parts	One-finger design can handle parts within wide size range.
d	Cam-operated hand	Easily manipulates heavy weights or bulky objects. The short distance between the center of gravity and wrist minimizes a heavy or bulky object's tendency to twist.
e	Wide-opening hand	May be useful if the part to be picked up is not always found in a constant orientation or at the same site. This hand sweeps part into its grasp as it closes. The hand develops low force when open and maximum force when closed.
f	Cam-operated hand with inside and outside jaws	A part is sometimes reoriented between the time when it is placed in a machine and when it is removed. This hand can grasp a part on the outer diameter by using the outer self-aligning pads, or the inner pads can grasp the inner diameter.
g	Special hand (one movable jaw)	Should be considered only when there is any access underneath a part, as on a rack. Simple design; one of the most economical hands.
h	Special hand (cartons)	Dual-jaw hand; opens wide to grasp inexactly located lightweight objects. Can lift and place cardboard cartons.
i	Special hand (modular gripper)	Has a pair of pneumatic actuators.
j	Special hand (glass tubes)	Grasps relatively short tubes securely. Pickup is effective even when tube length varies.
k	Special hand (chuck type)	A relatively simple mechanism (three fingers and a single actuator) to handle drums and similar large cylindrical parts. The actuator drives all three fingers simultaneously by means of a chain and sprockets.
l	Double hand	The hand can remove a finished part from a machine and replace it with an unfinished part by picking a part out of the chuck of a machine, swiveling, and placing a new part back in the chuck.
m	Vacuum gripper	Used to manipulate glass parts, such as cathode-ray–tube face.
n	Vacuum lifter	Often uses more than one cup for large pieces of glass, plastic, or metal.
o	Magnetic gripper	Good for flat pieces of ferrous metals.
p	Bladder gripper	Expandable, can be used to pick up large cylindrical pieces.

Figure key	Gripper	Comments
q	Tool gripper	Can manipulate grinders, sanders, or routers.
r	Spray-gun gripper	For spray painting.
s	Two-finger gripper	Ladles hot or hazardous materials.
t	Special heater end effector	Can be used to bake out foundry molds.
u	Tool gripper	Can operate drills, socket wrenches, and impact wrenches.
v	Spot-welding end effector	The single most common application of general-purpose robots.
w	Arc-welding end effector	A growing area of application for robots.
x	Stud-welding end effector	—

Computations and Derivations (Robot Mathematics)

14.1 Introduction

This chapter explains the mathematical methods that are commonly used to analyze and describe manipulator motions.

- Section 14.2 explains the concept of *homogeneous transformations* and shows how they are used to represent cartesian coordinate frames and the displacements and rotations of objects. Finally, it defines the 4×4 *Denavit-Hartenberg (DH) matrices* that are used to represent such transformations.

- Sections 14.3 and 14.4 discuss methods for representing the *displacement* and *orientation*, respectively, of a coordinate reference frame or an object.

- Section 14.5 presents the 3×3 matrices that represent *elementary rotations* about the individual axes of a reference frame.

- Section 14.6 explains how to multiply such matrices together to represent *successive rotations*.

- Section 14.7 shows how to represent rotations, translations, and combinations of the two in a single 4×4 homogeneous transformation matrix.

- Section 14.8 returns to the discussion of orientation and describes several *notational systems* that robot manufacturers use and the 4 × 4 matrices that correspond to them.

- Section 14.9 defines the *A matrices* that are commonly used to represent the joints and links of a manipulator, explains how to determine them by inspection of the shape of the manipulator, and presents the A matrices for several commonly encountered link shapes.

- Section 14.10 shows how to compute the position and orientation of the manipulator's wrist by multiplying the A matrices together to give the *T matrix.*

- Section 14.11 discusses ways of using the DH matrices to deal with problems such as *moving workpieces* and *tools* of unusual shapes.

- Section 14.12 discusses the role of the jacobian matrix in differential motion, force control, and transformation of force and torque between frames.

- Finally, Section 14.13 treats the difficult but crucial problem in manipulator control of the *arm solution*—computing the individual joint angles required to place the gripper or tool at a specified position with a specified orientation.

14.2   Homogeneous Transformations

14.2.1   Mathematical transformations in general

A "transformation" is a mathematical concept. In robotics we think of a transformation as a rule that changes any position in space into another position. For example, if we represent positions in a three-dimensional cartesian coordinate system by a column vector

$$\begin{bmatrix} X \\ Y \\ Z \end{bmatrix}$$

one transformation would be to change X into its absolute value, Y into Y squared, and any value of Z into 7.5.

$$\begin{bmatrix} X \\ Y \\ Z \end{bmatrix} \rightarrow \begin{bmatrix} |X| \\ Y \cdot Y \\ 7.5 \end{bmatrix}$$

This is not a very useful transformation in robotics, however. The most useful transformations are *displacement,* such as $X \rightarrow X + 3$, and *rotation,* such as

$$\begin{bmatrix} X \\ Y \\ Z \end{bmatrix} \rightarrow \begin{bmatrix} X\cos 30 - Y\sin 30 \\ -X\sin 30 + Y\cos 30 \\ Z \end{bmatrix}$$

which represents a 30° rotation about the z axis.

If you move the five or six joints of a typical robot, each joint generally displaces and rotates the hand in a different direction. Describing the final position of the hand thus would require very complex equations. This problem can be avoided by using certain mathematical transformations.

14.2.2 Homogeneous transformations

In the rest of this section, we write column vectors or matrices like those above in the horizontal form $(X \quad Y \quad \ldots)$ to save space. To simplify equations, computations, and computer programs for robot control, it is common practice to represent a position not as a set of three ordinary coordinates $(X \quad Y \quad Z)$ but as four "homogeneous" coordinates $(W \cdot X \quad W \cdot Y \quad W \cdot Z \quad W)$, where W is called the "scaling factor." All three coordinates $(X \quad Y \quad Z)$ have been scaled by W, so this is an example of a transformation. Because all three coordinates have been scaled in the same way (by W), this transformation is homogeneous (i.e., "the same everywhere"). The four new coordinates are called the homogeneous coordinates of the position $(X \quad Y \quad Z)$.

14.2.3 Uses of homogeneous transforms

Today, when we compute the location of a robot's hand, we first compute it in terms of homogeneous coordinates. Then we convert the homogeneous coordinates back to ordinary coordinates by dividing by the scaling factor. This sounds like more work, but in practice it is not, as we will see below. In fact, it makes the computer programs that do such computations much shorter and easier to write, and they are more generally useful.

There is another benefit from using homogeneous coordinates that will be increasingly important as industry begins to use computer graphics in robot programming and operation. Homogeneous coordinates are widely used in computer graphics software because they simplify the computations needed to show objects in perspective, to eliminate hidden lines and surfaces, and to display stereo pairs so that people can see three-dimensional objects.

The homogeneous transformation used in robotics is only one special case of transformations used in computer graphics. In robotics we always use a scaling factor W of either 0 or 1. In graphics, other values are used, too, for example, to produce perspective projections of objects. This is one reason that computations with four homogeneous coordinates are only a little more work than with three ordinary ones—any multiplications by scaling factors of 0 or 1 need not actually be carried out.

14.2.4   Points

Points are represented with a scaling factor of 1. Thus, the three ordinary coordinates of a point $(X \quad Y \quad Z)$ are transformed into the four homogeneous coordinates $(X \quad Y \quad Z \quad 1)$. To convert from one representation to the other is trivial.

Notice that $(X \quad Y \quad Z \quad W)$ can be thought of as a point in a four-dimensional space. Thus, all the points $(X \quad Y \quad Z)$ in ordinary three-space transform (or *map*) into the $W = 1$ plane in this four-space. $(X \quad Y \quad Z)$ and $(X \quad Y \quad Z \quad W)$ can also be thought of as three- or four-element vectors in their respective spaces.

14.2.5   Lines

Lines are useful in robotics for representing directions—especially the directions of the three axes of a coordinate reference frame. We represent a line by the point at infinity along that line.

For example, consider a point moving away along the $+x$ axis. In ordinary coordinates, it might start at $(1\ 0\ 0)$, then go to $(10\ 0\ 0)$, then $(1{,}000{,}000\ 0\ 0)$, and so on. We can represent these points in homogeneous coordinates by $(1\ 0\ 0\ 1)$, $(1\ 0\ 0\ \frac{1}{10})$, $(1 \quad 0 \quad 0 \quad \frac{1}{1{,}000{,}000})$, and so on. As W and X tend to the limit, we have the point at $(\infty \quad 0 \quad 0)$ represented conveniently by $(1 \quad 0 \quad 0 \quad 0)$, in which the scaling factor has gone to zero. This notation allows a computer to represent exactly the coordinates of a point at infinity, even though the computer represents each coordinate value internally with only a finite number of bits.

Notice that we have chosen to let the scaling factor vary so as to keep the magnitude of the x component equal to 1. We say that the resulting four-element vector is *normalized* to unit magnitude. So, in general, the first three elements of the homogeneous representation of a line will be the components of a "unit vector," i.e., $X \cdot X + Y \cdot Y + Z \cdot Z = 1$ for any line $(X \quad Y \quad Z \quad 0)$.

Each line transform is a point in a four-dimensional homogeneous coordinate space. Each one also represents an infinite number of points in ordinary three-dimensional space, according to the convention of representing a line by a point at infinity. So, it does not make sense to speak of converting the homogeneous representation of a line into ordinary coordinates. In practice there is no need to do so, and the four-element representation of a line is itself very useful in many important computations.

14.2.6 Summary

To summarize, we use a scaling factor of 0 to represent *lines* in homogeneous coordinates and a scaling factor of 1 to represent *points*. The x axis of a frame is represented by (1 0 0 0) in homogeneous coordinates, the y axis by (0 1 0 0), and the z axis by (0 0 1 0). The position $(X\ Y\ Z)$ is represented by $(X\ Y\ Z\ 1)$. In the following sections, we show how to represent a displacement in terms of a matrix, how to represent a rotation as a 3×3 matrix, and finally, how to represent a combined displacement and rotation as a single 4×4 homogeneous transform matrix.

14.3 Displacement

The displacement $(dX\quad dY\quad dZ)$ in ordinary coordinates is represented in homogeneous coordinates by the 4×4 matrix

$$
\text{trans}\,(dX\quad dY\quad dZ) =
\begin{bmatrix}
1 & 0 & 0 & dX \\
0 & 1 & 0 & dY \\
0 & 0 & 1 & dZ \\
0 & 0 & 0 & 1
\end{bmatrix}
$$

because, for any initial position $\mathbf{P}_1 = (X\ Y\ Z)$, the final position $\mathbf{P}_2$ is given in homogeneous coordinates by

$$
\mathbf{P}_2 = \text{trans}\,(dX\ dY\ dZ) \cdot
\begin{bmatrix}
X \\
Y \\
Z \\
1
\end{bmatrix}
$$

$$
=
\begin{bmatrix}
1 & 0 & 0 & dX \\
0 & 1 & 0 & dY \\
0 & 0 & 1 & dZ \\
0 & 0 & 0 & 1
\end{bmatrix}
\cdot
\begin{bmatrix}
X \\
Y \\
Z \\
1
\end{bmatrix}
$$

$$
=
\begin{bmatrix}
X + dX \\
Y + dY \\
Z + dZ \\
1
\end{bmatrix}
$$

14.4 Orientation

In robotics, we represent the orientation of an object as follows. First, we pretend that a cartesian coordinate frame is attached rigidly to the object. Sometimes it is called the "embedded frame" of the object, or just "the frame" of the object. After the object has rotated, we use the four-element homogeneous transforms of the lines along each axis of the frame to describe its new orientation.

The term "embedded" is a little misleading because the frame need not be located inside the object, and it can have any orientation we choose with respect to the object. People usually like to put the origin of the frame at some important place in the object, such as the working tip of a tool. They also like to align the $+z$ axis of the embedded frame with some special direction associated with the object. Some typical directions for the $+z$ axis are along the axis of a rotating joint, along the fingers of a robot hand, and along the vertical when the object is in its normal position.

Unfortunately, it is difficult to visualize the orientation of an object from a set of homogeneous transforms of the embedded frame axes. Until a robot can be completely programmed automatically by advanced manufacturing software, it will be necessary for its control system to describe orientations, and to accept orientation descriptions, in a form that people can understand easily.

The type of orientation description used in most robot control systems today is a sequence of angular rotations that take an object from a standard starting orientation to its actual orientation. These are sometimes called the "orientation angles" of the frame or object.

Most people can imagine an object rotating in different directions by different amounts, though it takes some practice to become proficient at it. Unfortunately, different manufacturers tend to use different sequences of rotations, and this is a source of great confusion to robot users.

Three of the rotation sequences in common use today to describe orientations are *Euler angles* ($\phi \quad \theta \quad \psi$), *roll, pitch, and yaw* (RPY), and *orientation, approach, and twist* (OAT). Unimation's VAL control system for their PUMA and 2000-series robots uses the OAT system. In a later section, we describe these different rotation sequences and develop the mathematics for them. Next, we develop the mathematics needed to describe a single rotation about a single axis.

14.5   Elementary Rotations

In this section we discuss elementary rotations—that is, rotations about a single coordinate axis.

14.5.1   Reference orientation

In the discussion of rotations, we assume that when an object is in its original, unrotated, or "reference" orientation, the x axis of its embedded frame is (1 0 0 0), the y axis is (0 1 0 0), and the z axis is (0 0 1 0).

14.5.2   3 × 3 rotation matrices

Because the last elements of the transforms mentioned above are all 0, we often put just the first three elements of each of these four vectors together to form a column of a 3 × 3 rotation matrix.

14.5.2.1   Identity rotation matrix.   For an unrotated object, the rotation matrix is just the 3 × 3 identity matrix

$$\mathbf{I} = \begin{bmatrix} 1 & 0 & 0 \\ 0 & 1 & 0 \\ 0 & 0 & 1 \end{bmatrix}$$

14.5.2.2   Rotation about *x* axis.   If the object is rotated only through an angle A about the x axis, the rotation matrix is

$$\text{rot } x(A) = \begin{bmatrix} 1 & 0 & 0 \\ 0 & \cos A & -\sin A \\ 0 & \sin A & \cos A \end{bmatrix}$$

14.5.2.3   Rotation about *y* axis.   Similarly, for a single rotation B about the y axis,

$$\text{rot } y(B) = \begin{bmatrix} \cos B & 0 & \sin B \\ 0 & 1 & 0 \\ -\sin B & 0 & \cos B \end{bmatrix}$$

14.5.2.4   Rotation about *z* axis.   For a single rotation C about the z axis,

$$\text{rot } z(C) = \begin{bmatrix} \cos C & -\sin C & 0 \\ \sin C & \cos C & 0 \\ 0 & 0 & 1 \end{bmatrix}$$

14.6   Rotation Sequences

14.6.1   Concatenation of rotations

When we discuss sequences of elementary rotations, we must specify the reference frame in which each rotation takes place. For example, consider the case in which an object is rotated 30° about the x axis and then 40° about the y axis. There is no ambiguity about the direction of the first rotation, but the second has not been completely specified: "about the y axis" could mean "about the original direction of the y axis." In that case, the correct mathematical expression would be

$$\mathrm{rot}\ y(40 \cdot \mathrm{rot}\ x(30)$$

in which we have *pre*multiplied by the matrix for the elementary y rotation. On the other hand, it could mean "about the direction in which the y axis points after rotating it around the x axis." Then we must *post*multiply by the rot y matrix

$$\mathrm{rot}\ x(30) \cdot \mathrm{rot}\ y(40)$$

Now, suppose this is the case and we rotate the frame or object a third time, by 50° around the direction of the z axis. If this means the *original* direction of the z axis, we *pre*multiply by rot $z(50)$

$$\mathrm{rot}\ z(50) \cdot \mathrm{rot}\ x(30) \cdot \mathrm{rot}\ y(40)$$

If, instead, this means the *rotated* direction of the z axis, we *post*multiply

$$\mathrm{rot}\ x(30) \cdot \mathrm{rot}\ y(40) \cdot \mathrm{rot}\ z(50)$$

Notice that the rotated direction of an axis is its direction after all preceding rotations have been performed.

Notice also that it is impossible to determine the sequence in which the rotations were performed by looking at the final expression. The expression above could represent a sequence of three rotations in which the first is about the x axis, the second is about the rotated y axis, and the third is about the rotated z axis (postmultiplying for each rotation). It could just as well represent a first rotation about the z axis, a second about the original y axis, and a third about the original x axis (premultiplying for each rotation).

In summary, then, we can represent any sequence of rotations as a product of the matrices for the individual rotations. We premultiply if the rotation is described in terms of the unrotated, or "base" axes, and postmultiply if it is described in terms of the rotated axes.

14.6.2   Noncommutativity of rotations

Matrix multiplication is not commutative, so multiplying matrices in a different order may give a different answer. This corresponds to the physical fact that applying a sequence of rotations in a different order generally results in a different final orientation.

14.6.3   Associativity of rotations

Matrix multiplication is associative, however. This represents the physical fact that a sequence of rotations can be replaced by a single rotation.

This fact can be used to increase computation efficiency. For example, if you must multiply several matrices together, it may be advantageous to group them in a particular way. For example, suppose $\mathbf{M}_i$, $i = 1, 2, 3, 4$, are four matrices such that $\mathbf{M}_2$ and $\mathbf{M}_3$ are constant while $\mathbf{M}_1$ and $\mathbf{M}_4$ vary. To evaluate $\mathbf{M}_1 \cdot \mathbf{M}_2 \cdot \mathbf{M}_3 \cdot \mathbf{M}_4$, then, it would be advisable to evaluate the product of $\mathbf{M}_2$ and $\mathbf{M}_3$ once (say, as $\mathbf{M}_{2,3} = \mathbf{M}_2 \cdot \mathbf{M}_3$), and then evaluate the product $\mathbf{M}_1 \cdot \mathbf{M}_{2,3} \cdot \mathbf{M}_4$ each time. This would reduce the number of matrix multiplications from three to two, for a 33 percent savings in computer time.

14.6.4 Rotation of a three-vector with a 3 × 3 matrix

A 3 × 3 rotation matrix like the ones shown above may be used as is to rotate an ordinary three-vector. To rotate three-vector $\mathbf{P}_1$ with a rotation matrix $\mathbf{M}$ to give a new vector $\mathbf{P}_2$, we use the equation

$$\mathbf{P}_2 = \mathbf{M} \cdot \mathbf{P}_1$$

Although this sort of computation is often useful in robotics, it is more common to use a 4 × 4 matrix to rotate the four-element homogeneous transformation of the three-vector. We show how to do this in the next section.

14.7 Frames

As noted above, an imaginary embedded coordinate frame is used to describe the location and orientation of an object. Thus, it is important to be able to represent such frames compactly in a computer program and to compute efficiently from them such things as relative positions, directions of motion, and velocities. Again, homogeneous transforms offer a solution.

Suppose a cartesian reference frame F is located so that its origin is at a position $(P_x \quad P_y \quad P_z)$ with respect to a reference frame that has axes x, y, and z. Suppose also that the unit vectors along the axes of frame F are $(X_x \quad X_y \quad X_z)$, $(Y_x \quad Y_y \quad Y_z)$, and $(Z_x \quad Z_y \quad Z_z)$, expressed in the reference frame. Then we can represent both the position and the orientation of frame F with a single 4 × 4 matrix

$$\mathbf{F} = \begin{bmatrix} X_x & Y_x & Z_x & P_x \\ X_y & Y_y & Z_y & P_y \\ X_z & Y_z & Z_z & P_z \\ 0 & 0 & 0 & 1 \end{bmatrix}$$

The first three columns are the homogeneous transforms of the lines that lie along the x, y, and z axes of frame F. The fourth column is the homogeneous transform of the position P of the origin of frame F. Note that the bottom row contains the scaling factors—0 for each axis line and 1 for the origin point.

Note also that the upper left 3×3 submatrix is just a rotation matrix like the ones we discussed earlier. Thus, we may think of such a 4×4 matrix as a way to represent both a displacement and a rotation. We may think of performing either the rotation or the displacement first. But, if we do the rotation first, we still have to displace the rotated frame by $(P_x \ \ P_y \ \ P_z)$ along the axes of the *reference* frame—not along the axes of the rotated frame.

In practice, there is no need to store the lower row of such a matrix, because it is always the same—(0 0 0 1). But, if we choose not to, then the routines that multiply a matrix by a matrix or a matrix by a vector must fill in the missing elements by themselves. However, because those missing elements are only either 0 or 1, this only means that some terms are either dropped or copied, respectively.

Suppose we were given a matrix like **F** above and asked to find the location of the origin of the corresponding frame F. Then all we would have to do would be to extract its position $(P_x \ \ P_y \ \ P_z)$ from the fourth column.

The simplest way to find the orientation of frame F would be to describe it in terms of unit vectors along its axes. Remember that because of the way we normalize the homogeneous transform of a line, each vector $(X_x \ X_y \ X_z)$, $(Y_x \ Y_y \ Y_z)$, and $(Z_x \ Z_y \ Z_z)$ will be a unit-length vector. Then we can simply extract the xyz components of the unit vector along the frame's x axis from the first column. The unit vector along the y axis is in the second column, and the third column contains the unit vector along the z axis.

14.8   "Understandable" Orientation Descriptions

In this section, we discuss the three most important representations used to describe the orientation of objects. These are *Euler angles,* the *roll-pitch-yaw* angles, and the *orientation-approach-twist* angles.

14.8.1   Euler angles

The Euler angle description of the orientation of a coordinate frame consists of three angles of rotation ϕ, θ, and ψ that turn the object from its unrotated or reference orientation to the orientation that is being described. Starting from the unrotated orientation, the embedded frame of the object (as well as the attached object) is

1. Rotated through angle ϕ about the frame's z axis, then

2. Rotated through angle θ about the frame's y axis (which now points in a different direction), and finally

3. Rotated through angle ψ, again about the frame's z axis, which is now pointing in a new direction

It is easier to apply such a description to decide how an object is oriented than it is to generate such a description for an arbitrary orientation of the object.

The net effect of the three rotations can be summed up in a single 3×3 matrix Euler(ϕ, θ, ψ). It can be computed by multiplying together the matrices for the elementary rotations described above

$$\text{Euler}(\phi, \theta, \psi) = \text{rot } z(\phi) \cdot \text{rot } y(\theta) \cdot \text{rot } z(\psi)$$

$$= \begin{bmatrix} \cos\phi\cos\theta\cos\psi - \sin\phi\sin\psi & -\cos\phi\cos\theta\sin\psi - \sin\phi\cos\psi & \cos\phi\sin\theta \\ \sin\phi\cos\theta\cos\psi + \cos\phi\sin\psi & -\sin\phi\cos\theta\sin\psi + \cos\phi\cos\psi & \sin\phi\sin\theta \\ -\sin\theta\cos\psi & \sin\theta\sin\psi & \cos\theta \end{bmatrix}$$

[after Paul (1981), p. 45]. The same rotation sequence can be represented in 4×4 form simply by augmenting the 3×3 with an extra row and column to represent a translation of zero distance.

$$\begin{bmatrix} & & & 0 \\ & \text{Euler}(\phi, \theta, \psi) & & 0 \\ & & & 0 \\ 0 & 0 & 0 & 1 \end{bmatrix}$$

14.8.2   Roll-pitch-yaw (RPY)

The angles used in the roll-pitch-yaw system are the same as those used to describe the attitude of an aircraft or boat (a "vehicle"). First we must embed a reference frame in the vehicle. Because the direction of motion of the vehicle may be considered to be special, we will align the z axis of the embedded frame with it. Then we will identify the x axis as being vertically upward and the y axis as being horizontal (to the right, or starboard). Then, starting from the reference orientation, the rotations are applied in the order in which they occur in the name of the representation—that is:

1. A *roll* rotation of R about the vehicle's roll $(+z)$ axis, then

2. A *pitch* rotation of P about the vehicle's new pitch $(+y)$ axis, then

3. A *yaw* rotation of Y about the vehicle's new yaw $(+x)$ axis

To produce the RPY rotation matrix we successively *post*multiply by the matrix for each individual rotation, because each of these rotations is about one of the vehicle's axes, rather than about one of the base frame's axes

$$\text{rot RPY}(R, P, Y) = \text{rot Z}(R) \cdot \text{rot Y}(P) \cdot \text{rot X}(Y)$$

$$= \begin{bmatrix} \cos R \cos P & \cos R \sin P \sin Y - \sin R \cos Y & \cos R \sin P \cos R + \sin R \sin Y \\ \sin R \cos P & \sin R \sin P \sin Y + \cos R \cos Y & \sin R \sin P \cos R - \cos R \sin Y \\ -\sin P & \cos P \sin Y & \cos P \cos Y \end{bmatrix}$$

(after Paul, p. 47).

As noted above, we can also interpret this matrix expression as having been generated by consecutively *pre*multiplying by each matrix. This would correspond to the rotation sequence of:

1. A *yaw* rotation of Y about the base frame's yaw $(+x)$ axis, followed by

2. A *pitch* rotation of P about the base frame's pitch $(+y)$ axis, and finally,

3. A *roll* rotation of R about the base frame's roll $(+z)$ axis.

14.8.3   Orientation-approach-twist (OAT)

Unimation's VAL controller describes hand (or tool) orientations to the user in terms of three angles called "orientation," "approach," and "twist." These angles actually describe the orientation of a reference frame which the user may embed in any desired position and orientation in the hand or tool, using the VAL command TOOL.

In discussions of hand orientation in the literature, people usually assume that the hand has the shape shown in Figure 14.1 and that the hand frame is embedded in it as follows:

1. The origin of the hand frame is midway between the fingertips.

2. The $+y$ axis lies along the line between the fingertips.

3. The $+z$ axis points in the direction of the fingers.

4. The $+x$ axis completes a right-handed coordinate system, at right angles to the plane in which the fingers lie.

The angles that describe the orientation of the PUMA's hand are:

- *Orientation (O).*   The "compass" direction in which the hand is reaching (measured counterclockwise from the $-y$ direction of the base frame, looking downward along $-z$). This is also the direction of the horizontal component of the hand or tool's $+z$, or "reach," axis.

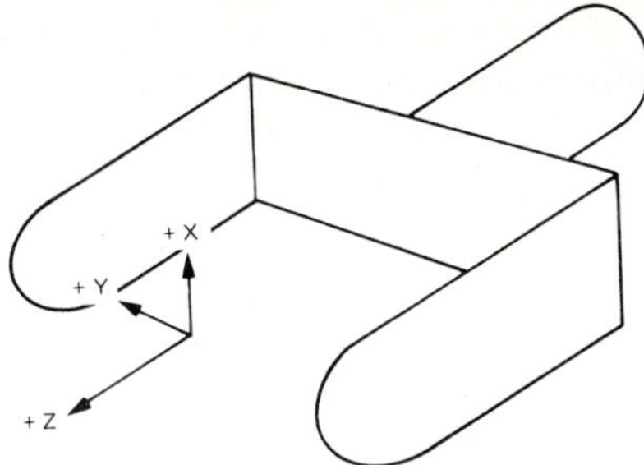

Figure 14.1 Conventional hand coordinate frame.

- *Approach (A).*   The angle of depression of the reach direction (positive if below the horizontal, negative if above).

- *Twist (T).*   The wrist rotation (measured clockwise looking along the reach direction, from an orientation in which the line between the fingertips is horizontal).

The following facts may help in visualizing hand orientation:

- When orientation angles O, A, and T are all zero, the fingers are horizontal and they point backward along the $-y$ axis. When the fingers point to the robot's right, $O = +90°$; to the left, $O = -90°$; forward, $O =$ either $+180°$ or $-180°$.

- When the fingers point vertically upward, $A = -90°$; vertically downward, $A = +90°$; horizontally, $A = 0°$.

- When $T = 0°$, the hand's $+y$ axis points from the right finger to the left finger looking in the direction in which the fingers point, and the line through the fingertips is horizontal. If the vertical component of the hand's $+x$ axis points downward, T is in the range -90 to $+90°$; upward, from -90 to $-180°$ or $+90$ to $+180°$.

Because a robot hand often has several kinds of symmetry, it can be difficult to determine the hand's orientation by looking at it. Students at Stanford University invented a clever solution to this problem: They mounted a small toy animal (a walrus) on the hand. It was then easy to tell that the walrus (hence, the hand) was upside down, for example, or that the walrus needed to turn more to its right. The up and down directions from the walrus's point of view were always along the hand's $+x$ and $-x$ directions, respectively. Similarly, the hand's $+y$ and $-y$ directions were always to the walrus's right and left, respectively. Some such method might be useful in industry, too.

The VAL control system displays O, A, and T to the user in response to the WHERE and HERE monitor commands, and the user can type in values for O, A, and T in order to specify an orientation explicitly. The

VAL controller also reports and accepts values for the six joint angles J_1 through J_6. However, the three wrist joint angles J_4, J_5, and J_6 are not the same angles as O, A, and T. In fact, O, A, and T depend in a very complicated way upon all six joint angles J_1 through J_6. For example, when all J_1 through J_6 are zero, the orientation angles are $O = 90°$, $A = -90°$, and $T = 0°$. The arm is then in the posture shown in Figure 14.2, and the orientation of the hand frame is the same as that of the base frame.

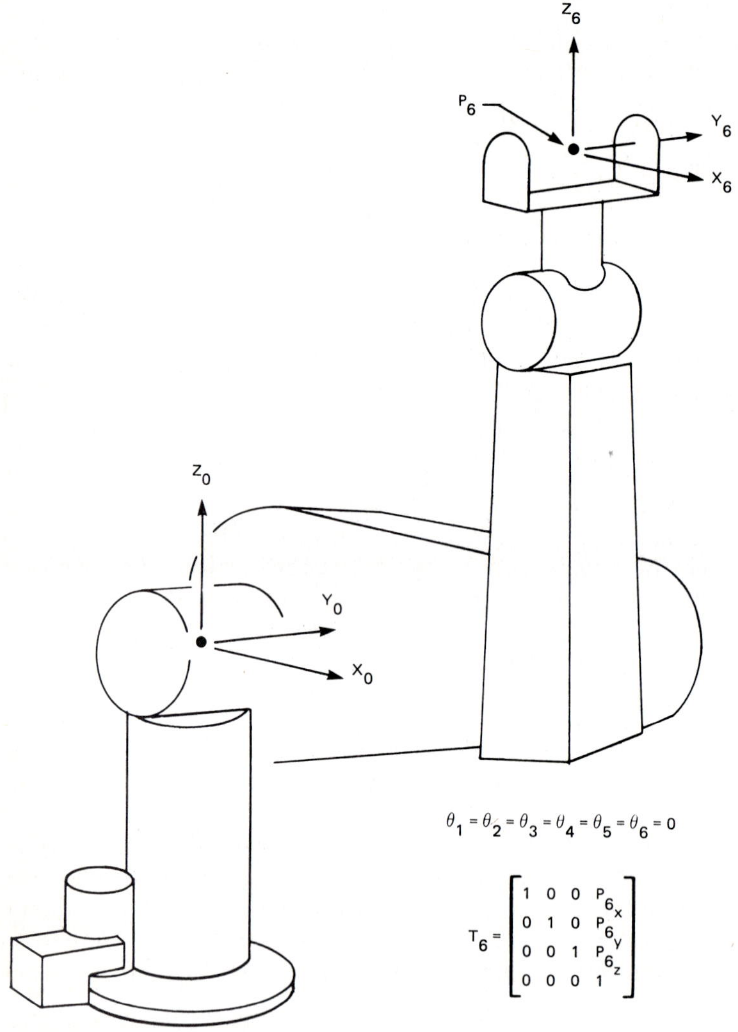

$$T_6 = \begin{bmatrix} 1 & 0 & 0 & P_{6_x} \\ 0 & 1 & 0 & P_{6_y} \\ 0 & 0 & 1 & P_{6_z} \\ 0 & 0 & 0 & 1 \end{bmatrix}$$

Figure 14.2 PUMA in zero-angle posture.

The posture in Figure 14.2 is called a "left-arm posture," because the elbow (joint 4) is bent in a way that only the elbow of a person's left arm could bend. The user can change a PUMA manipulator into a left or right arm under program control by issuing the VAL command LEFTY or RIGHTY before a MOVE instruction.

Similarly, notice that in Figure 14.2 the elbow lies *below* the line joining the shoulder (joint 2) and the wrist. The VAL controller also allows the user to select a posture in which the elbow lies either above or below that line by issuing the VAL command ABOVE or BELOW before a MOVE instruction.

The VAL command READY places the PUMA in a posture in which it reaches vertically upward so that it will be out of the way. This is not the position in which all joint angles are zero.

Below, we develop the 3×3 orientation matrix for the PUMA manipulator. We will do it step by step, according to the definitions published in the manual for the VAL language.

Unimation's orientation description is equivalent to yaw, pitch, and roll rotations as described above, except that the hand starts in a different orientation. In the RPY system, for $R = P = Y = 0°$ the hand frame has the same orientation as the base frame. In the OAT system, for $O = A = T = 0°$ the hand's $+x$ axis is along the base frame's $-z$ axis, its $+y$ axis is along the base frame's $+x$ axis, and its $+z$ axis is along the $-y$ axis of the base frame. To reach this orientation from the $R = P = Y = 0°$ orientation requires an initial rotation

$$\mathbf{R}_{\text{init}} = \begin{bmatrix} 0 & 1 & 0 \\ 0 & 0 & -1 \\ -1 & 0 & 0 \end{bmatrix}$$

as can be verified by inspection. Next, according to Unimation's definition of the approach angle O we rotate the hand by O about the base frame's $+z$ axis. We can represent this as

$$\text{rot } z(O) \cdot \mathbf{R}_{\text{init}}$$

Notice that we have *premultiplied* by rot $z(O)$ because it represents a rotation about the $+z$ axis of the *base* frame; if it were about the $+z$ axis of the *hand* frame, we would have *post*multiplied by that matrix.

Now, it happens that $\mathbf{R}_{\text{init}}$ has the effect of turning the hand so that its $+x$ axis lies along the base frame's $-z$ axis. So, we can also write the effect of rotating first by $\mathbf{R}_{\text{init}}$ and then by O as

$$\mathbf{R}_{\text{init}} \cdot \text{rot } x(-O)$$

Here we have postmultiplied by rot $x(-O)$ because it describes a rotation in the hand frame rather than the base frame. Either expression gives exactly the same 3×3 rotation matrix. We will use the latter expression in the following discussion. Next we rotate the hand by the approach angle A about its current pitch axis. This is the hand's $+y$ axis that runs from one fingertip to the other. To aid in visualizing this rotation, we note that the $+y$ axis will be horizontal at this stage and so will the plane of the fingers (although this fact is not important in the mathematical development). The matrix for this rotation is therefore rot $y(A)$, and because the y axis is the one in the hand frame, we must postmultiply by it.

$$\mathbf{R}_{\text{init}} \cdot \text{rot } x(-O) \cdot \text{rot } y(A)$$

Finally, we rotate the hand through the twist angle T about its current $+z$ axis. Again, we must *post*multiply, because it is the *hand's z axis*, not the base frame's z axis:

$$\mathbf{R}_{\text{init}} \cdot \text{rot } x(-O) \cdot \text{rot } y(A) \cdot \text{rot } z(T)$$

Expanding each matrix and multiplying them out, we obtain the 3×3 orientation matrix for the PUMA manipulator:

$$\text{rot OAT}(O, A, T) = \mathbf{R}_{\text{init}} \cdot \text{rot } x(-O) \cdot \text{rot } y(A) \cdot \text{rot } z(T)$$

$$= \begin{bmatrix} 0 & 1 & 0 \\ 0 & 0 & -1 \\ -1 & 0 & 0 \end{bmatrix} \begin{bmatrix} 1 & 0 & 0 \\ 0 & \cos O & \sin O \\ 0 & -\sin Q & \cos O \end{bmatrix} \begin{bmatrix} \cos A & 0 & \sin A \\ 0 & 1 & 0 \\ -\sin A & 0 & \cos A \end{bmatrix} \cdot \begin{bmatrix} \cos T & -\sin T & 0 \\ \sin T & \cos T & 0 \\ 0 & 0 & 1 \end{bmatrix}$$

$$= \begin{bmatrix} \cos Q \sin T - \sin O \sin A \sin T & \cos O \cos T + \sin O \sin A \sin T & \sin O \cos A \\ \sin O \sin T + \cos O \sin A \cos T & \sin O \cos T - \cos O \sin A \sin T & -\cos O \cos A \\ -\cos A \cos T & \cos A \sin T & -\sin A \end{bmatrix}$$

Exactly the same final result can be obtained by expanding and multiplying out

$$\text{rot OAT}(O, A, T) = \text{rot } z(O) \cdot \mathbf{R}_{\text{init}} \cdot \text{rot } y(A) \cdot \text{rot } z(T)$$

where we have considered the rotation by O to be about the $+z$ axis of the base frame instead of about the $-x$ axis of the hand frame. This is more in keeping with Unimation's own definition of angle O.

14.9   A Matrices

In this section, we describe how to represent the joints of a manipulator by 4×4 homogeneous transform matrices, traditionally called the "A matrices."

14.9.1 Contemporary manipulator design

With very few exceptions, all commercial and laboratory manipulators can be modeled as a number of rigid links with a joint between each successive pair of links. A few manipulators have been based on parallel arrangements of joints, but these are not in general use in industry, and other mathematical treatments may be more appropriate for analyzing them than the ones presented here.

14.9.2 Types of joints used in manipulators

The joints used in manipulators today are almost always "rotary" joints (i.e., "hinge" joints). Prab-Versatran robots, Unimate 2000 and 4000 series robots, Olivetti Sigma robots, and the IBM RS1 robot also have "prismatic" joints. These are sometimes called "telescoping," "sliding," or "linear" joints. Either type of joint has a single degree of freedom of motion, which is usually driven by a servo-controlled actuator.

14.9.3 Joint nomenclature

The *outboard* direction is from the base toward the wrist, while the *inboard* direction is toward the base. Each joint has an axis. For a rotary joint, it is the axis of rotation. For a prismatic joint, the axis direction is the direction in which the joint extends, but its location depends upon the joints that are outboard from it, as will be explained below.

The amount of rotation of a rotary joint is often represented by J, or by the Greek letter θ, and the distance that a prismatic joint extends by d. This number is called the "joint variable." Different manufacturers make different choices about how to represent a joint variable numerically—such as what joint position corresponds to a zero value. In this section we describe a uniform method for representing joint variables and joint types that is suitable for internal use by control software. If desired, the manipulator control system can use a completely different representation of joint variables for communication with a person.

14.9.4 Meaning of an A matrix

An A matrix describes the position and orientation of one link with respect to the preceding link. In the usual serial-joint manipulator, this description depends entirely upon the action of a single joint—the one between the two links. Thus, there is one A matrix per joint, and it is a function of that joint's variable.

14.9.5   Link reference frames

We describe the position and orientation of a link in terms of a coordinate reference frame embedded in it. Peiper (1968) suggested a particular way of embedding a frame in a link, which has become an industry standard. The method is as follows.

14.9.5.1   Link nomenclature.   The base of the manipulator is considered to be link number 0. A six-jointed arm will have another six links, a five-jointed arm five, and so on. Progressing from the base to the wrist we pass through link 0, joint 1, link 1, joint 2, and so on through joint 6 and link 6.

14.9.5.2   Virtual and physical links.   Most manipulators actually consist of only three or four physical links, because two or more joints are often combined. The missing links can be considered to be "virtual links" of zero length. Virtual links typically appear at the shoulder of an arm, where the axes of the first two or three joints often intersect, and at the wrist, where the axes of the last two or three joints often intersect. For example, in the wrists of the PUMA and the Milacron T3-746, the last three joint axes intersect at a point, and links 4 and 5 in those arms are virtual.

14.9.5.3   Link parameters.   The link that separates any successive pair of joints—rotary, prismatic, or one of each—can be completely described for most purposes by four numbers called the "link parameters," a_n, α_n, θ_n, d_n:

- a_n.   The distance between the axes of the joints at each end of the link n, measured along a line that is normal to both axes (the "common normal"). For prismatic joints, the location of the axis depends upon the outboard joints, as will be explained below, but is along the axis of rotation for rotary joints. The common normal may lie completely outside the physical body of the link, depending upon its shape. If the two joint axes happen to be parallel, as is often the case, there are an infinite number of common normals. In that case, we choose the one that intersects the next common normal (i.e., the one between joint $n + 1$ and joint $n + 2$).

- α_n.   The angle from the axis of joint n to that of joint $n + 1$, around the common normal.

- θ_n.   The variable of a rotary joint: the angle from the common normal between joint $n - 1$ and joint n, to the common normal between joint n and joint $n + 1$, measured around the axis of joint n.

- d_n. The variable of a prismatic joint: the distance between the two normals mentioned above, measured along the axis of joint n.

14.9.5.4   Standard embedding of link frames.

14.9.5.4.1   General principles. The origin of the frame embedded in link n is placed on the axis of the outboard joint in that link (joint $n + 1$), where the common normal of that link (from joint n to joint $n + 1$) intersects the axis of joint $n + 1$. This means that if the two axes in a link intersect, the origin for the link is placed at that point.

14.9.5.4.2   Special cases.

Links with parallel joint axes. If the axes of the link are parallel, there are an infinite number of normals for that link. In that case, we look ahead to the next outboard link and place the origin where the next outboard normal (from joint $n + 1$ to joint $n + 2$) crosses the axis of joint $n + 1$. If succeeding axes are also parallel, we must look further ahead until we reach either a nonparallel axis or the end of the arm before we can determine where to locate the origin. If the last joint axis is also parallel, then we may assign any convenient location on its axis as its origin. We might, for example, choose an origin to suit a particular hand or tool that the arm will carry.

Prismatic joints. When a link has a prismatic inboard joint, we must also look at the outboard links to decide where to place the origin of that link's embedded frame. It is placed on the next defined joint axis. If the next joint is another prismatic joint, then we must look at the next joint after that, and so on. We will reach either a rotary joint, whose axis is well defined, or the end of the arm. In the latter case, we assign any convenient axis and proceed.

14.9.6   Constructing the A matrix from the link parameters

Each link n of a robot manipulator can be thought of as a means for positioning the outboard joint of the link, joint $n + 1$, with respect to its inboard joint, joint n. Furthermore, we can imagine that it does this in four steps, starting from the position and orientation of the frame that is embedded in the inboard joint. Each step depends upon one of the four link parameters. We call each of these steps an "elementary link transformation."

14.9.6.1   Elementary link transformations. A link n performs four elementary link transformations:

1. It *rotates* the frame through an angle θ_n about the z axis of joint n.

2. It *translates* the frame a distance d_n along the z axis of joint n.

3. It *translates* the frame a distance a_n along the x axis of the rotated frame.

4. It *rotates* the frame through an angle α_n about that same axis.

Thus, we can write

$$\mathbf{A}_n = \text{trans } (0, 0, d_n) \cdot \text{rot } z(\theta_n) \cdot \text{trans } (a_n, 0, 0) \cdot \text{rot } x(a_n)$$

Notice that we have premultiplied rot z by trans $(0, 0, d_n)$ rather than postmultiplying by it, because the rotation is around the z axis of the unrotated frame, not the rotated one. In this case, it makes no difference whether we premultiply or postmultiply, because rot z does not change the orientation of the z axis of the frame. So, we can also write it as in Paul (1981), p. 53:

$$\mathbf{A}_n = \text{rot } z(\theta_n) \cdot \text{trans } (0, 0, d_n) \cdot \text{trans } (a_n, 0, 0) \cdot \text{rot } x(a_n)$$

This form of the expression has the advantage that it places the two pure translation matrices adjacent to one another, and it is particularly easy to multiply such matrices together. Just add up the entries in their right-hand columns.

Notice also that the first two of these elementary link transformations depend upon the inboard end of the link, while the last two depend upon the outboard end. If we expand and multiply out the matrices for each elementary link transformation, we obtain a closed-form symbolic expression for the A matrix of the link. This form is very convenient for use in internal computations by the control system.

14.9.6.2 Matrix for combined fundamental transformations. Expanding the elementary link transformation matrices, and combining the two translations into one matrix, as in Paul (1981), p. 53,

$$\mathbf{A}_n = \begin{bmatrix} \cos \theta_n & -\sin \theta_n & 0 & 0 \\ \sin \theta_n & \cos \theta_n & 0 & 0 \\ 0 & 1 & 0 & 0 \\ 0 & 0 & 1 & 0 \end{bmatrix} \cdot \begin{bmatrix} 1 & 0 & 0 & a_n \\ 0 & 1 & 0 & 0 \\ 0 & 0 & 1 & d_n \\ 0 & 0 & 0 & 1 \end{bmatrix} \cdot \begin{bmatrix} 1 & 0 & 0 & 0 \\ 0 & \cos \alpha_n & -\sin \alpha_n & 0 \\ 0 & \sin \alpha_n & \cos \alpha_n & 0 \\ 0 & 0 & 0 & 1 \end{bmatrix}$$

Multiplying these together gives the final expression

$$\mathbf{A}_n = \begin{bmatrix} \cos \theta_n & -\sin \theta_n \cos \alpha_n & \sin \theta_n \sin \alpha_n & a_n \cos \theta_n \\ \sin \theta_n & \cos \theta_n \cos \alpha_n & -\cos \theta_n \sin \alpha_n & a_n \sin \theta_n \\ 0 & \sin \alpha_n & \cos \alpha_n & d_n \\ 0 & 0 & 0 & 1 \end{bmatrix}$$

14.9.7 A matrices for common link shapes

In this section, we present the A matrices for several link shapes that are often found in robot manipulators. These shapes are

- *Prismatic joints.*   As in the Unimation 2000, Prab Versatran, Stanford Arm, seven-axis Kawasaki Unimate (which has two prismatic joints), the IBM RS1 (three), and the Olivetti Sigma robot (two sets of three)

- *Parallel rotary joints.*   As in the upper arm and forearm of the PUMA

- *Intersecting orthogonal rotary joints.*   As in the PUMA shoulder and wrist

- *Nonintersecting orthogonal rotary joints.*   As in the wrist of the Cincinnati Milacron T3

14.9.7.1 Prismatic joints.   For a prismatic joint n,

θ_n constant

d_n the joint variable

a_n 0

α_n constant

Then

$$\mathbf{A}_n = \begin{bmatrix} \cos\theta_n & -\sin\theta_n\cos\alpha_n & \sin\theta_n\sin\alpha_n & 0 \\ \sin\theta_n & \cos\theta_n\cos\alpha_n & -\cos\theta_n\sin\alpha_n & 0 \\ 0 & \sin\alpha_n & \cos\alpha_n & d_n \\ 0 & 0 & 0 & 1 \end{bmatrix}$$

Two special cases of prismatic joints often occur: In one the next joint axis is parallel to the prismatic joint's axis ($a_n = 0$ or 180°); in the other it is orthogonal ($a_n = 90$ or $-90°$).

14.9.7.1.1 Axis parallel to prismatic joint axis.   When the inboard joint of a link is prismatic and the axis of the next defined joint is parallel to it,

θ_n constant

d_n joint variable

a_n 0

α_n 0 or 180°

For $\alpha_n = 0°$,

$$\mathbf{A}_n = \begin{bmatrix} \cos\theta_n & -\sin\theta_n & 0 & 0 \\ \sin\theta_n & \cos\theta_n & 0 & 0 \\ 0 & 0 & 1 & d_n \\ 0 & 0 & 0 & 1 \end{bmatrix}$$

and for $\alpha_n = 180°$,

$$\mathbf{A}_n = \begin{bmatrix} \cos\theta_n & \sin\theta_n & 0 & 0 \\ \sin\theta_n & -\cos\theta_n & 0 & 0 \\ 0 & 0 & -1 & d_n \\ 0 & 0 & 0 & 1 \end{bmatrix}$$

14.9.7.1.2 Axis orthogonal to prismatic joint axis. When the inboard joint of a link is prismatic and the axis of the next defined joint is orthogonal to it,

θ_n constant

d_n joint variable

a_n 0

α_n 90 or $-90°$

For $\alpha_n = 90°$,

$$\mathbf{A}_n = \begin{bmatrix} \cos\theta_n & 0 & \sin\theta_n & 0 \\ \sin\theta_n & 0 & -\cos\theta_n & 0 \\ 0 & 1 & 0 & d_n \\ 0 & 0 & 0 & 1 \end{bmatrix}$$

and for $\alpha_n = -90°$,

$$\mathbf{A}_n = \begin{bmatrix} \cos\theta_n & 0 & -\sin\theta_n & 0 \\ \sin\theta_n & 0 & \cos\theta_n & 0 \\ 0 & -1 & 0 & d_n \\ 0 & 0 & 0 & 1 \end{bmatrix}$$

14.9.7.2 Parallel rotary joints. When a link contains two rotary joints whose axes are parallel,

θ_n joint variable

d_n 0

a_n distance between the axes

α_n 0 or $180°$

If a_n were zero, the two rotary joints would be collinear. This is an unlikely design, because the two could be replaced by a single rotary joint. For $\alpha_n = 0°$, the matrix $\mathbf{A}_n$ then reduces to

$$\mathbf{A}_n = \begin{bmatrix} \cos\theta_n & -\sin\theta_n & 0 & a_n\cos\theta_n \\ \sin\theta_n & \cos\theta_n & 0 & a_n\sin\theta_n \\ 0 & 0 & 1 & 0 \\ 0 & 0 & 0 & 1 \end{bmatrix}$$

and for $\alpha_n = 180°$,

$$\mathbf{A}_n = \begin{bmatrix} \cos\theta_n & \sin\theta_n & 0 & a_n\cos\theta_n \\ \sin\theta_n & -\cos\theta_n & 0 & a_n\sin\theta_n \\ 0 & 0 & -1 & 0 \\ 0 & 0 & 0 & 1 \end{bmatrix}$$

14.9.7.3 Intersecting orthogonal rotary joints. When a link contains two rotary joints whose axes intersect at right angles,

θ_n joint variable

d_n 0

a_n 0

α_n 90 or $-90°$

For $\alpha_n = 90°$,

$$\mathbf{A}_n = \begin{bmatrix} \cos\theta_n & 0 & \sin\theta_n & 0 \\ \sin\theta_n & 0 & -\cos\theta_n & 0 \\ 0 & 1 & 0 & 0 \\ 0 & 0 & 0 & 1 \end{bmatrix}$$

and for $\alpha_n = -90°$,

$$\mathbf{A}_n = \begin{bmatrix} \cos\theta_n & 0 & -\sin\theta_n & 0 \\ \sin\theta_n & 0 & \cos\theta_n & 0 \\ 0 & -1 & 0 & 0 \\ 0 & 0 & 0 & 1 \end{bmatrix}$$

14.9.7.4 Nonintersecting orthogonal rotary joints. When a link contains two rotary joints whose axes are at right angles, but do not intersect,

θ_n joint variable

d_n 0

a_n distance between the axes

α_n 90 or $-90°$

For $\alpha_n = 90°$,

$$\mathbf{A}_n = \begin{bmatrix} \cos\theta_n & 0 & \sin\theta_n & a_n\cos\theta_n \\ \sin\theta_n & 0 & -\cos\theta_n & a_n\sin\theta_n \\ 0 & 1 & 0 & 0 \\ 0 & 0 & 0 & 1 \end{bmatrix}$$

and for $\alpha_n = -90°$,

$$\mathbf{A}_n = \begin{bmatrix} \cos\theta_n & 0 & -\sin\theta_n & a_n\cos\theta_n \\ \sin\theta_n & 0 & \cos\theta_n & a_n\sin\theta_n \\ 0 & -1 & 0 & 0 \\ 0 & 0 & 0 & 1 \end{bmatrix}$$

14.10   T Matrices

In this section we show how to multiply together the A matrices for the joints of a manipulator to produce a "T matrix" that represents the entire manipulator.

$^i\mathbf{T}_j$ is the 4×4 matrix that represents the position of link j with respect to link i. Thus $^0\mathbf{T}_6$ represents the position of the wrist (link 6) with respect to the base of the manipulator (link 0). We note that by the definition of the A matrices,

$$\mathbf{A}_n = {}^{n-1}\mathbf{T}_n$$

and for any sequence of consecutive A matrices,

$$\mathbf{A}_n \cdot \mathbf{A}_{n+1} \cdot \cdots \cdot \mathbf{A}_{n+m} = {}^{n-1}\mathbf{T}_{n+m}$$

which represents the position of link $n + m$ with respect to link $n - 1$. In particular, for the whole arm the position of the wrist with respect to the base is

$$\mathbf{A}_1 \cdot \mathbf{A}_2 \cdot \mathbf{A}_3 \cdot \mathbf{A}_4 \cdot \mathbf{A}_5 \cdot \mathbf{A}_6 = {}^0\mathbf{T}_6$$

We will denote positions with respect to the base frame, $^0\mathbf{T}_n$, by $\mathbf{T}_n$, for any n.

To clarify these ideas, we will show how these computations apply to the Stanford Arm. For this manipulator, the link parameters are as shown in the table.

Link n	θ_n	d_n	a_n	α_n	Variable
1	θ_1	0	0	-90	θ_1
2	θ_2	d_2	0	90	θ_2
3	0	d_3	0	0	d_3
4	θ_4	0	0	-90	θ_4
5	θ_5	0	0	90	θ_5
6	θ_6	0	0	0	θ_6

Using the abbreviations $C_n = \cos\theta_n$, and $S_n = \sin\theta_n$, the A matrices for the Stanford Arm are (after Paul, 1981, p. 57),

$$\mathbf{A}_1 = \begin{bmatrix} C_1 & 0 & -S_1 & 0 \\ S_1 & 0 & C_1 & 0 \\ 0 & -1 & 0 & 0 \\ 0 & 0 & 0 & 1 \end{bmatrix}$$

$$\mathbf{A}_2 = \begin{bmatrix} C_2 & 0 & S_2 & 0 \\ S_2 & 0 & -C_2 & 0 \\ 0 & 1 & 0 & d_2 \\ 0 & 0 & 0 & 1 \end{bmatrix}$$

$$\mathbf{A}_3 = \begin{bmatrix} 1 & 0 & 0 & 0 \\ 0 & 1 & 0 & 0 \\ 0 & 0 & 1 & d_3 \\ 0 & 0 & 0 & 1 \end{bmatrix}$$

$$\mathbf{A}_4 = \begin{bmatrix} C_4 & 0 & -S_4 & 0 \\ S_4 & 0 & C_4 & 0 \\ 0 & -1 & 0 & 0 \\ 0 & 0 & 0 & 1 \end{bmatrix}$$

$$\mathbf{A}_5 = \begin{bmatrix} C_5 & 0 & S_5 & 0 \\ S_5 & 0 & -C_5 & 0 \\ 0 & -1 & 0 & 0 \\ 0 & 0 & 0 & 1 \end{bmatrix}$$

$$\mathbf{A}_6 = \begin{bmatrix} C_6 & -S_6 & 0 & 0 \\ S_6 & C_6 & 0 & 0 \\ 0 & 0 & 1 & 0 \\ 0 & 0 & 0 & 1 \end{bmatrix}$$

$$\mathbf{T}_6 = \begin{bmatrix} n_x & o_x & a_x & p_x \\ n_y & o_y & a_y & p_y \\ n_z & o_z & a_z & p_z \\ 0 & 0 & 0 & 0 \end{bmatrix}$$

This representation of the $\mathbf{T}_6$ matrix emphasizes the fact that each of the first three columns is a unit vector along one of the axes of the frame embedded in link 6 (the wrist). The **a** vector points along the axis of joint six, while the **n** and **o** vectors are normal to it. The fourth column p is the position of the origin of those axes, relative to the base frame.

Multiplying $\mathbf{A}_1$ through $\mathbf{A}_6$ together gives the following expressions for the elements of the $\mathbf{T}_6$ matrix (after Paul, p. 59).

$$n_x = C_1[C_2(C_4C_5C_6 - S_4S_6) - S_2S_5C_6] - S_1(S_4C_5C_6 + C_4S_6)$$

$$n_y = S_1[C_2(C_4C_5C_6 - S_4S_6) - S_2S_5C_6] + C_1(S_4C_5C_6 + C_4S_6)$$

$$n_z = -S_2(C_4C_5C_6 - S_4S_6) - C_2S_5C_6$$

$$o_x = C_1[-C_2(C_4C_5S_6 + S_4C_6) + S_2S_5S_6] - S_1(-S_4C_5S_6 + C_4C_6)$$

$$o_y = S_1[-C_2(C_4C_5S_6 + S_4C_6) + S_2S_5S_6] + C_1(-S_4C_5S_6 + C_4C_6)$$

$$o_z = S_2(C_4C_5S_6 + S_4C_6) + C_2S_5S_6$$

$$a_x = C_1(C_2C_4S_5 + S_2C_5) - S_1S_4S_5$$

$$a_y = S_1(C_2C_4S_5 + S_2C_5) + C_1S_4S_5$$

$$a_z = -S_2C_4S_5 + C_2C_5$$

$$p_x = C_1S_2d_3 - S_1d_2$$

$$p_y = S_1S_2d_3 + C_1d_2$$

$$p_z = C_2d_3$$

Note that the expressions for the a and p columns are simpler to compute than the n or o columns of $\mathbf{T}_6$. Since

$$\mathbf{n} = \mathbf{o} \times \mathbf{a} \qquad \text{and} \qquad \mathbf{o} = -\mathbf{n} \times \mathbf{a}$$

and the cross product is simpler to compute than either the n or o columns, one can save computation time by computing one of them that way.

14.11   Matrix Equations

In this section, we discuss matrix methods for dealing with tools of different shapes, workpieces being carried by a moving conveyor, and vision systems. We will use the equipment configuration shown in Figure 14.3 as an example. It consists of a work table, two manipulators, a camera, and a conveyor.

14.11.1   Transform graphs

We will use diagrams like the one in Figure 14.4 to analyze various control problems. This type of diagram is called a "transform graph" [Paul (1981), p. 123]. The points labeled W at the left and right of each line in the diagram represent the cartesian reference frame W of the worktable. For example, the position of the end effector of a manipulator is represented by a line running from the W on the left to the end of the arrow labeled $\mathbf{E}$ for end effector. The graph tells us that the expression for the

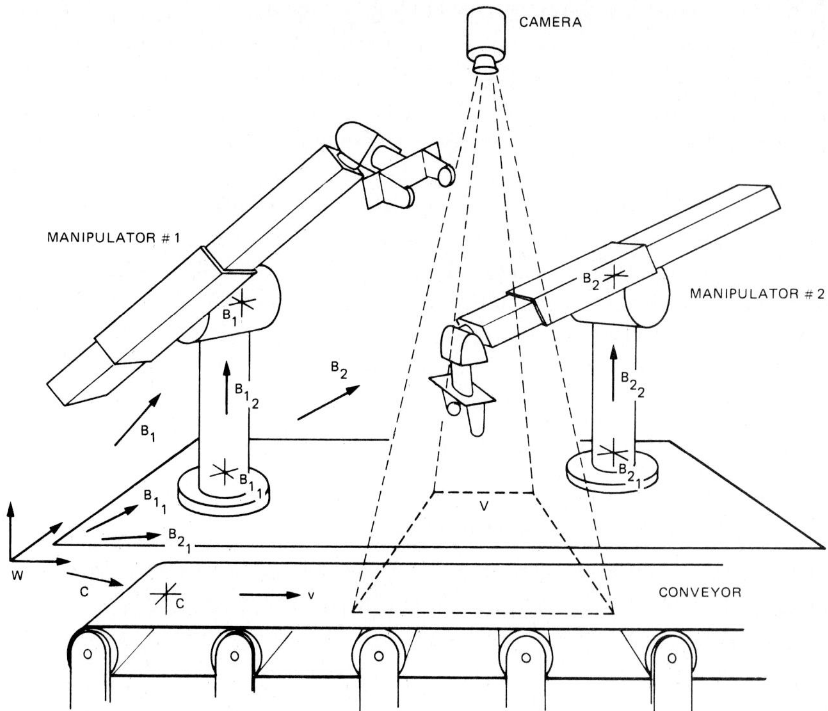

Figure 14.3   Equipment configuration.

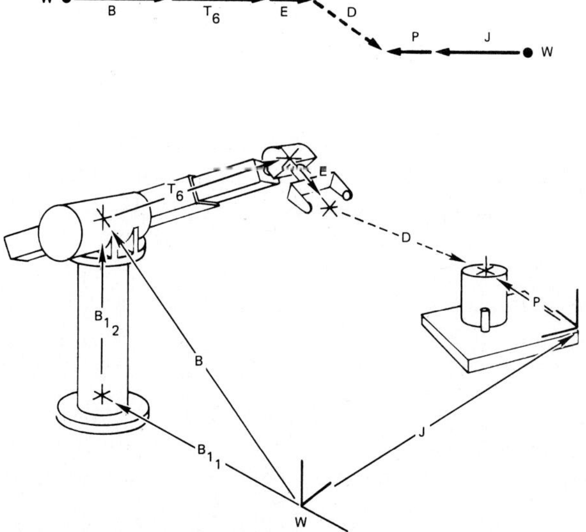

Figure 14.4   Transform graphs.

end effector position is $\mathbf{B} \cdot \mathbf{T}_6 \cdot \mathbf{E}$. The position of any other object in the work space is represented by a line running from the W on the right toward the left. Thus the position of the cylindrical workpiece in the square jig is given by $\mathbf{J} \cdot \mathbf{P}$, where $\mathbf{J}$ is the transform to a reference frame fixed in the jig, and $\mathbf{P}$ is the position of the workpiece relative to the jig.

Transform graphs make it easy to decide how to compute absolute and relative positions and orientations. For example, we may need to compute how to move the end effector so as to place it on the part. This movement is indicated by the dotted arrow $\mathbf{D}$ in Figure 14.4. To develop an expression for $\mathbf{D}$, we simply find a path from the tail of $\mathbf{D}$ to its head, then write down from left to right each transform we encounter along the path. However, if we encounter a transform that points backward along the path, we write down its *inverse*. Thus,

$$\mathbf{D} = \mathbf{E}^{-1} \cdot \mathbf{T}_6^{-1} \cdot \mathbf{B}^{-1} \cdot \mathbf{J} \cdot \mathbf{P}$$

14.11.2   Worktable coordinate system

The position of $\mathbf{W}$ with respect to itself is itself, so we can write $\mathbf{W}$ immediately as the 4×4 identity coordinate transformation matrix

$$\mathbf{W} = \begin{bmatrix} 1 & 0 & 0 & 0 \\ 0 & 1 & 0 & 0 \\ 0 & 0 & 1 & 0 \\ 0 & 0 & 0 & 1 \end{bmatrix}$$

14.11.3   Manipulator coordinate systems

The arrows labeled $\mathbf{B}_1$ and $\mathbf{B}_2$ represent the positions of the bases of manipulators number 1 and 2. Note that since we are using Stanford Arms in this example, the base reference frames are actually located *above the table* in the shoulder, rather than in the mounting plate that fastens to the table. It may be more convenient to break up each of these transformations into two consecutive transformations, say $\mathbf{B}_1 = \mathbf{B}_{1,1}\mathbf{B}_{1,2}$. $\mathbf{B}_{1,1}$ transforms from the $\mathbf{W}$ reference frame to the bottom of manipulator number 1, and $\mathbf{B}_{1,2}$ transforms from there up to the shoulder. $\mathbf{B}_{1,1}$ can be determined by measuring the position of the base plate on the table and will be different for each arm. $\mathbf{B}_{1,2}$ would be the same for each arm and could be determined either from measurements or from the blueprints for the arm.

14.11.4   End-effector transform

It is easy to deal with end effectors of unusual shape using transforms. In a Stanford Arm, for example, when the hand is closed, the fingertips lie on the axis of joint 6, and point along it away from the wrist. This is the usual position and orientation for a tool tip. If the fingertips are, say, 5 in long, then the transform from the wrist pivot point out to the fingertips is

$$\mathbf{E} = \begin{bmatrix} 1 & 0 & 0 & 0 \\ 0 & 1 & 0 & 0 \\ 0 & 0 & 1 & 5 \\ 0 & 0 & 0 & 1 \end{bmatrix}$$

Often, an end effector must be mounted at an angle to the axis of joint 6 or offset from that axis. The welding torch in Figure 14.5 is an example. Its tip is rotated through an angle of 60° around the y_6 axis, offset 7.5 in along the x_6 axis and 6 in along z_6 axis. The correct "tool matrix" for this tool is

$$\mathbf{E} = \begin{bmatrix} \cos 60° & 0 & \sin 60° & 7.5 \text{ in} \\ 0 & 1 & 0 & 0 \\ -\sin 60° & 0 & \cos 60° & 6.0 \text{ in} \\ 0 & 0 & 0 & 1 \end{bmatrix}$$

14.11.5 Conveyor coordinate system

The arrow $\mathbf{C}$ represents the transformation from the worktable frame W to a frame fixed in the conveyor belt. While frames B_1 and B_2 are constant, frame C changes with time as the conveyor travels. It may be convenient

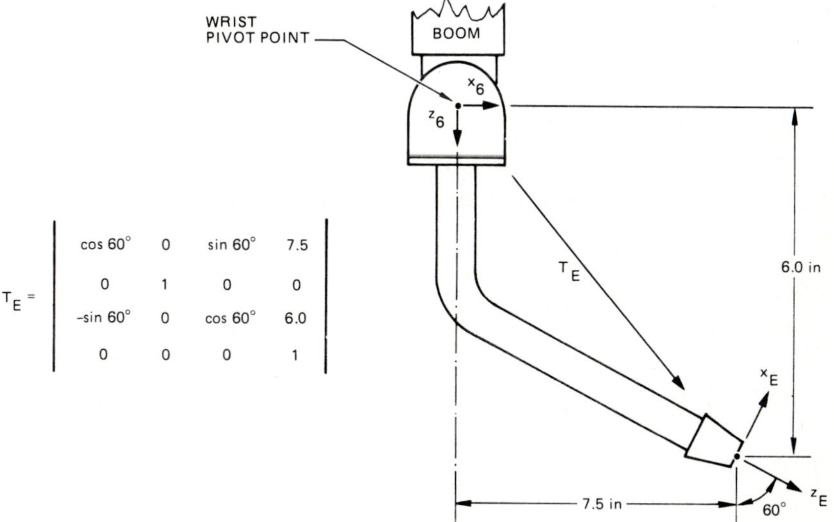

Figure 14.5 An end effector with offsets.

to choose an orientation for the conveyor frame so that the conveyor motion is along one of the coordinate axes of the frame. The z axis of the frame is the usual choice, in keeping with the way the z axis of a manipulator joint is aligned with its axis of motion. Then the 4×4 matrix for the conveyor will resemble the one for a prismatic joint

$$
\mathbf{C} = \begin{bmatrix} & & C_x \\ & [\mathbf{R}] & C_y \\ & & C_z \\ 0\ 0\ 0 & & 1 \end{bmatrix} \cdot \begin{bmatrix} 1 & 0 & 0 & 0 \\ 0 & 1 & 0 & 0 \\ 0 & 0 & 1 & V_c \cdot t \\ 0 & 0 & 0 & 1 \end{bmatrix}
$$

$$
= \begin{bmatrix} & & C_x + R_{13} \cdot V_c \cdot t \\ & [\mathbf{R}] & C_y + R_{23} \cdot V_c \cdot t \\ & & C_z + R_{33} \cdot V_c \cdot t \\ 0\ 0\ 0 & & 1 \end{bmatrix}
$$

The term $V_c \cdot t$ represents the motion of the belt as a function of time t, when it is moving at a constant velocity V_c. The submatrix $\mathbf{R} = [\mathbf{R}_{ij}]$ represents the orientation of the conveyor. C_x, C_y, and C_z are the coordinates of the belt's reference frame at $t = 0$. Note that the submatrix $\mathbf{R}$ can represent a belt that is inclined to the horizontal as easily as any other orientation.

A convenient method of calibrating the conveyor belt (that is, of filling in the numbers in the $\mathbf{C}$ matrix above) is to use the manipulators as coordinate measuring devices to train the belt position. A typical training procedure is shown in Figure 14.6. The steps are as follows:

1. Stop the belt, and place two marks 1 and 2 on it along a line approximately transverse to the belt direction and in positions that can be reached by the end effector of one of the manipulators. Mark 1 should be on the right side of the belt when facing in the direction of motion, and mark 2 should be on the left side, as shown in the figure. Put a calibrated pointer end effector on the arm. Place the pointer on each mark and record their positions as P_1 and P_2, respectively.

2. Move the belt until the first mark is almost out of reach of the end effector. Place the end effector on it and record its new position as P_3.

3. Define the origin of the belt frame to be at P_1. This is the fourth column of matrix $\mathbf{C}$.

4. Define the z axis of the belt frame to be along the vector $\mathbf{Q}_z = (P_3 - P_1)$. $\mathbf{Q}_z$ scaled to unit length is the third column of matrix $\mathbf{C}$.

5. Define the x axis of the belt frame to be along the vector $\mathbf{Q}_x = (P_3 - P_1)$. $\mathbf{Q}_x$ scaled to unit length is the first column of matrix $\mathbf{C}$.

6. Define the y axis of the belt frame to lie along $\mathbf{Q}_y = \mathbf{Q}_z \times \mathbf{Q}_x$, where

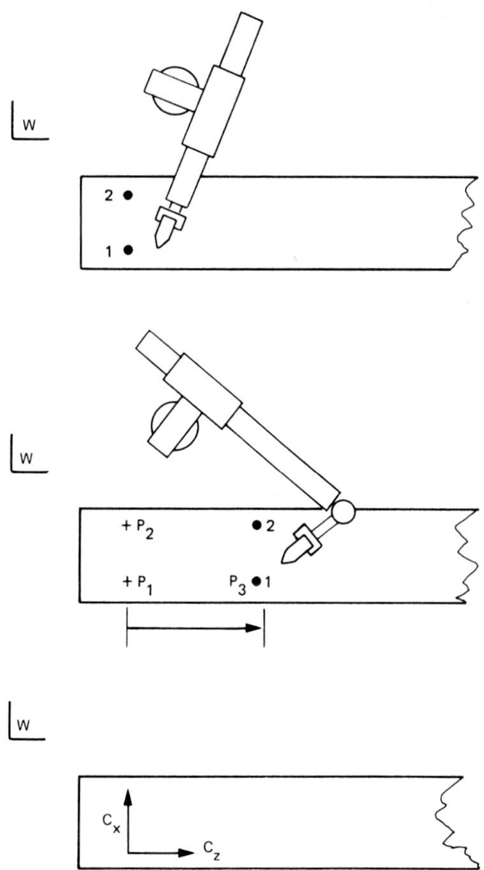

Figure 14.6   Calibrating a conveyor belt.

$\times$ is the vector cross product. $\mathbf{Q}_y$ scaled to unit length is the second column of matrix $\mathbf{C}$.

Defined in this way, the x, y, and z axes will form a right-handed coordinate system.

The x axis will lie in the plane of the belt, pointing from the right side of the belt to the left.

The y axis will be normal to the conveyor belt surface, pointing up.

The z axis will lie in the plane of the belt and point in the direction of conveyor motion.

As an example of how we can use the $\mathbf{C}$ matrix, we will compute the position in which to place the hand at each instant so as to track an object that is being carried by the conveyor. Assume that a workpiece is at position $\mathbf{F}$ relative to the frame $\mathbf{C}$ fixed in the conveyor. We want to keep the hand of the manipulator at a position $\mathbf{H}$ relative to a point $\mathbf{G}$ in the workpiece. Then the position of the hand $\mathbf{Q}$ is given by the matrix equation

$$\mathbf{Q} = \mathbf{B} \cdot \mathbf{T}_6 \cdot \mathbf{E}$$

For this tracking task, $\mathbf{Q}$ should also be positioned above the workpiece;

$$\mathbf{Q} = \mathbf{C}(t) \cdot \mathbf{F} \cdot \mathbf{G} \cdot \mathbf{H}$$

where the belt position $\mathbf{C}(t)$ depends on time t. Because these two equations represent the same hand position,

$$\mathbf{B} \cdot \mathbf{T}_6 \cdot \mathbf{E} = \mathbf{C}(t) \cdot \mathbf{F} \cdot \mathbf{G} \cdot \mathbf{H}$$

To control the arm, we need to solve for the wrist position $\mathbf{T}_6$, because our arm solution gives the joint angles in terms of the elements of that matrix

$$\mathbf{T}_6(t) = \mathbf{B}^{-1} \cdot \mathbf{C}(t) \cdot \mathbf{F} \cdot \mathbf{G} \cdot \mathbf{H} \cdot \mathbf{E}^{-1}$$

Here, we have shown that $\mathbf{T}_6$ must also be a function of time. To track the object, one must solve this equation and do an arm solution many times per second in order to obtain the joint position set points needed by the joint servos.

A variation of this method can be applied when the conveyor moves at an unpredictable velocity. Attach a position encoder to the belt and determine a calibration factor $K = $ distance/(encoder reading) so that the term $V \cdot t$ in the equation for matrix $\mathbf{C}$ can be replaced by a term of the form $K \cdot n$ where n is the encoder reading.

14.11.6   Camera coordinate system

The origin of a camera coordinate system should be in the plane of focus of the camera, not in the camera itself. This is because the camera produces information about the location of objects in that plane. Thus, in Figure 14.3 we showed the camera frame V in the surface of the table, not in the camera. It is necessary to convert from a position in the image to a position in the V frame. Image positions are usually reported in terms of the number of pixels U_v from the top of the image and U_h from the left side, since images are usually scanned from top to bottom and from left to right. Then to calibrate the camera it is necessary to

1. Locate the point in the work space that corresponds to the upper left corner of the image

2. Establish a scale factor for the camera

3. Develop a transformation to account for rotation of the camera with respect to the world frame axes

The theory behind the methods used to do these three things accurately is complex (Ballard and Brown, 1982, pp. 482–488). We will give a simple camera calibration procedure as an example. It requires a round object that appears to be about ⅛ of the width of the screen in size on the video monitor. A coin is a convenient object.

1. Place the object in one corner of the field of view (Figure 14.7*a*), taking care that it is completely visible. Measure the location of the center of the object $P_{1,h}$ $P_{1,v}$) in the image. This can be done by taking the average of the pixel coordinates of each pixel in the image of the object.

2. Measure the position of the center of the object $(P_{1,x}$ $P_{1,y}$ $P_{1,z})$ in work-space coordinates. Calibration marks on the table such as grid lines can be useful in making this measurement.

3. Move the object to the opposite corner of the field of view (Figure 14.7*b*) but so that it is still completely visible.

4. Measure $(P_{2,h}$ $P_{2,v})$ and $(P_{2,x}$ $P_{2,y}$ $P_{2,z})$.

A calibration matrix can be derived from these measurements that will transform pixel coordinates into a position in the worktable frame W. First compute the four differential quantities.

$$\frac{dX}{dU_h} = \frac{P_{2,x} - P_{1,x}}{U_{2,h} - U_{1,h}}$$

$$\frac{dX}{dU_v} = \frac{P_{2,x} - P_{1,x}}{U_{2,v} - U_{1,v}}$$

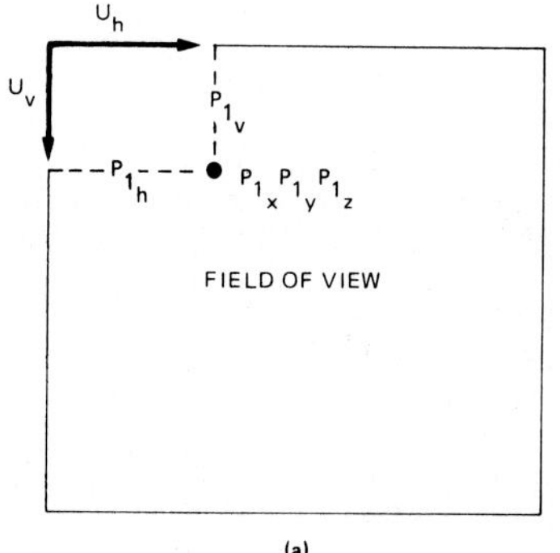

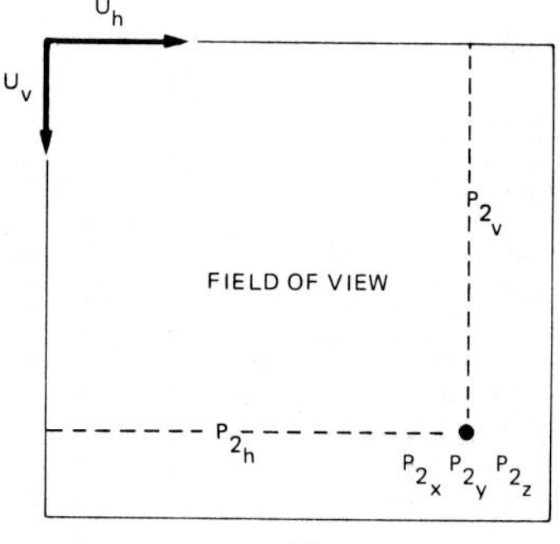

Figure 14.7   Calibrating a camera.

$$\frac{dY}{dU_h} = \frac{P_{2,y} - P_{1,y}}{U_{2,h} - U_{1,h}}$$

$$\frac{dY}{dU_v} = \frac{P_{2,y} - P_{1,y}}{U_{2,v} - U_{1,v}}$$

Then compute the location $(V_{0,x} \quad V_{0,y} \quad V_{0,z})$ of the origin of the camera frame at $U_h = U_v = 0$. Because the camera in this example is pointed vertically downward, the height of the origin is $V_0 = 0$ in the W frame

$$V_{0,x} = P_{1,x} - U_{1,h}\frac{dX}{dU_h} - U_{1,v}\frac{dX}{dU_v}$$

$$V_{0,y} = P_{1,y} - U_{1,h}\frac{dY}{dU_h} - U_{1,v}\frac{dY}{dU_v}$$

Finally, the camera calibration matrix is

$$\mathbf{T}_{cam} = \begin{bmatrix} dX/dU_h & dX/dU_v & V_{0,x} \\ dY/dU_h & dY/dU_v & V_{0,y} \\ 0 & 0 & V_{0,z} \\ 0 & 0 & 1 \end{bmatrix}$$

An object that appears at an image position $\mathbf{P}_{image} = (U_h \quad U_y)$ is then at the position $\mathbf{P}_W$ in the W frame given by

$$\begin{bmatrix} P_{W,x} \\ P_{W,y} \\ P_{W,z} \\ 1 \end{bmatrix} = \mathbf{T}_{cam} \cdot \begin{bmatrix} u_h \\ u_v \\ 1 \end{bmatrix}$$

Note that $\mathbf{T}_{cam}$ has four rows and three columns so that it can transform homogeneous two-vectors (image locations) into homogeneous three-vectors (spatial locations).

14.12   Jacobian Matrix

The jacobian matrix for a manipulator describes how *differential* (that is, small) motions of each joint combine to produce a net motion of the hand. The relationship between individual joint efforts and the total effort at the manipulator's hand can also be expressed simply in terms of the jacobian.

A jacobian matrix has six rows—one for each degree of freedom of the hand—and as many columns as the manipulator has joints. For example, a Stanford Arm has six joints, so its jacobian is a square 6×6 matrix.

Arms with more or fewer joints have nonsquare jacobians which are more difficult to work with (e.g., they cannot be inverted). The jacobian depends upon the instantaneous posture of the arm. That is, $\mathbf{J}$ is a function of the joint angles θ_k

$$\mathbf{J} = (\theta_1, \theta_2, d_3, \theta_4, \theta_5, \theta_6)$$

for a six-jointed Stanford Arm, for example.

We will first explain the jacobian as it relates to differential motion and then show how it can be used to relate forces and torques at the joints to net force and torque at the hand. Finally, we explain how a special kind of jacobian matrix is used to transform forces and torques from one coordinate frame to another.

14.12.1   Use of the jacobian in differential hand motion

Let $\mathbf{Q} = (q_1, \ldots, q_6)$ be a set of joint variables for a particular manipulator posture. Let $d\mathbf{P}$ be the six-element vector displacement and rotation of the hand caused by a set of differential, or infinitely small, displacements $d\mathbf{Q} = (dq_1, \ldots, dq_6)$ of the joints from their initial positions $\mathbf{Q}$. We express $d\mathbf{P}$ in terms of the coordinate system that is embedded in and rotates with the last link (i.e., in the frame represented by transform matrix $\mathbf{T}_6$).

$$d\mathbf{P} = (dX, dY, dZ, \delta_x, \delta_y, \delta_z) = (dP_1, dP_2, \ldots, dP_6)$$

where dX, dY, and dZ are differential *displacements* of the hand along the axes of the $\mathbf{T}_6$ frame and δ_x, δ_y, and δ_z are differential *rotations* of the hand about the axes of that frame.

Note that the orientation description that is used to describe small rotations in connection with the jacobian is essentially the roll-pitch-yaw description. That is, δ_x corresponds to a differential rotation about the x, or yaw, axis of the hand, δ_y about the y, or pitch, axis, and δ_z about the z, or roll, axis. These δ's do *not*, for example, correspond to Unimation's orientation, approach, and twist angles, or to Euler angles. This suggests that it may be advisable to adopt the roll-pitch-yaw orientation representation in robot control software in order to simplify any computations that involve jacobians. Otherwise, it will be necessary to develop appropriate transformation equations between roll-pitch-yaw and whatever representation is used.

We note in passing that it is well known that the order in which *differential* rotations are carried out makes no difference to the final orientation, whereas the order in which *finite* rotations are applied makes a great difference.

With these definitions, the element in row i column j of the jacobian matrix $\mathbf{J}$ is

$$J_{ij} = \frac{\partial P_i}{\partial q_j}$$

So, each element J_{ij} is just the contribution of the net hand motion in the i direction to one component dP_i per unit of motion dq_j in joint j. "Hand motion in the i direction" means *along* axis i of the hand frame for $i =$ 1, 2, or 3, and *around* axis 1, 2, or 3 for $i =$ 4, 5, or 6, respectively.

Thus, the jacobian matrix can be used to compute the net displacement and rotation of the hand, $^{T(6)}d\mathbf{P}$, expressed in the hand frame, for an arbitrary set $d\mathbf{Q}$ of differential joint motions as the matrix product

$$^{T(6)}d\mathbf{P}(dq_1, \ldots, dq_6) = \mathbf{J} \cdot d\mathbf{Q}$$

The same formula can be inverted to compute how far to move each joint in order to displace and rotate the hand by a specified amount.

$$d\mathbf{Q} = \mathbf{J}^{-1} \cdot d\mathbf{P}$$

These two formulas may also be used with good accuracy with *finite* joint motions if they are small (e.g., a few millimeters of displacement or a few degrees of rotation). The latter equation can be used in this way, for example, to implement compliance by moving the hand a short distance in the direction of any forces and torques that are felt by a sensor attached to the hand.

If $\mathbf{J}^{-1}$ does not exist, we say that the arm is "at a singularity." This means that it has lost some of its degrees of freedom of motion and cannot move in certain directions. This happens in the Stanford Arm, for example, whenever the axes of joints 4 and 6 become collinear. The hand is then unable to rotate about an axis normal to the axis of joint 5.

Arms with more or fewer than six joints have nonsquare jacobians, for which $\mathbf{J}^{-1}$ is not defined. Arms with fewer than six joints always have fewer than six degrees of freedom, and so there are always some directions in which they cannot move. The five-joint Unimate 2000, for example, is unable to rotate its hand about a vertical axis unless the last joint axis happens to be pointing straight up or down. For such arms, one can, however, compute a "pseudo inverse" of the Jacobian. Use of the pseudo inverse in the above equation then results in motion that is, loosely speaking, "as good as can be expected from that arm."

The opposite situation holds for "redundant" arms that have more than six joints. Not only can such an arm move its hand in any direction required, but it will be able to make certain motions in more than one way. That is, there will be more than one combination of joint motions that will produce certain hand motions. One can also compute a pseudo inverse for a redundant arm.

The usual methods for computing the pseudo inverse of a nonsquare A matrix result in a matrix $\mathbf{A}^{-1}$ for which $\mathbf{A} \cdot \mathbf{A}^{-1}$ is as close as possible to the identity matrix in a least-squared error sense. However, one could also define the pseudo inverse in other ways to produce different types of arm motion that might be better suited to a particular task. For example, in performing delicate assemblies with a redundant manipulator, one would prefer to make small motions with the fast outboard joints rather than with the slower inboard joints whenever possible. The jacobian for that arm could be tuned so that the outboard joints moved much more than the inboard joints.

14.12.2   Use of the jacobian in force control

The jacobian is also very useful in force control calculations. We will use the "principle of virtual work" to evaluate the contribution of the force or torque exerted by each joint to the net force or torque exerted by the hand. Then we will show how the jacobian appears naturally in these calculations.

In using the principle of virtual work, we imagine that the arm exerts a constant force and torque on some object while it makes a differential motion. Then the work done by the joints must equal the work done by the hand. The fact that this must be true for any force and torque and for any differential motion allows us to draw conclusions about the relation between the joint efforts and the net effort. Let $\mathbf{M}$ be a column vector of the joint efforts. For example, for the Stanford Arm,

$$\mathbf{M} = (\tau_1, \tau_2, f_3, \tau_4, \tau_5, \tau_6)$$

where the τ's are the torques exerted by the rotary joints and f_3 is the force exerted by prismatic joint number three. Then the virtual work done *by the joints* is

$$\delta W_{\text{joints}} = \mathbf{M}^T \cdot d\mathbf{Q}$$

where $\mathbf{M}^T$ is the transpose of $\mathbf{M}$, a row vector, so that δW is a scalar.

Let $^{\mathbf{T}(6)}\mathbf{F}$ be a column vector of the force and torque components exerted by the hand, expressed in the $\mathbf{T}_6$ frame. Then the virtual work done *by the hand* is

$$\delta W_{\text{hand}} = {}^{\mathbf{T}(6)}\mathbf{F}^T \cdot d\mathbf{P}$$

where again we have used the transpose to get a scalar value for the work.

We may assume an ideal frictionless manipulator, as we are analyzing only the effect of the geometry of the arm on force control. We may also assume that the arm is not experiencing any significant accelerations, as we are not analyzing the effects of arm dynamics on force control. Therefore, because the arm is a conservative energetic system, the work done *on* the arm by the joint motors must equal the work done *by* the arm on the imaginary object that it is touching with its hand

$$\mathbf{M}^T \cdot d\mathbf{Q} = {}^{T(6)}\mathbf{F}^T \cdot d\mathbf{P}$$

The differential motion $d\mathbf{P}$ can be expressed in terms of the jacobian as explained above.

$$\mathbf{M}^T \cdot d\mathbf{Q} = {}^{T(6)}\mathbf{F}^T \cdot \mathbf{J} \cdot d\mathbf{Q}$$

Because this relation must hold for any $d\mathbf{Q}$, we can eliminate it from both sides of this equation and solve for the joint torques

$$\mathbf{M}^T = {}^{T(6)}\mathbf{F}^T \cdot \mathbf{J}$$

Using the identity $(\mathbf{A}^T \cdot \mathbf{B})^T \equiv \mathbf{B}^T \cdot \mathbf{A}$,

$$\mathbf{M} = \mathbf{J}^T \cdot {}^{T(6)}\mathbf{F}$$

Thus we see that the transposed jacobian transforms the individual forces and torques exerted by each joint into the net force and torque exerted by the hand.

Space does not permit a complete example of how to compute a jacobian, but we can summarize the method given by Paul (1981, p. 103). He first defines the "differential coordinate transformation" $\mathbf{T}_{\mathrm{DCT},j}$ for a joint j to be the position of the hand frame with respect to the frame embedded in link $j - 1$. That is,

$$\mathbf{T}_{\mathrm{DCT},j} = {}^{j-1}\mathbf{T}_6$$

with columns n, o, a, and p. All the matrices ${}^{j-1}\mathbf{T}_6 = \mathbf{A}_j \cdot \cdots \cdot \mathbf{A}_6$ for $j = 1$ to 6 for the Stanford Arm are presented elsewhere in this book.

If the jth joint is rotary, Paul shows that the elements of the jth column of the jacobian can be computed from the columns n, o, a, and p of $\mathbf{T}_{\mathrm{DCT},j}$ as

$$J_{1,j} = -n_x p_y + n_y p_x$$

$$J_{2,j} = -o_x p_y + o_y p_x$$

$$J_{3,j} = -a_x p_y + a_y p_x$$

$$J_{4,j} = n_z$$

$$J_{5,j} = o_z$$

$$J_{6,j} = a_z$$

However, if joint j is prismatic, they are

$$J_{1,j} = n_z$$

$$J_{2,j} = o_z$$

$$J_{3,j} = a_z$$

$$J_{4,j} = J_{5,j} = J_{6,j} = 0$$

Thus the jacobian can be computed by multiplying together the $\mathbf{A}_j$ matrices for each joint i to form the six matrices $^{j-1}\mathbf{T}_6$ and by inserting elements from them into the above expressions according to the type of joint.

14.12.3   Use of the jacobian to transform forces and torques

Forces and torques must often be transformed from one coordinate system to another. For example, the equation above gives the joint efforts $\mathbf{M}$ in terms of the required force and torque measured in the T_6 frame. Often, however, they are given in the wrong frame. For example, a force/torque sensor attached between the wrist and the hand will report forces and torques in its own coordinate system. These must be transformed into the T_6 frame before the equivalent joint torques can be computed in the way described in the previous section. They must be transformed into the A frame in the fingertips to determine what force and torque the hand is exerting on whatever it is touching.

In this section we will explain how to transform forces and torques between coordinate systems using the example in Figure 14.8. In this figure a force $\mathbf{F}_E$ and torque τ_E are applied to the hand of a Stanford Arm at the origin of a frame E located between the fingertips. It is desired to compute the force and torque at the origin of the T_6 coordinate frame, which is at the wrist pivot point.

If we consider the portion of the arm from the wrist pivot point to the fingertips to be a rigid body at rest and in equilibrium, elementary statistics tells us that the sum of the forces dend the sum of the torques acting on the body are both zero. If we assume that the only points of contact with that portion of the arm are the wrist pivot point and the fingertips, then there is only one additional force and torque to consider. Let $\mathbf{F}_6$ be the net force acting at the wrist pivot point, and let τ_6 be the net torque acting there. These quantities, expressed in the T_6 frame, are what we are seeking.

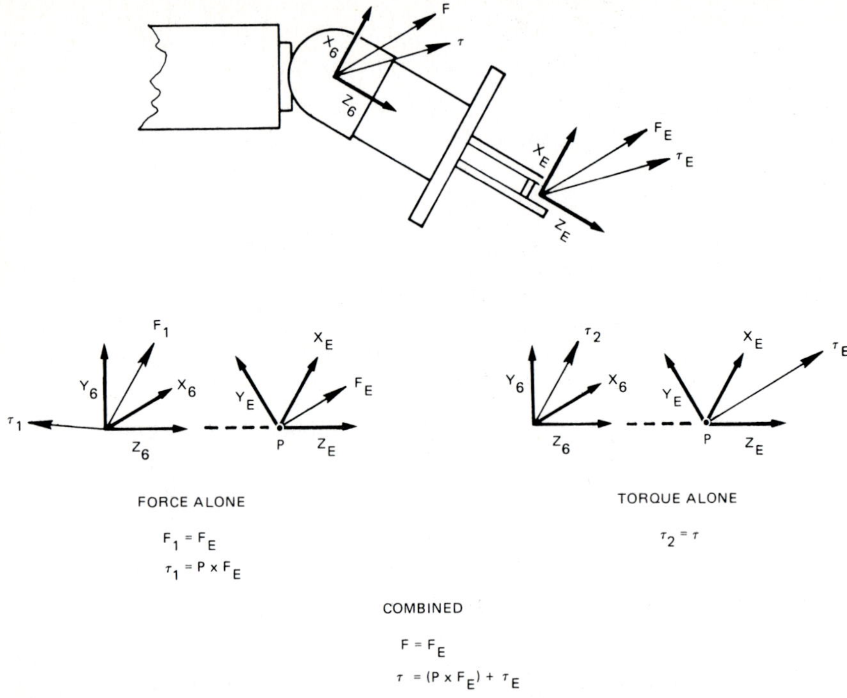

Figure 14.8 Transformation of force and torque to a different coordinate system.

Let the tool transform be

$$E = \begin{bmatrix} n_x & o_x & a_x & p_x \\ n_y & o_y & a_y & p_y \\ n_z & o_z & a_z & p_z \\ 0 & 0 & 0 & 1 \end{bmatrix}$$

The force $\mathbf{F}_E$ applied at the fingers will be felt as an equal force $\mathbf{F}_1$ at the wrist pivot point plus a torque τ_1. $\mathbf{F}_1$ is just $\mathbf{F}_E$ rotated into the T_6 frame with the rotation portion of the $\mathbf{E}$ transform matrix (matrix $\mathbf{R}$, consisting of the upper-left 3×3 submatrix of $\mathbf{E}$).

$$\mathbf{F}_1 = \mathbf{R} \cdot {}^{E}\mathbf{F}_E$$

τ_1 is the vector cross product of the fingertip position relative to the desired frame and the force vector expressed in the desired frame.

$$\tau_1 = \mathbf{p} \times {}^{T(6)}\mathbf{F}_E = \mathbf{p} \times (\mathbf{R} \cdot {}^{E}\mathbf{F}_E)$$

The torque applied to the fingertips will be felt at the wrist pivot point as an equal torque τ_2 in the same direction, so

$$\tau_2 = {}^{T(6)}\tau_E = \mathbf{R} \cdot {}^{E}\tau_E$$

Paul (1981, pp. 95–103, 219) gives a matrix representation for these computations that may be more convenient for implementation. He defines a matrix $\mathbf{J}$ which he calls "the jacobian." This matrix is *not* the same as the manipulator jacobian that he describes on p. 103 and that we have described in the preceding section. Nevertheless, it is a jacobian in the general sense because it shows how to convert a differential motion from one coordinate system to another. The manipulator jacobian converts a differential motion of an arm from joint coordinates to cartesian coordinates. The other jacobian converts a differential motion of an arbitrary object from one cartesian coordinate system to another. To avoid confusion, we will call this latter kind of jacobian $\mathbf{J}_{\text{cart}}$.

Let a vector torque $\mathbf{m}$ and a vector force $\mathbf{f}$ be given in a reference frame and let matrix

$$
\mathbf{C} = \begin{bmatrix} n_x & o_x & a_x & p_x \\ n_y & o_y & a_y & p_y \\ n_z & o_z & a_z & p_z \\ 0 & 0 & 0 & 1 \end{bmatrix}
$$

be the transformation from the reference frame to the frame in which the force and torque are to be expressed. Then

$$
\begin{bmatrix} {}^c m_x \\ {}^c m_y \\ {}^c m_z \\ {}^c f_x \\ {}^c f_y \\ {}^c f_z \end{bmatrix} = \mathbf{J}_{\text{cart}} \begin{bmatrix} m_x \\ m_y \\ m_z \\ f_x \\ f_y \\ f_z \end{bmatrix}
$$

where

$$
\mathbf{J}_{\text{cart}} = \begin{bmatrix}
n_x & n_y & n_z & (\mathbf{p} \times \mathbf{n})_x & (\mathbf{p} \times \mathbf{n})_y & (\mathbf{p} \times \mathbf{n})_z \\
o_x & o_y & o_z & (\mathbf{p} \times \mathbf{o})_x & (\mathbf{p} \times \mathbf{o})_y & (\mathbf{p} \times \mathbf{o})_z \\
a_x & a_y & a_z & (\mathbf{p} \times \mathbf{a})_x & (\mathbf{p} \times \mathbf{a})_y & (\mathbf{p} \times \mathbf{a})_z \\
0 & 0 & 0 & n_x & n_y & n_z \\
0 & 0 & 0 & o_x & o_y & o_z \\
0 & 0 & 0 & a_x & a_y & a_z
\end{bmatrix}
$$

So,

$$
{}^c m_x = \mathbf{n} \cdot [(\mathbf{f} \times \mathbf{p}) + \mathbf{m}]
$$

$$
{}^c m_y = \mathbf{o} \cdot [(\mathbf{f} \times \mathbf{p}) + \mathbf{m}]
$$

$$
{}^c m_z = \mathbf{a} \cdot [(\mathbf{f} \times \mathbf{p}) + \mathbf{m}]
$$

$$
{}^c f_x = \mathbf{n} \cdot \mathbf{f}
$$

$${}^{c}f_y = \mathbf{o} \cdot \mathbf{f}$$

$${}^{c}f_z = \mathbf{a} \cdot \mathbf{f}$$

$\mathbf{J}_{\text{cart}}$ also converts differential displacements and rotations from one coordinate system to another. If $\mathbf{d}$ is the vector of differential displacements along the reference frame's axes and δ is the vector of differential rotations around them,

$$
\begin{bmatrix}
{}^{c}d_x \\
{}^{c}d_y \\
{}^{c}d_z \\
{}^{c}\delta_x \\
{}^{c}\delta_y \\
{}^{c}\delta_z
\end{bmatrix}
= \mathbf{J}_{\text{cart}}
\begin{bmatrix}
d_x \\
d_y \\
d_z \\
\delta_x \\
\delta_y \\
\delta_z
\end{bmatrix}
$$

(See Paul, 1981, p. 95).

14.13   Arm Solution

In Section 14.10, we discussed the problem of determining the position and orientation of the wrist for any given set of joint angles or extensions. In this section we discuss the inverse problem—finding a set of joint angles or extensions that will place the wrist in a specified position and specified orientation. We will first explain why there is usually more than one set of joint coordinates that will accomplish this. Then we will show (with a simple example) some of the difficulties that the existence of multiple solutions can cause in trying to make smooth motions. Finally, we will explain the methods used in practice to compute the joint coordinates from the wrist position and orientation. We will use the Stanford Arm as an example throughout.

14.13.1   Nomenclature

There is currently some confusion in the industry as to which problem should be called "the arm solution," and which should be called "the backward arm solution." We will arbitrarily call the problem of finding the wrist position the *back solution,* and the more difficult problem of finding the joint angles the *arm solution.*

14.13.2   Origin of multiple arm solutions

For a conventional manipulator whose joints are all in series, there is a unique wrist position and orientation for each set of joint angles. (This may not be true if the joints are kinematically in parallel, but that is an advanced topic not treated in this book.) The position can be calculated in a straightforward way by simply multiplying together the A matrices for each successive joint from the base of the arm to its wrist, as described in Section 14.10. Thus, the back solution is unique for any common manipulator.

For a conventional serial-joint manipulator, however, the arm solution is very often not unique. There may be two, four, eight, or more different sets of joint angles that will place the wrist at exactly the same position and orientation.

The equations that relate cartesian coordinates to joint coordinates for arms with rotary joints inevitably require trigonometric functions and squared terms (corresponding to squares of distances). The periodicity and symmetry of these functions give rise to multiple arm solutions. As an example of the effect of periodicity,

$$\tan \theta = \tan (\theta + 180)$$

so if the arm solution equations determine only the tangent of an angle, then that angle can have two values 180° apart. Similarly, if the arm solution equations determine only the cosine of an angle, then, because

$$\cos \theta = \cos (-\theta)$$

that angle can have two values with opposite signs.

Finally, the squared terms in the equations can determine a prismatic joint's extension as the roots of a polynomial. Each real root results in another arm solution.

14.13.3   Example of multiple arm solutions

To illustrate how there can be multiple arm solutions for a given end-effector position and orientation, Figures 14.9 and 14.10 show how the Stanford Arm can reach for an object in *four* different ways. Each way corresponds to a different set of joint angles and a different answer to the arm solution problem. For clarity, we have shown an asymmetric hand as the end effector and have shaded portions of the wrist mechanism to destroy its symmetry.

The two positions shown in Figure 14.9 are both right-arm postures, because the configuration of the shoulder joints resembles that of the right arm of a human being. The shoulder joint angles θ_1 through θ_3 are identical in both postures. Only the wrist joint angles θ_4 through θ_6 differ. θ_4 in Figure 14.9b is rotated by 180° from its position in Figure 14.9a. The same is true of θ_6. If we take the midposition of joint 4 as the position corresponding to a joint angle of $\theta_4 = 0°$, then θ_4 has opposite signs in the two postures shown in Figure 14.9.

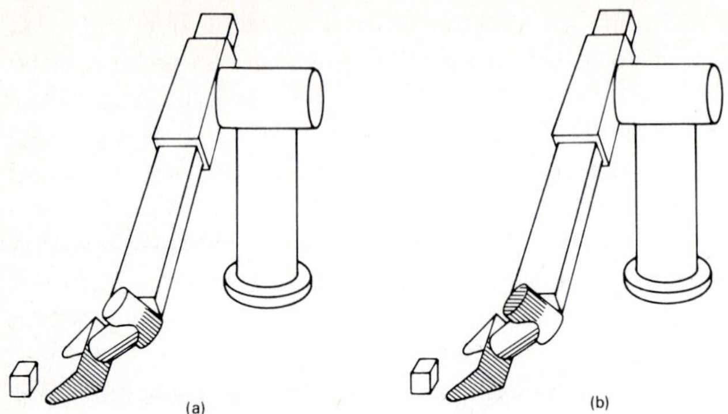

(a)   (b)

Figure 14.9 Multiple arm solutions—right arm postures. (*a*) Solution 1; (*b*) solution 2.

Figure 14.10 shows the second pair of arm solutions for the Stanford Arm. These are left-arm solutions. It is obvious that the wrist joint angles shown in this figure are not related to those in the preceding figure in any simple way. With a little thought, however, it becomes apparent that, as shown in Figure 14.11, the shape of the arm inboard of the wrist is the reflection in the (vertical) plane through the wrist and the axis of joint 1 of its position in the other figure. Thus, the extension d_3 of joint 3 is the same in all four postures, the sign of θ_2 in Figure 14.10 is opposite to that in Figure 14.9, and the axis of joint 2 makes the same angle ϕ with the plane of symmetry in both figures.

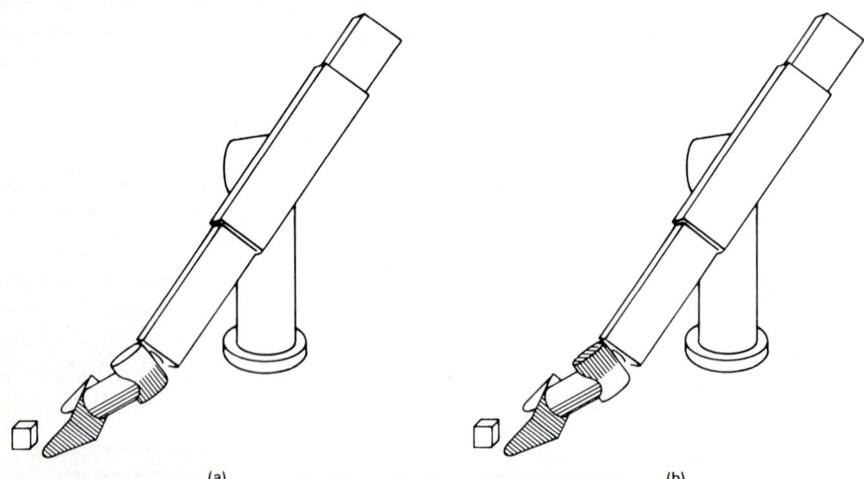

(a)   (b)

Figure 14.10 Multiple arm solutions—left arm postures. (*a*) Solution 3; (*b*) solution 4.

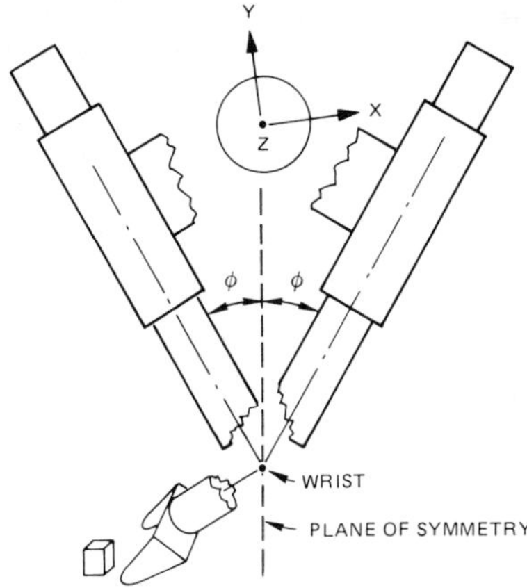

Figure 14.11 Inboard of the wrist, left- and right-angle solutions are reflections in the vertical plane.

14.13.4 Control problems resulting from multiple arm solutions

Note that the Stanford Arm cannot convert from a right-arm to a left-arm posture while keeping its wrist (i.e., joint 4) in one position. In fact, to make the conversion the whole arm must swing through quite a large portion of the work space. The arm can convert from one wrist posture to the other without moving its wrist, but it must still swing the end effector quite drastically about the wrist. Such configuration changes may be required during a task. They must be programmed with great care, because they can easily make the arm collide with objects in the work space.

The practical significance of multiple arm solutions is that the control software cannot simply use *any* arm solution. It must select the one that is best depending upon what the arm is supposed to do next and on what it is doing at the moment.

To illustrate this principle, suppose that a Stanford Arm is supposed to spray paint an object or lay down a bead of sealant or glue on it during an assembly process. Then obviously the hand must move smoothly along some path, maintaining its orientation. Also, let us assume that the control software operates in the usual way, that is, it repeatedly solves for the joint angles for successive hand positions spaced every few millimeters along the path. It is obvious from the previous example that for each position along the path there will be four radically different sets of joint angles that could be sent to the joint servos. What would happen if the control software were to simply select one of the four arm solutions at random each time—perhaps a right-arm posture at one position on the path and a left-arm posture for the next? Obviously, the arm would wave about violently every time it changed its configuration, the hand would not follow the required path at all, and paint or glue would be sprayed in all directions.

Problems with these smooth or continuous-path motions are much less frequent in assembly, but they are just as serious. Gross transportation or repositioning motions can usually be performed as point-to-point motions because it is rarely necessary to hold the end effector or workpiece in any particular orientation as it is being moved. However, any sort of fine motion for fitting, fastening, or sensor-servoed motions must be performed as a continuous-path motion, because any configuration change would be disastrous.

14.13.5   Continuous-path control with multiple arm solutions

The proper way to make continuous-path motions is to *always use the same arm solution*. In other words, if a Stanford Arm is in a left-arm configuration when it starts to spray a workpiece or insert a pin into a hole, it must remain a left arm until it can turn off the spray gun or release the pin. Although the Stanford Arm's wrist postures do not have such obvious names as "left" and "right," the wrist must also remain in whichever of its two postures it is in when the activity started.

For continuous-path control, the control software must therefore have some consistent way of classifying the various arm solutions, be able to realize when a motion will require a configuration change, and be able to decide how best to change configurations when necessary. In this regard, it is instructive to consider the thought experiment of mapping the six-dimensional space of joint coordinates J^6 onto the six-dimensional space C^6 of wrist positions and orientations expressed in some cartesian reference frame. Because the arm solution is complicated, this will obviously require some stretching and turning of local regions in J^6. But, because the solution is also multivalued, as we have seen, it will also require some "folding" of J^6 space, because different points in it must arrive at the same point in C^6. For any point in C^6 corresponding to some wrist position and orientation, each point from J^6 that maps into it represents one of the possible arm postures that will place the wrist there.

Although we cannot draw a six-dimensional diagram of this mapping, we can give some idea of what is going on with a simplified example in two dimensions.

14.13.6   Mapping diagrams

The two-joint manipulator shown in Figure 14.12 has two arm solutions which we will call "right arm" and "left arm" as follows:

$$0 < \theta_2 < 180 \qquad \text{right arm}$$

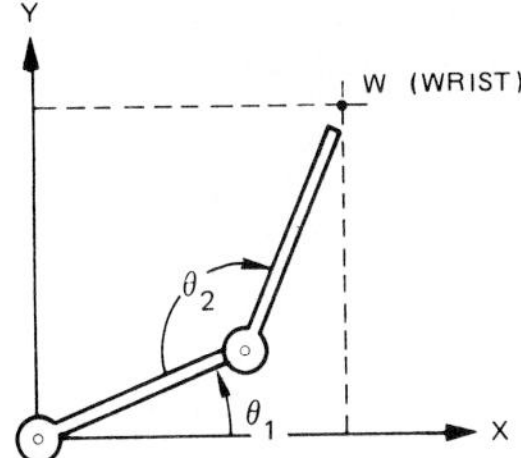

Figure 14.12 A two-joint manipulator.

and $\qquad\qquad -180 < \theta_2 < 0 \qquad$ left arm

Figure 14.13 shows several stages in the mapping of the $\theta_1\theta_2$ space J^2 into the cartesian XY space C^2. The top diagram shows the rectangular region of joint space corresponding to all possible sets of joint angles. The bottom diagram shows the circular region in cartesian space that the arm can reach. The intervening diagrams show how the former can be deformed and folded onto the latter. It is apparent that this mapping entails only a single fold, because there are two layers of J^2 in the bottom diagram. This of course results in the required mapping of two different sets of joint angles into each cartesian wrist position. Note that the line $\theta_2 = 0$ in J^2 becomes the single point at the origin in C^2. The vertical dimension in the bottom diagram does not correspond to a physical coordinate: It represents the choice of arm solution.

For any commercial manipulator arm with six joints, the corresponding diagram would be six-dimensional rather than planar. For an IBM RS1 manipulator, the joint space would be folded once, because it has two arm solutions as a result of its wrist design. For a Stanford Arm, the joint space would be folded twice to produce its four arm solutions. A PUMA manipulator has eight different arm solutions, so its joint space would be folded back on itself three times.

The second diagram from the bottom best illustrates the situation that confronts the control software. At any instant the arm must be in one posture or the other. If we plot the location of the wrist in the XY plane as shown, then the point directly above it on the upper sheet corresponds to the right-arm posture, while the one below it on the lower sheet corresponds to the left-arm posture. Thus, the control software must simply remember on which sheet the arm is currently operating in order to remember its configuration. If it loses this information for some reason (for example, because it was turned off), it has only to check θ_2 to know on which sheet the (simple) arm is currently operating.

In the case of a real six-joint manipulator, the control software would have to make one test for each fold in the mapping to know on which sheet

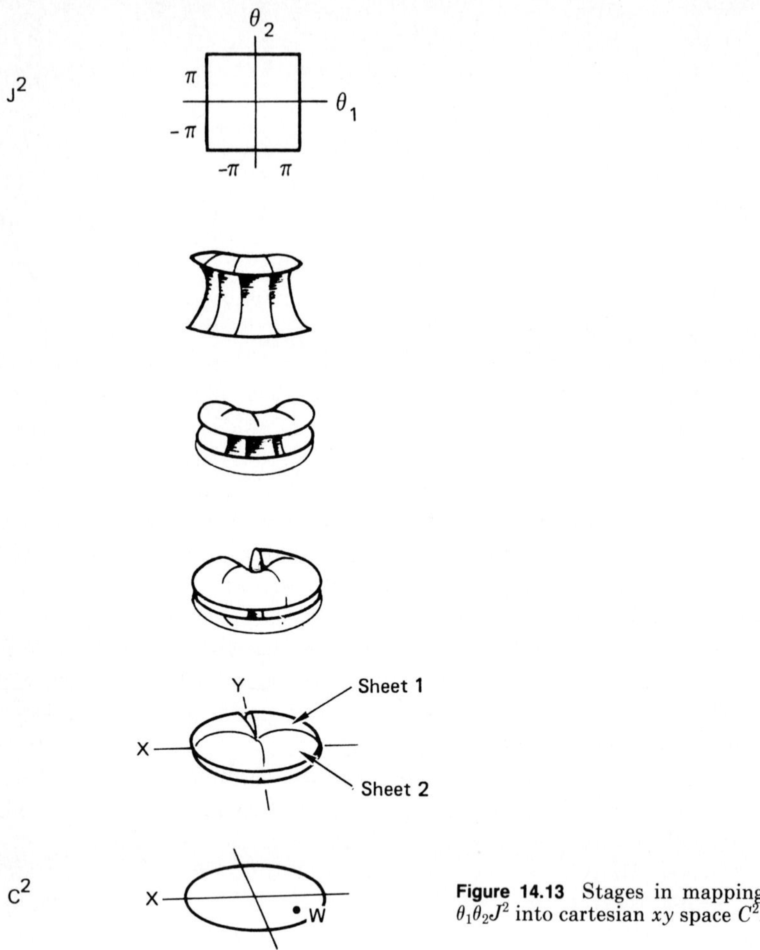

Figure 14.13 Stages in mapping $\theta_1\theta_2 J^2$ into cartesian xy space C^2.

the arm is currently operating. As noted above, a PUMA has eight possible sheets or solutions, so three tests are needed. These tests determine

- Whether the arm is in the "righty" or "lefty" configuration
- Whether the elbow is pointing up or down
- Whether the wrist is in the flipped or unflipped orientation

In most commercial manipulators, the tests required are quite simple. They usually involve measuring a joint coordinate and testing whether it is in a certain range. The result of each test can be stored in one bit of data. Thus, the control software needs to maintain only one bit of posture information for an RS1 manipulator, two for a Stanford Arm, and three

for a PUMA in order to be able to make continuous-path motions
properly.

To illustrate how the control software might make use of the posture
information, suppose that the arm has to make a continuous-path movement in a straight line from position A to position B in Figure 14.14. If
the arm arrives at A in a right-arm configuration, as shown, it must stay
in that configuration until the wrist reaches point B. That means that the
control software must use the arm solution for which θ_2 is positive all along
the path. That is, the path of the arm in the joint space should stay on
the upper sheet as shown. If, however, the arm arrived at point B in a left-
arm configuration, it would have to stay in that configuration all the way
to point B, the control software would have to use the solution for which
θ_2 was negative, and the solution point in joint space would have to move
from A to B along the lower sheet.

Such diagrams are also useful in thinking about configuration changes.
For example, suppose the arm is at position A in a right-arm configuration, as shown in Figure 14.14. How can it change to a left-arm configuration? The general rule is that the operating point of the arm cannot
"jump" from one sheet to another (because then some joint would have
to change its position instantaneously). It can only cross over from the
right-arm sheet to the left-arm sheet at places where the two sheets meet.

The diagram in Figure 14.13 shows that the two sheets meet at two
different places—at the perimeter of the reachable region and at the origin. Therefore, the diagram shows us that the arm can change configurations in two very different ways. One way is to extend itself fully and then
retract in the left-arm configuration. Alternatively, it can fold up until the

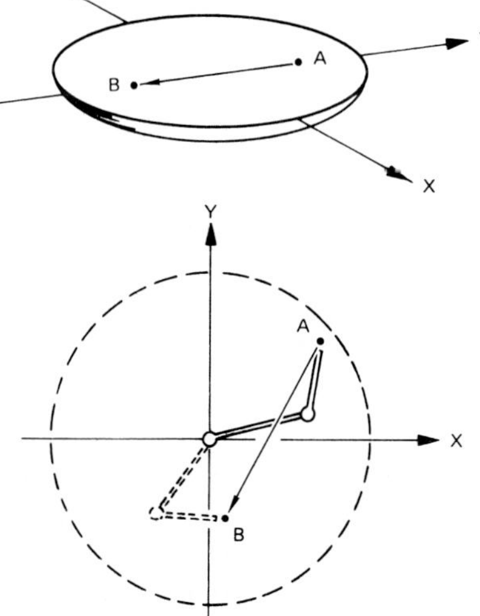

Figure 14.14 Solution for a continuous-path straight-line movement from A to B.

wrist reaches the shoulder joint at the origin and then unfold as a left arm. We can see immediately that if the arm has to change its configuration, it has no choice but to retract or extend itself fully. This is useful information for the control software. It would allow the software to choose the quickest way to switch configuration, for example. Or, if the path to the perimeter should be blocked by an obstacle, the control software could select the other path through the origin.

14.13.7   Effects of joint motion limits

A mapping diagram like this can also represent the effect of limitations on the motion of individual joints. For example, if joint 2 in the example were only able to rotate over a range of, say,

$$-90 < \theta_2 < 180$$

then the lower quarter of the rectangular region in the top diagram would be missing. Consequently, in the bottom diagram the outer portion of the bottom sheet would be missing. That would indicate that wrist positions greater than a certain distance from the origin could only be reached with the arm in a right-arm configuration. With a more severe constraint, say

$$0 < \theta_2 < 180$$

then there would effectively be only one arm solution. Although this is certainly one way to simplify the control problem, it is not very good, because it greatly reduces the arm's dexterity. It is useful to be able to reach for an object in different ways—to avoid obstacles, for example.

The kinematic equations for an arm do not represent joint motion limits. Therefore, the usual method of finding an arm solution is to first find all possible solutions, ignoring the limits. Then each individual solution is checked to see if it violates any of the limits. Solutions that do must be discarded, because the arm cannot be put into the corresponding positions.

The joint motion limits greatly complicate the manipulator control problem. For example, it could be impossible to make a simple straight-line motion with an end effector, even though every point on the path was well within reach, if one of the joints would be driven into one of its limit stops along the way.

This is an important problem. When movements are partly determined by sensory information in a program, it becomes impossible to predict arm movements exactly in order to discover problems with limit stops. In some tasks it may be possible to predict the limits of variation that will be possible in a given motion trajectory. For example, if a fixed camera locates a part to be picked up, then any trajectories to that part will have to terminate in the field of view of the camera. Other tasks may inherently be too variable to permit useful predictions. Simulation and thorough testing may be useful in reassuring the programmer that limit-stop errors will at least be acceptably infrequent. This can be expensive and time-consuming, however, and cannot guarantee that limit-stop problems will never happen during production.

The major cause of limit-stop errors is unpredictable orientation of the parts to be handled. The end effector must often be oriented to match the part's orientation. The inboard joints of a manipulator usually have little or no control over the orientation of the end effector, so variations in workpiece orientation usually require large variations in the wrist joint angles. This makes it very probable that a wrist joint will hit a limit stop in trying to orient the end effector to pick up or put down an object.

There are no good remedies for the problems introduced by limit stops. Continuously rotatable joints (i.e., those that have *no* limit stops) would solve the problem, of course. But such a wrist would probably be very expensive and heavy. Any external cables to the outboard joints or to the end effector would defeat the purpose of such a wrist, because they would get wound up tightly and break. A continuous-rotation wrist joint would require internal slip couplings to carry electric or pneumatic power to the end effector, and power and feedback connections to the outboard joints. A three-roll wrist design, such as the one on the Cincinnati Milacron T-746 arm, might be a good starting point for designing a continuous-rotation wrist. In that design the wrist actuators are all located inboard of the wrist, and hence no slip couplings would be required, at least for joint power and feedback signals. Nevertheless, end-effector connections would still pose a problem.

14.13.8 State of the art

Existing software does not make use of representations such as mappings, although such representations clearly could provide insight into control problems. At most, existing software can stay on a particular solution sheet. It generally does not "reason" about the best way to cross from one sheet to another.

Unimation's VAL controller, for example, checks only the beginning and end points of a continuous-path motion to decide whether the arm will be able to make the motion without a configuration change. However, it is quite possible that a motion may require *two* configuration changes during the motion, ending up with the arm in the original configuration. This will also make the motion impossible, but the VAL controller will not notice the problem. When the arm reaches the point at which the first configuration change is required, the VAL controller will simply stop the arm and abort execution of the program. It is unable to recover by itself.

We have used mapping diagrams merely as an aid to exposition. It is obvious that in commercial arms the mapping is far more complicated, and the control problem is correspondingly more difficult. It would be possible to write a computer program to display the full six-dimensional mapping for a particular arm design. However, the resulting picture might not provide much insight, because only two- to four-dimensional (using time as a fourth axis) cross sections of the surfaces could be shown at any one time. To our knowledge, this has not been attempted, and there are probably some good opportunities for creative approaches.

Advanced manipulator control software could well make use of a computer representation of the mapping—not necessarily related to any graphics approaches—to improve control over current practice. However, this is an advanced topic that cannot be treated here in further detail.

14.13.9 Computational approaches to obtaining arm solutions

Peiper (1968) provided the first comprehensive treatment of the arm-solution problem. He observed that if a manipulator had three consecutive joints with intersecting axes, then it was relatively easy to obtain closed-form symbolic expressions for the joint variables in terms of the link parameters and the wrist position and orientation. He then outlined this method of solution for all possible arm designs that met this criterion. The arm solution for poorly designed arms without intersecting joints requires other methods, e.g., linearization of the equations and successive approximation, or finding roots of a polynomial. The former method is slow; in extreme cases the polynomial can have 64,000 roots to sort through. Thus it is desirable to have intersecting joint axes in an arm design.

To use Peiper's geometric solution technique it is necessary to have some ability to visualize the arm and to think in three dimensions. Paul et al. (1981*a; b*) presented a more systematic method of obtaining an arm solution based on matrix algebra that relies much less on such skills. Using this method, they have been able to obtain arm solutions for each of the popular commercial manipulators in a matter of hours.

The computer programs that compute the arm solution for the Stanford Arm, the model 2000 Unimate, and other popular designs typically require calculation of only a dozen or so trigonometric functions and a few dozen multiplications. Solutions typically run in about 10 ms in integer arithmetic on rather slow 16-bit computers such as a DEC LSI-11. Thus, real-time control of an arm by a computer is practical and cost-effective.

It should be realized that so far no one has developed an algorithm that will *automatically* produce a minimal set of equations for computing joint angles from wrist position, given a particular kinematic design for an arm. Although such an algorithm may be possible, it is not clear that one is needed; new arm designs are not developed very often, and a solution can be developed rapidly with existing techniques.

We first illustrate below the derivation of an arm solution using the geometric approach propounded by Peiper. Then we will derive the same solution using Paul's matrix-algebra approach. We use the Stanford Arm as an example.

14.13.9.1　Geometric derivation of Stanford Arm solution.　At first it may appear that the Stanford Arm will be difficult to analyze because of the 6-in horizontal offset at its shoulder, between the axes of joints 1 and 3. In fact, some arm designers have tried to eliminate this offset, partly in order to simplify the solution and partly to balance the arm a little better. (The Maker robot from U.S. Robots is a notable example of such a design.) A second cause of concern is the complex way in which the three wrist joints interact with one another to produce the final end-effector orientation. A third cause of concern is how to deal with the end effector, which may be of different lengths and may even have a bend in it.

In fact, these problems are not serious. The first step is to remove the effect of the end effector. First we represent the transformation from the pivot point in the wrist to the tip of the end effector by a 4×4 matrix $\mathbf{A}_{\text{tool}}$ called the "tool transform." We can either compute this matrix from blueprints or measurements of the end effector, or we can deduce the transform by positioning the tool tip at a known point.

The arm solution is generally stated in terms of finding joint angles, given the wrist position $\mathbf{T}_6$ rather than the tool tip position $\mathbf{P}$ (see Figure 14.15). However,

$$\mathbf{T}_6 \cdot \mathbf{A}_{\text{tool}} = \mathbf{P}$$

so,

$$\mathbf{T}_6 = \mathbf{P} \cdot \mathbf{A}_{\text{tool}}^{-1}$$

Using this formula, we compute $\mathbf{T}_6$ as the first step in any arm solution.

For the Stanford Arm, the origin of the T_6 frame is at the center of the mechanism of joint 5, where the axes of joints 4, 5, and 6 intersect. Furthermore, the origins of frames T_3, T_4, and T_5 also coincide at this point. We will refer to this point in the following discussions as vector $\mathbf{P}_3$, the fourth column of the $\mathbf{T}_3$ matrix, and we note that it runs from the origin of the base frame to the wrist pivot point at the outboard end of the boom.

The Stanford Arm is an example of a simple arm design, because it has three intersecting axes at its wrist. Because these joints are all rotary, they form an effective ball joint there. This simplifies the arm solution; the arm can be split conceptually into two pieces:

- Everything from the base out to the center of the ball joint
- The wrist

The effect of the inboard joints 1 to 3 is to position this ball joint, which

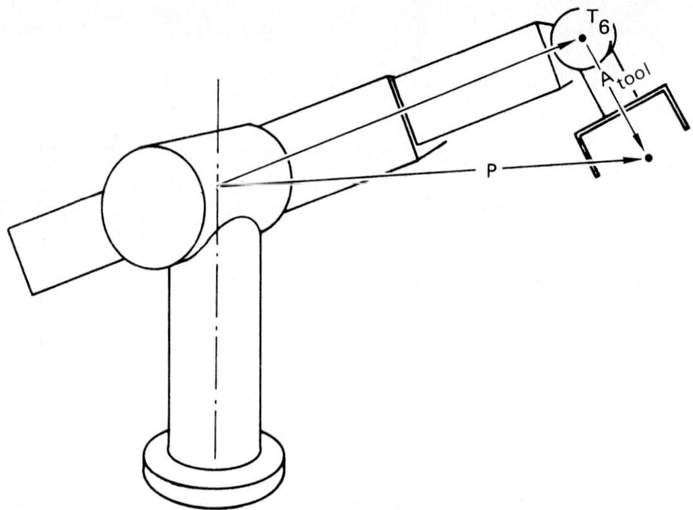

Figure 14.15 Solving for wrist position.

is carried at the end of the boom, and the effect of the wrist joints 4 to 6 is to orient the end effector with respect to the boom.

In brief, the method of solution is as follows. Considering first the portion of the arm inboard of the wrist, we choose the angles of rotary joints 1 and 2 and the extension of linear joint 3 so as to position the end of the boom correctly. We can then use these joint variables in the expressions for the rotational portion of the $\mathbf{T}_3$ matrix to obtain the orientation of the boom. From that and the orientation of joint 6, we can compute the correct angles for rotary joints 4 to 6 in the wrist. The following sections show how these steps are carried out.

14.13.9.1.1 Computing the inboard joint angles (θ_1, θ_2, and d_3). θ_1, θ_2, and d_3 can all be computed from the coordinates of $\mathbf{P}_3$.

Computing d_3. Figure 14.16 shows how to compute d_3. Note that the following three points form a right triangle:

- P_1, at the point of intersection of the axes of joint 1 and 2
- P_2, where the axes of joints 2 and 3 intersect
- P_3

(The plane of this triangle is not necessarily horizontal.) Because P_1 is at the origin, we can compute d_3 using the Pythagorean theorem:

$$d_3 = \sqrt{|P_3|^2 - (6 \text{ in})^2}$$

where $|P_3|$ is the magnitude of P_3, and 6 in is the distance from P_1 to P_2 (the Stanford Arm's shoulder offset).

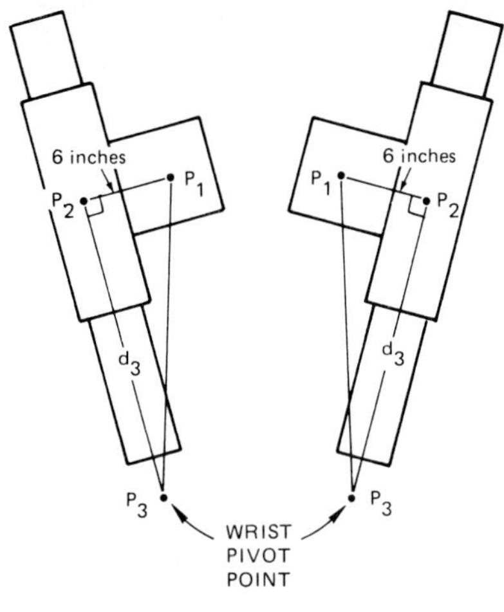

$$d_3 = |\vec{P_3} - \vec{P_2}|$$

$$d_3 = \sqrt{|P_3|^2 - (6\ \text{inches})^2}$$

Figure 14.16 Methods for computing d_3.

At this point, we can sometimes determine that no solution exists. The construction of the arm makes it impossible for the wrist to enter a vertical cylinder of radius 6 in. So, with P_{xy} the horizontal distance of the wrist point from the axis of joint 1, if the desired position and orientation of the particular end effector would require

$$P_{xy} = \sqrt{P_{3,x}^2 + P_{3,y}^2} < 6\ \text{in}$$

we need proceed no further, as the arm solution is infeasible.

The expression for d_3 shows how multiple solutions can arise and how joint motion limits can reduce the number of such solutions that need to be considered. The sign of the root can be either $+$ or $-$, which would give rise to two sets of arm solutions. In the case of the Stanford Arm, however, d_3 must always be positive, because the hand cannot retract completely through the square sleeve of joint 3 and emerge out the back. So here a joint limit eliminates half of the possible arm solutions. If there were no limit on d_3, the Stanford Arm would have eight, not four, different arm solutions!

Computing θ_1.   We can compute θ_1 in terms of d_3 just derived and the two angles α and β shown in Figure 14.17:

$$\theta_1 = \alpha \pm \beta$$

where $\alpha = \text{arctan2}\ (-P_{3,x}, P_{3,y})$
$\quad\quad \beta = \text{arctan2}\ (d_{3,xy},\ 6\ \text{in})$
$d_{3,xy} = \sqrt{d_3^2 - P_{3,z}^2} = $ horizontal component of boom extension

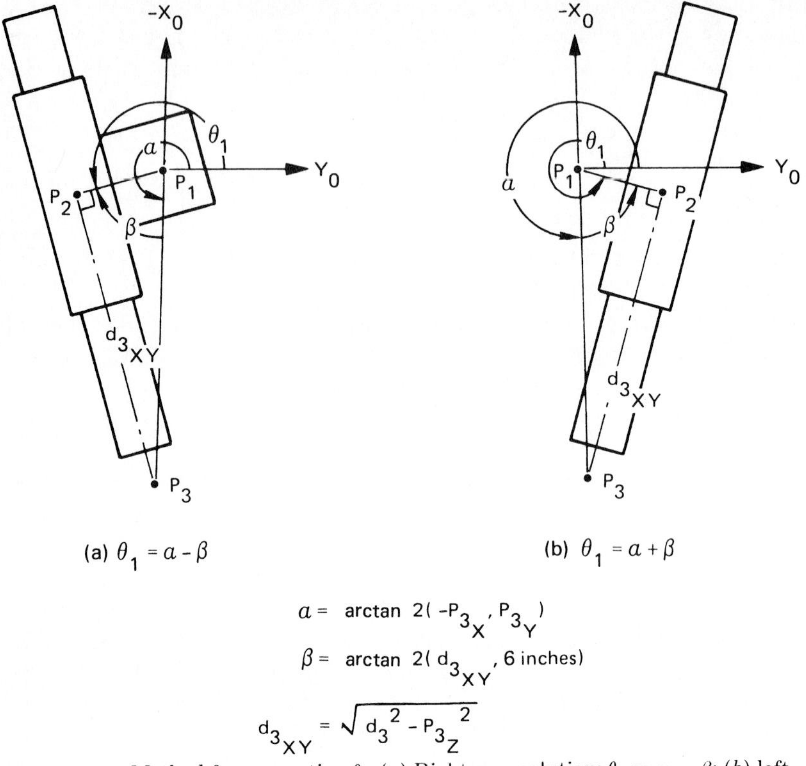

$$a = \text{arctan } 2(-P_{3_X}, P_{3_Y})$$

$$\beta = \text{arctan } 2(d_{3_{XY}}, 6 \text{ inches})$$

$$d_{3_{XY}} = \sqrt{d_3{}^2 - P_{3_Z}{}^2}$$

Figure 14.17 Method for computing θ_1. (a) Right-arm solution: $\theta_1 = \alpha - \beta$; ($b$) left-arm solution: $\theta_1 = \alpha + \beta$.

Again, different sets of arm solutions result from each of the two possible values for θ_1. Adding angle β results in a left-arm configuration; subtracting it gives a right-arm configuration.

These equations also illustrate an important technique: Always use the computer arctangent function of *two* arguments to compute an angle rather than an arcsine, arccosine, or arctangent. One reason is that the latter functions do not return values over the whole range of -180 to $+180°$ in which the angles to be computed may lie. Furthermore, the arcsine is inaccurate around $\pm 90°$, and the arccosine is inaccurate around 0 and 180°. The arctan2 function is defined by the relation

$$\tan^{-1} \frac{Y}{X} = \text{arctan2 } (Y, X)$$

for angles in the domain of the $\tan^{-1}$ function (usually $\pm 90°$). The arctan2 function, however, can also determine when the angle lies outside that range, by checking the signs of X and Y. Furthermore, arctan2 will not cause a divide-by-zero error when the value of X goes to zero, but attempting to use the $\tan^{-1}$ function would.

Computing θ_2.   The angle of joint 2 as shown in Figure 14.18 is measured from the position in which the boom points straight up, and is

$$\theta_2 = \pm \arctan2\,(d_{3,xy},\, P_{3,z})$$

Here, the $+$ sign ($\theta_2 > 0$) is correct for a left-arm configuration, and the $-$ sign ($\theta_2 < 0$) for a right-arm configuration.

14.13.9.1.2 Computing the outboard joint angles θ_4, θ_5, and θ_6. The outboard joint angles are computed by first deriving the orientation of the axis of joint 5 and, from it, the matrix $\mathbf{T}_4$. Then we compute each of the wrist joint angles by calling the arctan2 function with appropriate unit vectors from matrixes $\mathbf{T}_3$, $\mathbf{T}_4$, $\mathbf{T}_5$, and $\mathbf{T}_6$ as arguments.

Because of the way the Stanford Arm's wrist is constructed, the unit vector $\mathbf{a}_4$ along the axis of joint 5 is always normal to the axes of joint 4 and joint 6. Therefore, $\mathbf{a}_4$ must be $\pm$ the cross product of the unit vectors $\mathbf{a}_3$ along axis 4 and $\mathbf{a}_5$ along axis 6. These unit vectors are the third columns of the matrices $\mathbf{T}_3$ and $\mathbf{T}_5$, respectively.

There is no need to work out $\mathbf{T}_5$, however, since its third column is the same as that of $\mathbf{T}_6$, which we derived earlier, because both are unit vectors along the axis of joint 6. For the Stanford Arm, $\mathbf{T}_3 = \mathbf{A}_1 \cdot \mathbf{A}_2 \cdot \mathbf{A}_3$, or

$$\mathbf{T}_3 = \begin{bmatrix} n_{3,x} & o_{3,x} & a_{3,x} & p_{3,x} \\ n_{3,y} & o_{3,y} & a_{3,y} & p_{3,y} \\ n_{3,z} & o_{3,z} & a_{3,z} & p_{3,z} \\ 0 & 0 & 0 & 0 \end{bmatrix}$$

Multiplying out $\mathbf{A}_1$ through $\mathbf{A}_3$,

$$\mathbf{T}_3 = \begin{bmatrix} C_1C_2 & -S_1 & C_1S_2 & C_1S_2d_3 - S_1d_2 \\ S_1C_2 & C_1 & S_1S_2 & S_1S_2d_3 + C_1d_2 \\ -S_2 & 0 & C_2 & C_2d_3 \\ 0 & 0 & 0 & 1 \end{bmatrix}$$

We only need to compute the elements of the y_3 column in this matrix, which is quite simple. Once we have done that,

$\mathbf{a}_4$	$\mathbf{a}_3 \times \mathbf{a}_6$
θ_4	angle from $\mathbf{o}_3$ to $\mathbf{a}_4$ about $\mathbf{a}_3$
θ_5	angle from $\mathbf{a}_3$ to $\mathbf{a}_6$ about $\mathbf{a}_4$
θ_6	angle from $\mathbf{a}_4$ to $\mathbf{o}_6$ about $\mathbf{a}_6$

One way to implement these computations is to write a function subroutine that takes three unit vectors as parameters and returns the angle. This saves space but costs a little time to pass parameters and transfer control. Alternatively, one can write in-line code for each of these com-

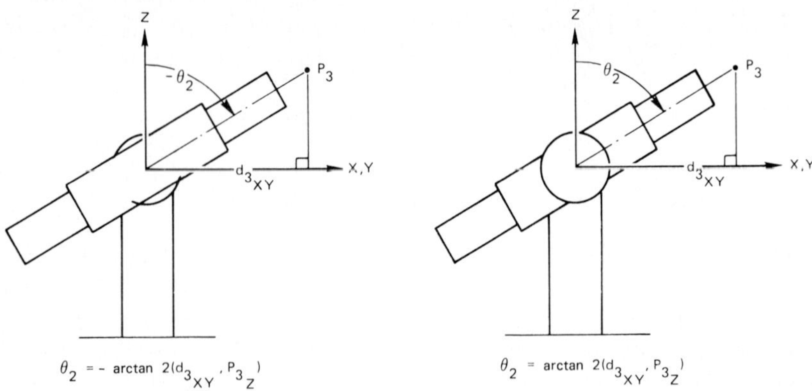

Figure 14.18 Method for computing θ_2.

putations, trading greater memory requirements for higher speed. The function itself can be written in terms of the dot product ($\cdot$), cross product ($\times$), and the arctan2 function

$$\text{angle } (f, t, a) = \text{arctan2} \, (\mathbf{t} \cdot (\mathbf{a} \times \mathbf{f}), \mathbf{t} \cdot (\mathbf{a} \times \mathbf{f} \times \mathbf{a}))$$

where $\mathbf{f}$ = "from" unit vector
    $\mathbf{t}$ = "to" unit vector
    $\mathbf{a}$ = "about" unit vector

and the dot and cross products in the above expressions can themselves be implemented either as subroutines or in-line code.

Note that this function will return the correct value of the angle throughout the range -180 to $+180°$, as long as neither the $\mathbf{f}$ nor the $\mathbf{t}$ vectors are collinear with the $\mathbf{a}$ vector. This will only happen when computing θ_6 for a posture in which the axes of joints 4 and 6 are closely aligned. Such postures correspond to an inherent singularity in the Stanford Arm's wrist and should be avoided when moving the arm.

14.13.9.2 Paul's matrix analysis method for deriving an arm solution.

Briefly stated, the matrix analysis method of obtaining an arm solution is to move forward along the arm from the base to the last link, writing a matrix equation in the frame of each link. Each matrix equation isolates all the inboard joint variables on one side of the equals sign and all the outboard joint variables on the other side. Each equation represents a set of 12 different algebraic equations. One inspects each set to see if one of them can be solved easily for the last inboard joint variable. If so, one then looks for equations that can be solved easily for the first outboard joint variable, using the result obtained for the preceding joint variables.

This method works because the equations are written as if the link in the arm currently under consideration were the base of the manipulator.

The joint at either end of that link then appears to be the first joint in an arm. Because the equations involving the first joint are always simpler than the others, we can solve for that joint variable easily. To illustrate the method, we will find the arm solutions for the Stanford Arm.

The starting point of the method is

$$\mathbf{T}_6 = \mathbf{A}_1 \cdot \mathbf{A}_2 \cdot \mathbf{A}_3 \cdot \mathbf{A}_4 \cdot \mathbf{A}_5 \cdot \mathbf{A}_6$$

$\mathbf{T}_6$ is treated as a known quantity, because it can be computed either numerically or symbolically from the tool transform and the desired position and orientation of the tool, as described above. The problem is to solve for the joint variables, all of which appear in this equation on the right side. So we first transform the equation so that the first joint variable appears only on the left side by multiplying both sides by $\mathbf{A}_1^{-1}$. This gives us our first "stage equation."

$$\mathbf{A}_1^{-1} \cdot \mathbf{T}_6 = \mathbf{A}_2 \cdot \mathbf{A}_3 \cdot \mathbf{A}_4 \cdot \mathbf{A}_5 \cdot \mathbf{A}_6$$

This equation corresponds to a situation in which we are pretending that link 1 is the base of a manipulator. The joints attached directly to this new "base" are joint 1 and joint 2. So, we may expect this equation to help us find θ_1 and θ_2.

In the following, we will use the notation

$${}^n\mathbf{T}_6 = \mathbf{A}_n \cdot \mathbf{A}_{n+1} \cdot \cdot \cdot \cdot \cdot \mathbf{A}_6$$

so that

$$\mathbf{A}_1^{-1} \cdot \mathbf{T}_6 = {}^1\mathbf{T}_6$$

The left side of the stage equation, $\mathbf{A}_1^{-1} \cdot \mathbf{T}_6$ can be written as

$$\mathbf{A}_1^{-1} \cdot \mathbf{T}_6 = \begin{bmatrix} C_1 & S_1 & 0 & 0 \\ 0 & 0 & -1 & 0 \\ -S_1 & C_1 & 0 & 0 \\ 0 & 0 & 0 & 1 \end{bmatrix} \cdot \begin{bmatrix} n_x & o_x & a_x & p_x \\ n_y & o_y & a_y & p_y \\ n_z & o_z & a_z & p_z \\ 0 & 0 & 0 & 0 \end{bmatrix}$$

when $\mathbf{n}$, $\mathbf{o}$, and $\mathbf{a}$ are the unit vectors along the axes of the reference frame embedded in link 6, and $\mathbf{p}$ is the position of the origin of that frame in base coordinates. The next step is to rewrite this equation in terms of three functions $\mathbf{f}_{1i}(v)$, $i = 1, 2, 3$ of a vector $\mathbf{v} = (v_x \quad v_y \quad v_z)$ such that

$$\mathbf{A}_1^{-1} \cdot \mathbf{T}_6 = \begin{bmatrix} f_{11}(n) & f_{11}(o) & f_{11}(a) & f_{11}(p) \\ f_{12}(n) & f_{12}(o) & f_{12}(a) & f_{12}(p) \\ f_{13}(n) & f_{13}(o) & f_{13}(a) & f_{13}(p) \\ 0 & 0 & 0 & 0 \end{bmatrix}$$

By inspection, it is obvious that

$$f_{11}(v) = C_1 v_x + S_1 v_y$$

$$f_{12}(v) = -v_z$$

$$f_{13}(v) = -S_1 v_x + C_1 v_y$$

The next step is to obtain the expression for the right side of the stage equation. For the Stanford Arm, the right side of the first stage equation is

$$
{}^1\mathbf{T}_6 =
\begin{bmatrix}
C_2(C_4C_5C_6 - S_4S_6) - S_2S_5C_6 & -C_2(C_4C_5S_6 + S_4C_6 + S_2S_5S_6C_2C_4S_5 + S_2C_5 & S_2d_3 \\
S_2(C_4C_5C_6 - S_4S_6) + C_2S_5C_6 & -S_2(C_4C_5S_6 + S_4C_6) - C_2S_5S_6S_2C_4S_5 - C_2C_5 & -C_2d_3 \\
S_4C_5C_6 + C_4S_6 & -S_4C_5S_6 + C_4C_6 & S_4S_5 & d_2 \\
0 & 0 & 0 & 1
\end{bmatrix}
$$

Note that in this equation d_2 is a constant (6.0 in), while we must solve for d_3, the extension of joint 3, and the angles for each joint n that appear in the C_n and S_n terms.

Because the stage equation is a matrix equation, corresponding elements can be equated to give 16 algebraic equations. Four of these, the four that arise from equating elements in the fourth row on each side, are trivial. They just say $0 = 0$ or $1 = 1$. The remaining 12 are meaningful. We look for one that can be solved easily for θ_1. The equation that results from equating the elements in row 3, column 4

$$f_{13}(p) = d_2$$

or
$$-S_1 p_x + C_1 p_y = d_2$$

is particularly simple because the only unknown is θ_1. Paul (1981, p. 74) recommends the following transformation to polar coordinates r and α in order to solve an equation of this form:

$$p_x = r \cos \alpha$$

$$p_y = r \sin \alpha$$

where $r = +\sqrt{p_x^2 + p_y^2}$ and $\alpha = \arctan2\,(p_y, p_x)$. r corresponds to the line from P_1 to P_3 in the geometric diagram in Figure 14.17; α is the same as angle α in that diagram.

Making the substitution in the equation obtained from the matrix,

$$\sin \alpha \cos \theta_1 - \cos \alpha \sin \theta_1 = \frac{d_2}{r}$$

Using the trigonometric identity for the difference of two angles,

$$\sin (\alpha - \theta_1) = \frac{d_2}{r}$$

This difference is the complement of the angle β in the geometric construction diagram in Figure 14.17. Because the range of the sine is from -1 to $+1$, and because $r > 0$ and $d_2 > 0$, we obtain the condition

$$r > d_2$$

which corresponds to the feasibility requirement mentioned above in deriving the arm solution by the geometric method. Finally, we can compute θ_1 as follows:

$$\theta_1 = \alpha - (\alpha - \theta_1)$$

$$= \text{arctan2} \,(p_y,\, p_x) - \text{arctan2} \,(\sin (\alpha - \theta_1),\, \cos (\alpha - \theta_1))$$

$$= \text{arctan2} \,(p_y,\, p_x) - \text{arctan} \, 2 \left(\frac{d_2}{r},\ \pm \sqrt{1 - \left(\frac{d_2}{r}\right)^2} \right)$$

The left side of the stage equation can now be considered to be a known quantity, and we can look for elements on the right side that are simple functions of the next joint variable only. The next joint variable is θ_2, and it happens that no elements are functions of it alone. The simplest expressions are to be found in the elements in row 1, column 4, and in row 2, column 4. These only involve θ_2 (being sought), θ_1 (already found), and d_3, the extension of joint 3 (still unknown). Equating those elements to the corresponding elements from the left side, we obtain two equations

$$f_{11}(p) = S_2 d_3$$

$$f_{12}(p) = -C_2 d_3$$

Substituting in the definitions of f_{11} and f_{12},

$$C_1 p_x + S_1 p_y = S_2 d_3$$

$$-p_z = -C_2 d_3$$

Now it is obvious that we can eliminate d_3. However, this elimination will be invalid if $d_3 = 0$. This gives us another insight that was not so obvious in the geometric solution method, namely, there is a singularity at zero extension. That is, θ_2 could then have any value. This is not a problem in practice, because the Stanford Arm has a physical limit stop in joint 3 that prevents d_3 from being less than about 12 in. Dividing one equation by the other to obtain $\tan \theta_2$, and inverting,

$$\theta_2 = \text{arctan2}\,(C_1 p_x + S_1 p_y, p_z)$$

The procedure for determining the remaining four joint variables proceeds in a similar manner. We will not show it in detail here for lack of space but instead will outline it. The next stage equation is

$$\mathbf{A}_2^{-1} \cdot \mathbf{A}_1^{-1} \cdot \mathbf{T}_6 = {}^2\mathbf{T}_6$$

$${}^2\mathbf{T}_6 = \begin{bmatrix} C_4C_5C_6 - S_4S_6 & -C_4C_5S_6 - S_4C_6 & C_4S_5 & 0 \\ S_4C_5C_6 + C_4S_6 & -S_4C_5S_6 + C_4C_6 & S_4S_5 & 0 \\ S_5S_6 & S_5S_6 & C_5 & d_3 \\ 0 & 0 & 0 & 1 \end{bmatrix}$$

and the $f_{2i}(v)$ equations are

$$f_{21}(v) = C_2(C_1 v_x + S_1 v_y) - S_2 v_z$$

$$f_{22}(v) = -S_1 v_x + C_1 v_y$$

$$f_{23}(v) = S_2(C_1 v_x + S_1 v_y) + C_2 v_z$$

Equating the 3,4 elements yields

$$d_3 = S_2(C_1 p_x + S_1 P_y) + C_2 p_z$$

The next useful stage equation is the fourth:

$$\mathbf{A}_4^{-1} \cdot \mathbf{A}_3^{-1} \cdot \mathbf{A}_2^{-1} \cdot \mathbf{A}_1^{-1} \cdot \mathbf{T}_6 = {}^4\mathbf{T}_6$$

in which

$${}^4\mathbf{T}_6 = \begin{bmatrix} C_5C_6 & -C_5S_6 & S_5 & 0 \\ S_5C_6 & -S_5S_6 & -C_5 & 0 \\ S_6 & C_6 & 0 & 0 \\ 0 & 0 & 0 & 1 \end{bmatrix}$$

and the $f_{4i}(v)$ equations are

$$f_{41}(n) = C_4[C_2(C_1 v_x + S_1 v_y) - S_2 v_z] + S_4[-S_1 v_x + C_1 v_y]$$

$$f_{42}(n) = -S_2(C_1 v_x + S_1 v_y) - C_2 v_z$$

$$f_{43}(n) = -S_4[C_2(C_1 v_x + S_1 v_y) - S_2 v_z] + C_4[S_1 v_x + C_1 v_y]$$

This stage equation happens to yield solvable equations for all three of the wrist joint angles, because the last three joint axes in the Stanford Arm intersect at the same point. Equating the 3,3 elements yields

$$\theta_4 = \text{arctan2} \pm ([-S_1a_x + C_1a_y], \pm[C_2(C_1a_x + S_1a_y) - S_2a_z])$$

where corresponding signs are taken, leading to two angles 180° apart. Each solution corresponds to one of the two possible wrist configurations found in the geometric solution method. Furthermore, in this expression it is possible for both of arctan2's arguments to go to zero simultaneously. Geometrically, this corresponds to the singularity that is observed whenever $\theta_5 = 0$, which aligns the axes of joints 4 and 6. It is possible to determine that this happens when $\theta_5 = 0$ entirely from the matrices, without any need for geometrical insight. Examining the 1,3 and 2,3 elements of the stage equation involving ${}^2\mathbf{T}_6$ reveals that the above expression is equivalent to

$$\theta_4 = \text{arctan2}\ (C_4S_5,\ S_4S_5)$$

which obviously has no unique solution when $S_5 = 0$, that is, when $\theta_5 = 0$ or 180°. When this happens, θ_4 can have any value. As we shall see, choosing a value for θ_4 will determine θ_6. This equation also warns us to beware of roundoff errors in calculating θ_4 when S_5 is small, especially if the software performs its calculations in a scaled-integer representation. A floating-point representation may be necessary, because its precision will be constant over a larger dynamic range for a given number of bits, although it will be lower.

Equating the 1,3 and 2,3 elements of the same stage equation (the one in ${}^4\mathbf{T}_6$ above) yields

$$\theta_5 = \text{arctan2}\ (C_4[C_2(C_1a_x + S_1a_y) - S_2a_z] + S_4[-S_1a_x + C_1a_y],$$

$$S_2(C_1a_x + S_1a_y) + C_2a_z)$$

We can also derive the expression for θ_6 from the fourth stage equation

$$f_{43}(n) = S_6$$

$$f_{43}(o) = C_6$$

$$\theta_6 = \text{arctan2}\ (f_{43}(n), f_{43}(o))$$

Paul (1981, pp. 77–78) points out an important practical consideration. It will often be more convenient in manipulator control software to store only two of the first three columns of a 4 × 4 matrix, because one of them is redundant and can be easily recomputed by means of a cross product when required. Except for θ_6, which makes use of the first, or n, column of $\mathbf{T}_6$, the Stanford Arm solution so far has used only columns 2, 3, and 4 (the o, a, and p columns). So for Stanford Arm control it may be advantageous to omit the n column. Paul suggests that if this is the case, a solution for θ_6 that does not involve the n column would be preferable, because it would require one less cross product. Such a solution can be obtained from the fifth stage equation:

$$\mathbf{A}_5^{-1} \cdot \mathbf{A}_4^{-1} \cdot \mathbf{A}_3^{-1} \cdot \mathbf{A}_2^{-1} \cdot \mathbf{A}_1^{-1} \cdot \mathbf{T}_6 = {}^5\mathbf{T}_6$$

for which the $f_{5i}(v)$ functions are

$$f_{51} = C_5\{C_4[C_2(C_1v_x + S_1v_y) - S_2v_z] + S_4(-S_1v_x + C_1v_y)\}$$
$$+ S_5[-S_2(C_1v_x + S_1v_y) - C_2v_z]$$

$$f_{52} = -S_4[C_2(C_1v_x + S_1v_y) - S_2v_z] + C_4(-S_1v_x + C_1v_y)$$

$$f_{53} = S_5\{C_4[C_2(C_1v_x + S_1v_y) - S_2v_z] + S_4(-S_1v_x + C_1v_y)\}$$
$$+ C_5[S_2(C_1v_x + S_1v_y) + C_2v_z]$$

and the right side of the stage equation is

$$
{}^5\mathbf{T}_6 = \begin{bmatrix}
C_6 & -S_6 & 0 & 0 \\
S_6 & C_6 & 0 & 0 \\
0 & 0 & 1 & 0 \\
0 & 0 & 0 & 1
\end{bmatrix}
$$

The sixth joint angle can then be found by

$$\theta_6 = \arctan2\,(S_6,\ C_6)$$

where expressions for S_6 and C_6 in terms of known quantities can be obtained by equating the 1,2 and 2,2 elements of the left and right sides of the fifth stage equation

$$S_6 = f_{51}(o)$$

$$C_6 = f_{52}(o)$$

and vector $\mathbf{o}$, the second column of the $\mathbf{T}_6$ matrix, is known from the tool transform and desired tool position and orientation.

As mentioned above during the discussion of the solution for θ_4, alignment of the axes of joints 4 and 6 in the Stanford Arm makes its solution degenerate. We see that even when this happens, the resulting equations will assign a correct value to θ_6 for whatever value the control software assigns to θ_4. In continuous-path control, for example, the control software would assign whatever the initial value of θ_4 was as the arm approached the singular position, in order to prevent a configuration change.

Research Thrusts

15.1 Robots of the Future

The next generation of robots will be perceptibly different in their appearance and behavior. Robots of the future will have much lighter structures, and the structures will be noticeably more flexible. The limbs will deform under loads and during motion. Robot motion will not be awkward and jerky, but will be considered and graceful. Manipulators will take advantage of the flexing of their limbs rather than fighting the strange motions. They will use flexibility to accomplish tasks not even considered for earlier robots such as throwing and hammering.

These future manipulators will handle loads heavier than themselves, and they will not have to be anchored to the floor to do so. They will be modular so that a different limb can be inserted for a different task and so that repairs can be made easily. The manipulators will ride on light, mobile vehicles. Because of their light weight, they will be cost-effective; partly because of their flexibility, they will be safer in their interactions with people and other equipment and collisions will not be catastrophic. These robots will be found working together with people, on construction and even on routine maintenance, as well as in dangerous and high-payoff situations.

15.2 Limitations of Current Robotic Manipulators

1. Present manipulators have a low payload-to-weight ratio. Figure 15.1 shows a histogram of commercially available manipulators according to catalog information on maximum payload-to-weight ratios. Regardless of size, most manipulators fall in the less-than-10-percent category, i.e., they have a payload that is small compared to their weight. Low payload-to-weight ratios lead to robots that are expensive, heavy, and dangerous. These three negative features have limited robots to the factory environment, where they can be fed work regularly and, therefore, have an economic payoff; where they can be set on a substantial foundation; and where they can be isolated from human workers.

2. Present manipulators must be bolted to a solid foundation. Their control systems are unable to accommodate vibrations from softness in the base and must be "detuned" if these vibrations occur. Whereas manipulators have nearly been excluded from mobile applications on the basis of payload-to-weight ratio, the need for a solid foundation makes mobile application even more unlikely because mobile bases are inevitably softer.

3. Manipulators are designed to have rigid limbs. Rigidity further exacerbates the safety problem, because impacts are less distributed in time. Furthermore, the reliability problem is made worse, because the manipulator is more likely to cause damage to others or to damage itself.

It is of interest to see how these limitations came into being. Early work in robotics concentrated on bringing existing equipment under computer control and adding sensory capabilities. Although there have been advances, e.g., increasing use of electrical control and the SCARA configuration, the mechanical aspects of robots have lagged behind advances in vision and artificial intelligence.

15.3 The Machine Tool Tradition

Robotics evolved from the machine tool tradition. Machine tools are characterized by the ability to hold position independently of force. Humans have a limited ability to hold the position of a tool to cut metal to a predetermined shape; a person will yield when the forces become great. Machine tools, with their solid beds and strong carriages, provided a means to machine precisely shaped parts out of rough castings or forgings. The key feature of this advance was that machine tools could hold their position regardless of the forces applied to them.

Automatic controls were applied to automate machine tools. Sensors

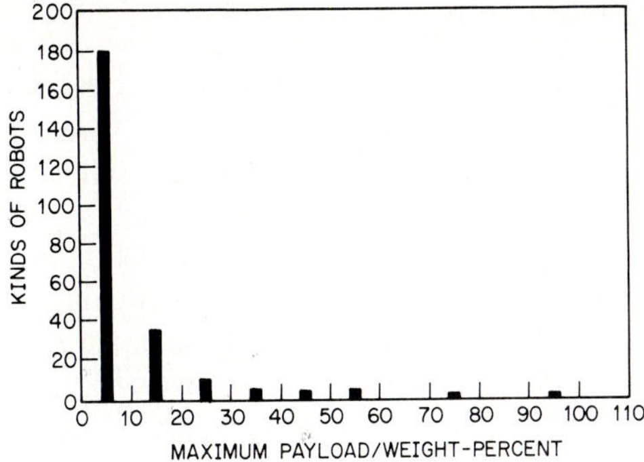

Figure 15.1 Payload-to-weight ratios of available industrial manipulators.

were used to feed back the machine positions for comparison with the desired positions. At first the desired position was conveyed to the numerical control (NC) machines by punched tape. More recently, this information has been conveyed directly by computers; this is called computer numerical control (CNC).

The first manipulators were point-to-point devices without position control (and constitute a large part of the total number of manipulators currently being used). The earliest robotic manipulators capable of going to specified positions had control systems that were direct descendants of the NC machines. The manipulators had rigid limbs whose positions relative to each other were measured and compared to a commanded position using NC-type controllers.

Although the control has become more sophisticated through methods that consider the activity of other joints and that compensate for dynamics, the concept of position control remains. One might even say that it has become embedded in the thinking of those who deal with robotics, a consequence of robotics having evolved from the machine tool tradition.

Both the accuracy and frequency design procedures for the machine tool type of manipulator discussed in Section 3.4 lead to a manipulator that is big and rigid. [Tesar (1985) points out manipulators are not really rigid because they have low structural frequencies compared to most automated equipment, but the manipulators are rigid from a control viewpoint.] They tend to have low payload-to-weight ratios.

The obvious solution is to allow deflections and to compensate for them by replacing manipulator mass with computer capability. This is reasonably straightforward for static deflections, but much more difficult when dynamics must be considered, especially for devices such as manipulators where the dynamics are nonlinear. Several solutions have been offered for particular linear cases (Bauldry et al., 1983; Schaecter, 1982; Cannon and Schmitz, 1984), but the general case is considered to be beyond the state of the art (Firshein and Georgeff, 1985).

15.4   Trends in Manipulator Development

The need for compliance (i.e., for force control) was discussed in Section 2.2. Compliance can be located in the end effector (as with the RCC), in the joints (as with the SCARA), in the limbs, or in the base. Compliance is usually built into the end effector; in this way a conventional, rigid, position-controlled robot can be made to behave with force control. However, it is preferable to use available compliance in the components of the manipulator rather than to have to add it later. Compliance is usually undesirable in the base because there is too much mass between the manipulator endpoint and the base for the compliance to be effective in dynamic situations. The ideal location for compliance is in the joints and the limbs.

Figure 15.2 shows the combinations of compliance in the joints and in the limbs. As mentioned before, the earliest robots had evolved from the machine tool tradition; they had rigid limbs and stiff joints (non-backdrivable), either because they were hydraulic or because they were highly geared. (The reference to compliance in the joint refers to deliberate compliance rather than to natural compliance as from a spring in a gear train.) Most robots still fall into this category. The next step historically has been to use back-drivable joints so that compliance can be electronically controlled. The extreme of this technology is the direct-drive device; the motor is connected directly across the joint with no gearing. This type of manipulator has the advantage of being fast but typically has an even lower payload-to-weight ratio because of the large motors required. Robots which have a combination of rigid limbs and compliant joints are a rapidly growing class; the SCARA robots belong to this class.

Research on compliant limbs began with Book (1975) and Dubowsky (1977). The study of compliant limbs has become an increasingly popular research topic (Truckenbrodt, 1979, 1980; Balas, 1978; Fukuda and Yutaka, 1983; Futami et al., 1983; Judd and Falkenberg, 1983; Hudgins and Perreira, 1982; Hardt and Zalucky, 1982; Kamiya et al., 1981; Mendelson and Rinderle, 1984; Alberts et al., 1985). Researchers are concerned with the deflection of limbs due not only to heavy loads but also to rapid motion. The stiffest of robots becomes flexible when it is moving rapidly. Often structural resonance problems are initiated because of the compli-

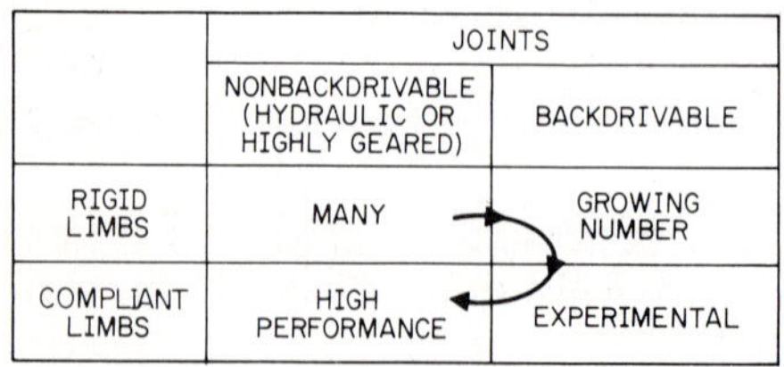

Figure 15.2 Compliance combinations. Arrow indicates trend.

ance in the mounting. Several research manipulators have been built where the flexibility has been exaggerated for ease of study. Usually researchers have used joints with electronically controlled compliance.

There are only a few robots whose joints are stiff but whose limbs are compliant, the final category. An example is Professor Book's flexible limb manipulator (Book, 1986) that has hydraulic actuators. In this manipulator, a servo loop around the joint allows the actuator to behave as if it were back-drivable.

It is likely that future high-performance (at least in terms of payload-to-weight) manipulators will have their compliance in the limbs. In this way, both the limbs and the joints are selected to have the highest individual performance. Highly geared joints have a higher payload-to-weight ratio than direct-drive motors; limbs that are allowed to deflect similarly have a higher performance.

15.5 Control

Manipulator control for the most part has been predicated on position control and, in particular, on the separation of planning and execution, where execution has been assumed to be primarily of commanded positions. Most research on the control of manipulators with flexible limbs accepts these conditions.

15.5.1 Approaches in the literature

Several of the flexible robot control strategies described in the literature take similar approaches. The manipulator links are modeled as a Euler-Bernoulli beam, the first few vibration modes of which are considered important (Book, 1975, 1979; Cannon and Schmitz, 1984; Fukuda and Yutaka, 1983, 1984; Futami et al., 1983). The additional state variables which describe the bending modes are included with those which describe the rigid link motion of the manipulator, forming a more complete model of the manipulator. The governing differential equations are typically high-order and coupled. Small deflection assumptions and other simplifications are used to simplify and linearize. A modern control theory approach is often taken to design a suitable feedback control law (Book et al., 1975; Book, 1979; Balas, 1978; Cannon and Schmitz, 1984; Fukuda and Yutaka, 1984; Futami et al., 1983; Truckenbrodt, 1979).

The control laws typically require full state feedback (information on the bending modes of the beams). Most manipulators are equipped only with position and velocity sensors colocated with the joint actuators. Additional sensing necessary to provide the bending information must be added. Researchers have taken different approaches to obtaining bending information.

Cannon and Schmitz (1984) advocate the use of independent endpoint sensing, an optical sensor to follow a light on the tip of the manipulator. With such an approach, it is necessary to use state estimators (observers) in order to provide all the necessary feedback information. This approach requires an accurate model of the manipulator dynamics.

More direct state information on the bending of a link can be obtained by using strain gauges (Fukuda and Yutaka, 1984). One measurement is required for each bending mode considered, and it is best to locate gauges at a bending node for that particular mode. Balas (1978) argues that it is often difficult to locate the sensors at the precise locations desired because of structural considerations and the accuracy required. He suggests that bandpass filters be used to prefilter the sensor data and to discern information on specific modes. He recommends phase-locked loops for the filters since they can lock onto a particular frequency even if it is not exactly known.

Futami et al. (1983) used feedback from accelerometers located on the moving links of the three position axes of a five-axis commercial manipulator. This research builds upon an existing classical control design which has a current feedback loop internal to a velocity feedback loop which in turn is internal to a position feedback loop. Acceleration feedback is used to effectively provide viscous damping of the bending of the links in a velocity control loop and thereby improve the overall positioning performance of the manipulator.

Of the research described above, Cannon and Schmitz, Fukuda and Yutaka, and Futami et al. include experimental results (tests on actual manipulators) in their work. In each of these three cases, significant improvement in manipulator performance was achieved by including flexibility effects in the control law. Cannon and Schmitz used a single-link manipulator, which lacks the dynamic complexity of a manipulator in general. Fukuda and Yutaka have demonstrated what could be considered control of a true flexible manipulator, in this case a two-axis serial-link device with three significant bending modes in each link. Futami et al. applied their approach to a manipulator of commercial design which displayed fairly good behavior under normal operation even without the acceleration feedback.

Cannon and Schmitz, and Book warn that inaccuracies in modeling the manipulator flexibility can seriously degrade manipulator performance even to the point where ignoring the flexibility would provide better performance. This fact is important since, in general, manipulators would be difficult to accurately model and might also grasp payloads of unknown inertia which could significantly affect their dynamics. Fukuda and Truckenbrodt suggest that adaptive control can be used to compensate for such inaccuracies. Unfortunately, the practical application of adaptive control to robotic manipulators, even rigid manipulators, is still in its infancy (Dubowsky and Kornbluh, 1984).

A quite different route to manipulator control is the "variable structure system with sliding modes" (Utkin, 1971, 1972). The basic concept of variable structure control is the use of more than one control law, none of which may be by itself stable. Switching from one control law to another takes place under sensor guidance in various parts of the phase plane to ensure that a stable combination of control modes is used. Although a sliding mode has not been proposed specifically for a flexible manipulator, it has been used in manipulator control (Young, 1978; Morgan and Ozguner, 1985; Klein, 1979). Although there is no reason to think that sliding mode control would be advantageous (allowing for more rapid motion) for the flexible robot, it is an applicable approach.

15.5.2   An alternative approach

The benefits of feedback control are the ability to cope with unexpected disturbances and to respond to errors and eventually reach a desired endpoint. If one is for the moment willing to give up these advantages, other strategies for control become available that have other advantages. In particular, prescribing joint torques allows one to eliminate frequencies that excite resonances. Elimination of feedback removes the controller bandwidth to structural frequency consideration.

Consider an operator who has the task of programming a robot to do a particular repetitive task in automated production. The operator programs the robot with several positions that are automatically smoothed into a path by the addition of several points. At first the path program is executed slowly for verification. Then it is run increasingly faster as the operator brings the speed up to the "line speed" to coordinate with other equipment. Further speed increase will be required if the line speed is increased.

To achieve the desired coordinated motion of robot joints specified by the programmer through path selection, joint actuators often must apply rapidly changing torques. As speed is increased, the torques become more pronounced and at some point reach the limits of the actuator. Furthermore, as speed is increased, the torque spikes are accentuated, stimulating vibration in transmissions, limbs, and foundation. Because of interactions with the control system, the actuator torques not only change scale but become more ragged. Increases in speed amplify the problem.

To achieve further speed increases, the operator must rework the program using the limited variables available—more careful path selection and perhaps control system adjustments (gains). What the operator is doing is attempting to smooth the actuator torques by a better path selection. However, since the path must be differentiated twice to obtain torque and since differentiation roughens the torque, a smooth torque is difficult to obtain.

A more sensible approach would appear to be to deal with the joint torques directly. Smoothing might be done by eye (when displayed) or by filtering the signal by any of a number of algorithms. One method, for example, might be to take the Fourier transform of the signal, eliminate undesired frequencies (those close to resonances), and then take the inverse transform. The modified torque functions would then be played back to the robot open loop so that the programmer can observe the behavior.

The robot is unlikely to go to the desired motion or process endpoint. The torque functions must be adjusted over repeated trials until the exact process endpoint is reached. It should be noted that this is using the information in a trial to affect the choice of torque in the next trial. It is not feedback in the usual sense, but it is feedback around the process as suggested in Figure 15.3.

One possible method of adjusting the torque is to add torque time functions that can be described by a limited number of variables. For example, a cycle of a sine function is completely described by its amplitude, period, and starting time. A set of values describing the torque addition can be used to alter the torque in well defined ways in tests so that the influence of changing values can be determined. Best values can be chosen so that the final position is achieved to arbitrary accuracy. An algorithm that is applicable to six joints with at least two flexible limbs should be possible. In a sense, the manipulator through repeated motions could learn to adjust the right values.

The chosen torque could then be used in a feedforward mode, together with the path that it produced, as a position-control override in a conventional feedback control system as shown in Figure 15.4. The open-loop control may be the control method for athletes doing high-performance activities. Athletes learn a groove, a coordinated sequence of muscle firings. They adjust their grooves to their condition on a particular day through the process of warming-up.

15.6   Kind of Compliance

Introduction of compliance into the limbs introduces a set of variables. Compliance will involve not just how much a limb deflects under force or torque, but *how* it deflects as well. As pointed out in Section 2.2, the deflection from a force or torque may not be just in the direction of the

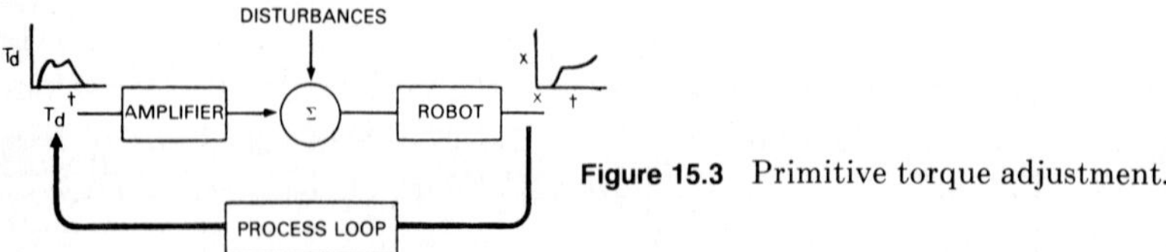

Figure 15.3 Primitive torque adjustment.

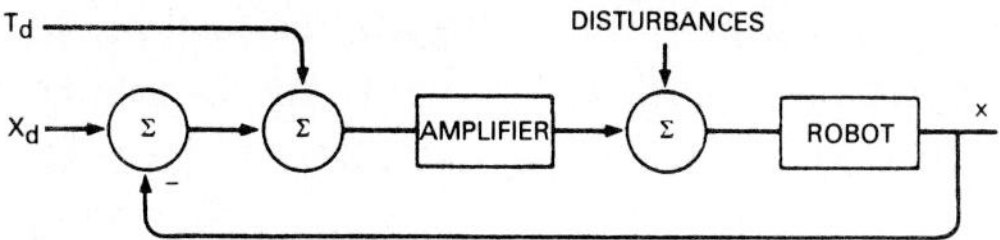

Figure 15.4   Dual control.

application. Force in one direction can produce deflection in the orthogonal direction and twists.

When the manipulator is following a surface, the incorrect beam compliance can cause a jam. For example, a jam will occur if the manipulator hits a step and deflects into the surface being followed rather than away from it. The jam could be detected actively by the control system, and perhaps damage could be prevented. However, passive compliance that would prevent jamming in the first place would be a superior design choice.

The two beam designs shown in Figure 15.1 illustrate another kind of compliance issue. The different beam designs have very different rotations in response to the same applied couple and moment. Limbs made by the two beams would have to be treated differently by the manipulator control system to correct for the position of other limbs. Thus the introduction of compliance has introduced the entirely new issue of the nature of compliance into manipulator research.

15.7 Tasks

The introduction of compliance in the limbs opens up applications that have been difficult or not even considered for manipulators before.

Pushing.   Pushing would seem to be possible by controlling either the force or the position. However, pushing by position can lead to high forces in the event of jams, and pushing by force alone can lead to run-

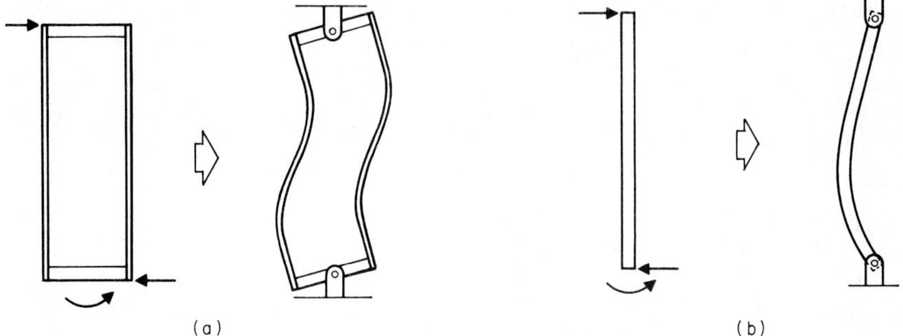

Figure 15.5   Pinned beam response. (*a*) Parallel element construction; (*b*) ordinary beam.

aways. Some compliance is clearly desirable. Pushing is the robot manipulation that is intuitively more satisfactorily achieved with force rather than with position control.

Hammering.   Hammering is a momentum-to-impulse exchange that needs to take place independently of position. That is, the part being hammered may deform or not. The amount of that deformation should not have to be anticipated to allow hammering; therefore, freedom of motion to the extent of the uncertainty must be allowed.

Contour following.   Contour following implies that a stiffness step is generated in resistance to motion into the cam being followed at the contour surface. The forces involved can go from small values to very large values for small positional differences without compliance.

Constrained motion.   The constrained-motion problem is similar to contour following; large forces are generated for small position errors in the constrained directions. Compliance is required in the constrained direction to keep forces within machine tolerances reasonable. Examples of the constrained-motion problem are turning a doorknob, opening a door, opening a drawer, turning a crank, pushing a button, or throwing a switch.

Shaking hands.   "Shaking hands" is the term given to the problems of machine cooperation. If two manipulators attempt to pick up a part together, the position of the part is overconstrained. Each robot performs a constrained-motion problem relative to the other. The problem is reduced to the constrained motion, and the need for compliance is clear.

Throwing.   There are two reasons that rigid robots are not well suited to throwing. The first is an impedance matching issue. This is best illustrated by the casting rod, the means to match the impedance between a person fishing and the lightweight lure. The second reason is more complex and has to do with the process. Throwing can be thought of as releasing an object at a given position at a given velocity vector. In reality, the end result can be achieved from an infinite variety of positions with the corresponding velocity vector. Control of position is not the most desirable way to describe a throwing motion, because it restricts positions and velocity vectors which might be chosen and which may be superior for a variety of other reasons, including limb compliance.

References and Bibliography

Agin, G. J., 1979: "Real-Time Robot Control with a Mobile Camera," *Robotics Today,* Fall, p. 35.

Alberts, T. E., S. L. Dickerson, and W. J. Book, 1985: "Modeling and Control of Flexible Manipulators," *Robots9 Conference Proceedings,* June, Detroit, vol. 1, pp. 1-59 to 1-73.

Albus, J. S., 1975*a:* "A New Approach to Manipulator Control: The Cerebellar Model Articulation Controller (CMAC)," *J. Dynamic Systems, Measurement and Control, Trans. ASME,* **97**(3), 220–227.

Albus, J. S., 1975*b:* "Data Storage in the Cerebellar Model Articulation Controller (CMAC)," *J. Dynamic Systems, Measurement and Control, Trans. ASME,* **97** (3), pp. 228–233.

Allan, R., 1985: "Nonvision Sensors," *Electronic Design,* **33:**(15), 103–115.

Asada, H., T. Kanade, and I. Tateyama, 1982: "Control of Direct-Drive Arm," Carnegie-Mellon University Robotics Institute Publications, prepared for the Office of Naval Research, NTIS ADA 144969.

Bahr, A. B., 1986: "Electromagnetic Sensor Arrays—Experimental Studies," in *Review of Progress in Quantitative Nondestructive Testing,* vol. 5*a,* D. O. Thompson and D. E. Chimenti, eds., Plenum, New York, pp. 691–698.

Balas, M., 1978: "Feedback Control of Flexible Systems," *IEEE Trans. Automatic Control,* **AC-23,** 673–679.

Ballard, D. H., and C. M. Brown, 1982: *Computer Vision,* Prentice-Hall, Englewood Cliffs, New Jersey.

Bauldry, R. D., J. A. Breakwell, et al., 1983: "A Hardware Demonstration of Control for a Flexible Offset-Feed Antenna," *J. Astronautical Sciences,* **31:**(3), 455–470.

Bejczy, A. K., 1974: "Robot Arm Dynamics and Control," NASA-JPL Technical Memorandum 33-669, Jet Propulsion Laboratories, Pasadena, California, February.

Belforte, G., N. D'Alfio, and A. Romiti, 1982: "A Self-Adaptive Guided Assembler (SAGA)," *Robots VI,* p. 544.

Belytschko, T., 1974: "Transient Analysis," in *Structural Mechanics Computer Programs,* W. D. Pilkey, K. Saczalski, and H. Schaeffer, eds., University Press of Virginia, Charlottesville, pp. 254–276.

Bernin, Victor M., 1976: "Positional Transducer Utilizing Magnetic Elements," U.S. Patent 3,958,203, May 18.

Binford, T. O., et al., 1977: "Exploratory Study of Computer Integrated Assembly Systems," Progress Report 4, Stanford Artificial Intelligence Laboratory, Computer Science Department, Stanford University, Stanford, California, June.

Blanchard, M., 1976: "Digital Control of a Six-Axis Manipulator," Working Paper 129, MIT Artificial Intelligence Laboratory, Cambridge, Massachusetts.

Bolles, R., and R. P. Paul, 1973: "The Use of Sensory Feedback in a Programmable Assembly System," Report AIM-220, Stanford Artificial Intelligence Laboratory, Computer Science Department, Stanford University, Stanford, California, October.

Book, W. J., 1979: "Analysis of Massless Elastic Chains with Servo Controlled Joints," *J. Dynamic Systems, Measurement and Control, Trans. ASME,* **101,** (3), 193–200.

Book, W. J., 1986: "Modeling and Control Problems in Light-Weight, High-Speed Robot Arms," Presentation at *Symposium on Kinematics, Dynamics and Control of Mechanisms and Manipulators,* Rensselaer Polytechnic Institute, Troy, New York, June.

Book, W. J., O. Maizza-Neto, and D. E. Whitney, 1975: "Feedback Control of Two Beam, Two Joint Systems with Distributed Flexibility," *J. Dynamic Systems, Measurement, and Control, Trans. ASME,* **97:**(2), 424–431.

Bowman, Joseph, Henry Eikelberger, Jeffrey Rand, and Arnold M. Miller, 1978: "Contactless Linear Position Sensor," U.S. Patent 4,088,977, May 9.

Buckingham, Earle, 1949; *Analytical Mechanics of Gears,* McGraw-Hill Book Company, Inc., New York, New York.

Buckley, S., 1985: "Automation Sensing Mimics Organisms," *Sensors,* June.

Campbell, P. and K. Wasson, 1986: "Plastic Ferrite Magnets Ease Assembly and Cost Constraints for Motors and Encoders," *Powerconversion & Intelligent Motion,* Intertec Communications, Inc., Ventura, California, July.

Cannon, R. H. Jr., and E. Schmitz, 1984: "Initial Experiments on the End-Point Control of a Flexible One-Link Robot," *International J. Robotics Research,* **3**(3), 62–75.

Carmichael, Colin, ed., 1950: *Kent's Mechanical Engineer's Handbook,* Design and Production Volume, 12th ed., John Wiley & Sons, Inc., New York.

Chironis, N., 1960: "Harmonic Drive," *Product Engineering,* **31,** February 8, pp. 47–51.

Cushing, I. B., 1985: "A New High Accuracy Angular Position Transducer," *Powerconversion & Intelligent Motion,* Intertec Communications, Inc., Ventura, California, September.

Drake, S., 1977: "High Speed Robot Assembly of Precision Parts Using Compliance Instead of Sensory Feedback," *Proceedings, 7th International Symposium on Industrial Robots,* October 19–21, Japan Industrial Robot Association, Tokyo, pp. 87–98.

Drake, S. H., R. M. Spencer, and S. N. Simunovic, 1979: "Using Compliance in Assembly—An Engineering Approach to Float," Technical Paper MS79-873, Computer and Automated Systems Association of SME, Dearborn, Michigan.

Drives & Controls International, 1982: **2**(1), February.

Dubowsky, S., 1981: "On the Adaptive Control of Robotic Manipulators: The Discrete-Time Case," *Proceedings, 1981 Joint Automatic Control Conference,* Charlottesville, Virginia, June.

Dubowsky, S., and D. DesForges, 1979: "The Application of Model-Referenced Adaptive Control to Robotic Manipulators," *J. Dynamic Systems, Measurement and Control, Trans. ASME,* **101**(3), 193–200.

Dubowsky, S., and T. N. Gardner, 1977: "Design and Analysis of Multi-Link Flexible Mechanism with Multiple Clearance Connections," *ASME J. Engineering for Industry,* **99**(1), pp.88–96.

Dubowsky, S., and R. Kornbluh, 1984: "On the Development of High Performance Adaptive Control Algorithms for Robotic Manipulators," *Proceedings of the 2nd International Symposium on Robotics Research,* Kyoto, Japan, MIT Press.

Eckerle, J. S., and P. M. Newgard, 1977: "A Noninvasive Transducer for the Continuous Measurement of Arterial Blood Pressure," *Proceedings, AAMI 12th Annual Meeting,* Association for the Advancement of Medical Instrumentation, March 13–17, Arlington, Virginia.

Edel, M., and D. Jolly, 1982: "Conception and Realization of an Intelligent and Autonomous End Effector," *Robots VI,* p. 552.

Engelberger, J. F., 1980: *Robotics in Practice: Management and Applications of Industrial Robots,* Kogan Page in association with Aveberry Publishing Company, London.

Farrand Controls, 1981*a: Inductosyn Precision Linear and Rotary Position Transducers,* Farrand Controls, 99 Wall Street, Valhalla, New York.

Farrand Controls, 1981*b: The Farrand High Gain Inductosyn Transducer,* Farrand Controls, 99 Wall Street, Valhalla, New York.

Finkel, R., et al., 1975: "An Overview of AL, A Programming Language for Automation," *Proceedings, Fourth International Joint Conference on Artificial Intelligence,* Tbilisi, Georgia, U.S.S.R., pp. 758–765.

Firschein, O., and M. P. Georgeff, 1985: "Review of Automation Technologies Applicable to the National Space Program," *AIAA/NASA Symposium on Automation, Robotics and Advanced Computing for the National Space Program,* Washington, D.C., September.

Fukuda, T., and K. Yutaka, 1983: "Precise Positioning Control of Flexible Arms with Reliable Control System," *Proceedings of the International Conference on Advanced Robotics,* Science University of Tokyo, September 12–13, pp. 237–244.

Fukuda, T., and K. Yutaka, 1984: "Flexibility Control of Elastic Robotic Arms and Its Application to Contouring Control," *Proceedings of the International Conference on Robotics,* Science University of Tokyo, March 13–15, pp. 540–545.

Futami, S., N. Kyura, and S. Hara, 1983: "Vibration Absorption Control of Industrial Robots by Acceleration Feedback," *IEEE Trans. on Industrial Electronics,* **IE-30**(3), 229–305.

Genz, Earl James, 1978: "Linear Position Sensor," U.S. Patent 4,121,185, October 17.

Girod, G. F., 1982: "Utilization of Microprocessors in Robot Automation Systems," *Robots VI,* p. 565.

Guckel, H., and D. W. Burns, 1984: "Planar Processed Polysilicon Sealed Cavities for Pressure Transducer Arrays," IDEM-84, December, p. 223.

Hackwood, S., G. Beni, L. A. Hornak, R. Wolfe, T. J. Nelson, 1983: "A Torque-Sensitive Tactile Array for Robotics," *International J. Robotics Research,* **1**(2), 46–61.

Harmonic Drive Division, 1979: *Responsyn Fine Angle Stepping Motors and Digital Drive Systems,* USM Corporation, November.

Heginbotham, W. B., M. Dooner, and K. Case, 1979: "Rapid Assessment of Industrial Robots Performance by Interactive Computer Graphics," *Proceedings of the 9th International Symposium on Industrial Robots,* March 13–15, Washington, D.C.

Highway Safety Research Institute, 1977: *Modular Program Development for Vehicle Crash Simulation,* vol. I: *Summary Report, Test Results and Theory,* prepared for the U.S. Department of Transportation, Contract DOT-HS-4-00855, Michigan University, Ann Arbor, Michigan, August.

Hollerbach, J. M., 1980: "A Recursive Lagrangian Formulation of Manipulator Dynamics and a Comparative Study of Dynamics Formulation," *IEEE Trans. Systems, Man and Cybernetics,* **SMC-10**(11), November, 730–736.

Hooker, W. W., and G. Marguillies, 1965: "The Dynamic Attitude Equations for an n-body Satellite," *J. Astro. Sci.,* **12,** 123–128.

Hurty, W. C., 1965: "Dynamic Analysis of Structural System Using Component Modes," *AIAA Journal,* **3**(4), 678–685.

IRI, 1982*a: Industrial Robots International,* Technical Insights, Inc., Fort Lee, New Jersey, May 24, p. 4.

IRI, 1982*b: Industrial Robots International,* Technical Insights, Inc., Fort Lee, New Jersey, June 14, p. 4.

IRI, 1982*c: Industrial Robots International,* Technical Insights, Inc., Fort Lee, New Jersey, June 28, pp. 4, 5, 8.

IRI, 1982*d: Industrial Robots International,* Technical Insights, Inc., Fort Lee, New Jersey, September 13, pp. 2, 3.

IRI, 1983*a:* "Visiting Robot Labs, You May Notice a Second Commercial Tactile Sensor," *Industrial Robots International,* Technical Insights, Inc., Fort Lee, New Jersey, November 28.

IRI, 1984: "Small, Cool Direct Drive," *Industrial Robots International,* Technical Insights, Inc., Fort Lee, New Jersey, June 25.

IRI, 1984*a:* "Tactile Sensors on Chips," *Industrial Robots International,* Technical Insights, Inc., Fort Lee, New Jersey, July 9.

IRI, 1985: "Tactile Sensor for Robot Palm," *Industrial Robots International,* Technical Insights, Inc., Fort Lee, New Jersey, August.

Jacobsen, S. C., E. K. Iversen, D. F. Knutti, R. T. Johnson, and K. B. Biggers, 1986: "Design of the Utah/M.I.T. Dexterous Hand," *Proceedings of IEEE International Conference on Robotics and Automation,* San Francisco, **3,** 1520–1532.

Jones, Franklin D., ed., 1930: *Ingenious Mechanisms for Designers and Inventors,* vols. 1–4, Industrial Press Inc., New York, New York.

Judd, R. P., and D. R. Falkenburg, 1983: "Control Strategy for Flexible Robot Mechanism," *Proceedings of the Fifteenth Southeastern Symposium on System Theory,* Oakland University, March 28–29, pp. 278–281.

Kinnucan, Paul, 1981: "Robots, Developers Aim for 3-D Vision, Sense of Touch, Better Dexterity, and Rudimentary Reasoning," *High Technology,* September/October, p. 37.

Klein, C. A., and J. J. Maney, 1979: "Real-Time Control of a Multiple-Element Mechanical Linkage with a Microcomputer," *IEEE Trans Industrial Electronics and Control Instrumentation,* **IECI-26**(4), 227–234.

Korn, G. A., 1978: *Digital Continuous System Simulation,* Prentice-Hall, Englewood Cliffs, New Jersey.

Lane, J. D., 1979: "Applications of the Remote Center Compliance Device to Assembly Operations," SME Paper MS79-872.

Lechtman, H. et al., 1982: "Connecting the PUMA Robot with the MIC Vision System and Other Sensors," *Robots VI,* p. 447.

Luh, J. Y. S., 1983: "Conventional Controller Design for Industrial Robots—A Tutorial," *IEEE Trans. on Systems, Man, and Cybernetics,* **SMC-13**(3), 298–316.

Luh, J. Y. S., and E. Kinne, 1965: "On the Stability of a Sampled-Data Control System with High-Frequency Resonant Poles," *IEEE Trans. Automatic Control,* January, 107–116.

Luh, J. Y. S., M. W. Walker, and R. P. C. Paul, 1980: "Resolved-Acceleration Control of Mechanical Manipulators," *IEEE Trans. Automatic Control,* **AC-25**(3), June, 468–474.

Luneau, J. R., 1986: "A New Feedback Device for Brushless DC Servo Systems," *Powerconversion & Intelligent Motion,* Intertec Communications, Inc., Ventura, California, September.

Machine Design, 1978: "Motors," **50**(11), May 18, pp. 55–60.

Machine Design, 1984: "Motor Improves Robot Performance," **56**(16), 42.

Maiorca, Philip P., and Norman H. MacNeil, 1979: "A Closed-Loop System for Smoothing and Matching Step Motor Responses," *Hewlett-Packard Journal,* Hewlett-Packard Company, Palo Alto, California, February.

Makino, H., and N. Furuya, 1980: "Selective Compliance Assembly Robot Arm," *Proceedings of the First International Conference on Assembly Automation,* IFS Publications Ltd., Bedford, pp. 77–86.

Martin, Harold C., 1966: *Introduction to Matrix Methods of Structural Analysis,* McGraw-Hill Book Company, pp. 314–315.

Mason, M., 1981: "Compliance and Force Control for Computer Controlled Manipulators," *IEEE Trans. Systems, Man, and Cybernetics,* **SMC-11**(6), June, 418–432.

McCammon, I., 1984: "Design of a Conformal Tactile Sensing Array," *SPIE Intelligent Robots and Computer Vision,* **512,** 296.

Mendelson, J. J., and J. R. Rinderle, 1984: "Design of a Compliant Robotic Manipulator," in *Proceedings, Robotics Research, The Next Five Years and Beyond,* August 14–16, Bethlehem, Pennsylvania. (Available as Robotics International Technical Paper MS84-495, Soc. Mfg. Engineers, Dearborn, Michigan.)

Meyer, Hans Ulrich, 1974: *Electrical Length Measuring System,* United Kingdom Patent 1,366,284, The Patent Office, 25 Southhampton Buildings, London, England, September 11.

Miyahara, Nobufumi, and Takashi Uchida, 1982: "Rotary Magnetic Sensor," U.S. Patent 4,334,166, June 8.

Morgan, R. G., and U. Ozguner, 1985: "A Decentralized Variable Structure Control Algorithm for Robotic Manipulation," *IEEE Journal of Robotics and Automation,* **RA-1**(1), 57–65.

Nerozzi, A., and G. Vassura, 1980: "Study and Experimentation of a Multifinger Robot," *10th International Symposium on Industrial Robots/5th International Conference on Industrial Robot Technology,* Milan, Italy, March 5–7.

Nevins, J. L., and D. E. Whitney, 1973: "The Force Vector Assembler Concept," *Proceedings of the First CISM-IFTOMM Symposium on Theory and Practice of Robots and Manipulators,* September.

Nishimura, T., and K. Ishizuka, 1986a: "Laser Rotary Encoders, Part 1," *Motion, J. Electronics Motion Control Association,* Orange, California, July/Aug.

Nishimura, T., and K. Ishizuka, 1986*b:* "Laser Rotary Encoders, Part 2," *Motion, J. Electronics Motion Control Association,* Orange, California, Sept./Oct.

Nitzan, David, Alfred E. Brain, and Richard O. Duda, 1977: "The Measurement and Use of Registered Reflectance and Range Data in Scene Analysis," *Proceedings of the IEEE,* **65,**(2), February, 206–220.

Ong, M. C., 1981: "An Approach to the Mechanical Design of a Sheep Shearing Robot," Australian Wool Corporation, *Second National Wool Harvesting Research and Development Conference Proceedings,* Sydney, Australia.

Paden, Bradley E., 1986: "Kinematics and Control of Robot Manipulators," Memorandum No. UCB/ERL M86/5, Electronics Research Laboratory, College of Engineering, University of California, Berkeley, California (also a Ph.D. thesis).

Paul, R., 1972: "Modeling, Trajectory Calculation and Servoing of a Computer Controlled Arm," Report AIM-177, Stanford Artificial Intelligence Laboratory, Computer Science Department, Stanford University, Stanford, California.

Paul, R., 1977: "The Mathematics of Computer-Controlled Manipulation," *Proceedings, 1977 Joint Automatic Control Conference,* July, pp. 124–131.

Paul, R., 1979: "Manipulator Cartesian Path Control," *IEEE Trans. Systems, Man, and Cybernetics,* **SMC-9**(11), November, 702–711.

Paul, Richard P., 1981: *Robot Manipulators: Mathematics, Programming, and Control,* The MIT Press, Cambridge, Massachusetts.

Paul, R., and B. Shimano, 1976: "Compliance and Control," *Proceedings, 1976 Joint Automatic Control Conference,* pp. 694–699.

Paul, R., B. Shimano, and G. E. Mayer, 1981*a:* "Differential Kinematic Control Equations for Simple Manipulators," *IEEE Trans. Systems, Man, and Cybernetics,* **SMC-11**(6), 456–460.

Paul, R., B. Shimano, and G. E. Mayer, 1981*b:* "Kinematic Control Equations for Simple Manipulators," *IEEE Trans. Systems, Man, and Cybernetics,* **SMC-11** (6), 449–455.

Peiper, D. L., 1968: "The Kinematics of Manipulators under Computer Control," Report A172/CS116, Report AIM-177, Stanford Artificial Intelligence Laboratory, Computer Science Department, Stanford University, Stanford, California.

Pennywitt, K. E., 1986: "Robot Tactile Sensing," *BYTE,* January, 177–200.

Phelan, Richard M., 1970: *Fundamentals of Mechanical Design,* McGraw-Hill Book Company, New York.

Prahl, B. W., and P. M. Tracey, 1986: "Pressure/Tactile Sensing with Intrinsic Fiber-Optic Sensors," *Sensors,* **3**(8), pp. 48–52.

Raibert, M., 1976: "A State Space Model for Sensorimotor Control and Learning," Memo 351, National Institutes of Health Grant NIH-5-T01-6MO-1064-14, Office of Naval Research Contract N00014-75-C-0634, MIT Artificial Intelligence Laboratory, Cambridge, Massachusetts, January.

Raibert, M. H., and J. J. Craig, undated: "Hybrid Position/Force Control of Manipulators," Project Report, NASA Contract NAS-7-100, Jet Propulsion Laboratory, Pasadena, California.

Rothbart, Harold A., ed., 1964: *Mechanical Design and Systems Handbook,* McGraw-Hill Book Company, New York.

Salisbury, K. J., 1980: "Active Stiffness Control of a Manipulator in Cartesian Coordinates," *Proceedings, 19th IEEE Conference on Decision and Control,* December.

Salisbury, K., 1984: "Interpretation of Contact Geometries from Force Measurements," *Proceedings of Robotics 84,* Atlanta, pp. 240–247.

Salisbury, K., D. Brock, and S. Chiu, 1985: "Integrated Language, Sensing and Control for a Robot Hand," Preprints of the *Third International Symposium of Robotics Research,* Gouvieux, France, October, pp. 54–61.

Schaecter, D. B., 1982: "Hardware Demonstration of Flexible Beam Control," *J. Guidance and Control,* **5**(1), 48–53.

Scheinman, V. D., 1969: "Design of a Computer Controlled Manipulator," Report AIM-92, Stanford Artificial Intelligence Laboratory, Computer Science Department, Stanford University, Stanford, California, August.

Seffernick, G. H., 1982: "Spray Oil-Cooled Motors for High Speed Industrial Applications," *Drives & Controls International,* Intertec Communications, Inc., Oxnard, California, June/July.

Seltzer, D. S., 1982: "Tactile Sensory Feedback for Difficult Robot Tasks," *Robots VI,* p. 467.

Sensor Technology, 1985: "Tactile Sensor Employs Optics to Increase Resolution," Technical Insights, Englewood, New Jersey, June, p. 7.

Sesak, J. R., P. W. Likins, and T. Coradetti, 1979: "Flexible Spacecraft Control by Model Error Sensitivity Suppression," *J. Astro. Sci.,* **XXVII**(2), April/June, 131–156.

Shimano, B., 1978: "The Kinematic Design and Force Control of Computer Controlled Manipulators," Report AIM-33, Stanford Artificial Intelligence Laboratory, Computer Science Department, Stanford University, Stanford, California.

Shrader, E., R. Kornbluh, and G. B. Andeen, 1986: "A Thermal Slip Detector," *Proceedings Sensors Exposition,* Chicago, vol. 1, September, pp. 365–388.

Sidor, Edward F., 1976: "Positional Transducer Utilizing Magnetic Elements Having Improved Operating Characteristics," U.S. Patent 3,598,202, May 18.

Skinner, Frank, 1975: "Designing a Multiple Prehension Manipulator," *Mechanical Engineering,* September, pp. 30–37.

Smith, David E., 1980: "Electronic Distance Measurement for Industrial and Scientific Applications," *Hewlett-Packard Journal,* **31**(6), Hewlett-Packard Company, Palo Alto, California, June.

Soroka, B. I., 1980: "Debugging Robot Programs with a Simulator," *Proceedings SME Autofact West,* November 17–20, Anaheim, California.

Sunada, W. H. and S. Dubowsky, 1983: "On the Dynamic Analysis and Behavior of Industrial Robotic Manipulators with Elastic Members," *J. Mechanical Design, Trans. ASME* **105** (1), 42–51.

Takase, K., R. Paul, and E. Berg, 1981: "A Structured Approach to Robot Programming and Teaching," *IEEE Trans. Systems, Man, and Cybernetics,* **SMC-11**(4), 274–289.

Tella, R., R. Kelley, and J. Birk, 1980: "A Contour Adapting Vacuum Gripper," *Proceedings 10th International Symposium on Industrial Robots,* University of Rhode Island, Department of Electrical Engineering, Kingston, Rhode Island, IFS Publications Ltd., Bedford, England, p. 175.

Tesar, D., 1985: "Next Generation of Technology for Robotics," *AIAA/NASA Symposium on Automation, Robotics and Advanced Computing for the National Space Program,* Washington, D.C., September.

Truckenbrodt, A., 1979: "Dynamics and Control Methods for Moving Flexible Structures and Their Application to Industrial Robots," *Proceedings 5th World Congress on Theory of Machines and Mechanisms,* vol. 1, ASME, New York, pp. 831–834.

Truckenbrodt, A., 1980: *Bewegungsverhalten und Regelung hybrider Meehrkorpersystem mit Anwendung auf Industrieroboter (Vibration damping and control of hybrid many body systems with application to industrial robots),* Verlag, Dusseldorf.

Uicker, J. J., 1969: "Dynamic Behavior of Spatial Linkages," *J. Engineering for Industry, Trans. ASME,* **91,** 251–258.

Utkin, V., 1971: "Equations of Sliding Mode in Discontinuous Systems, Part I," *Automation and Remote Control,* **32**(12), 1897–1907.

Utkin, V., 1972: "Equations of Sliding Mode in Discontinuous Systems, Part II," *Automation and Remote Control,* **33**(2), 211–219.

Vasilash, Gary S., 1981: "General Electric: Bringing the Factory of Tomorrow to Life," *Manufacturing Engineering,* **87**(4), 72–75.

Vranish, J. M., 1982: "Magnetoelastic Force Feedback Sensors for Robots and Machine Tools," *Robots VI,* p. 492.

Warnecke, H. J., and D. Haaf, 1981: "Components for Programmable Assembly—Grippers, Sensors, Conveying Systems for an Industrial Robot Engaged Together with Fixed Automation," *Assembly Automation,* IFS Publications Ltd., Bedford, England, May.

Whitney, D. E., 1969: "Resolved Motion Rate Control of Manipulators and Human Prostheses," *IEEE Trans. Man-Machine Systems,* **MMS-10,** October, 47–53.

Whitney, D. E., and J. L. Nevins, 1979: "What Is the Remote Center Compliance and

What Can It Do?" paper presented at the *9th International Symposium and Exposition on Industrial Robots (ISIR)*, Washington, D.C., March 13–15. (Also published as CSDL Report P-728, November 1978.)

Wiener, Norbert, 1961: *Cybernetics: or Control and Communication in the Animal and the Machine*, 2d ed., The MIT Press, Cambridge, Massachusetts, p. 212.

Woo, L. W., and F. Freudenstein, 1971: "Dynamic Analysis of Mechanisms Using Screw Coordinates," *J. Engineering for Industry, Trans. ASME*, **93**, 273–276.

Wu, C-H., and R. P. Paul, 1980: "Manipulator Compliance Based on Joint Torque Control," *Proceedings, 19th IEEE Conference on Decision and Control*, December, pp. 88–94.

Wyllie, C. E., 1982: "Are Actuation Systems Ready for All-Electric Aerospacecraft?" *Drives & Controls International*, **2**(3), Intertec Communications, Inc., Oxnard, California, April/May, 36–37.

Youcef-Toumi, K., and H. Asada, 1986: "The Design of Open-Loop Manipulator Arms with Decoupled and Configuration-Invariant Inertia Tensors," *Proceedings 1986 IEEE International Conference on Robotics and Automation*, San Francisco, **III**, April, 2018–2026.

Young, K-K. D., 1978: "Controller Design for a Manipulator Using Theory of Variable Structure Systems," *IEEE Trans. on Systems, Man, and Cybernetics*, **SMC-8**(2), 101–109.

Zienkiewicz, O. C., 1977: *The Finite Element Method*, 3d ed., McGraw-Hill Book Company.

Ac motor, **10.11**
Accommodation:
  active, **2.7**
  passive, **2.7**
Accuracy, **2.1, 2.2, 2.4** to **2.6, 2.14, 2.20,
    3.16, 3.24, 3.26, 5.3, 6.2, 12.2, 12.4,
    12.8** to **12.10, 12.13, 12.26**
Adaptive control, **3.22, 6.20, 6.21, 15.6**
AL control system, **2.27, 2.28**
AML control system, **2.27, 2.28**
Approach, **2.22, 14.6, 14.10, 14.12**
Arrays, tactile sensor, **2.15, 6.3, 12.18,
    12.21**
Assembly, **2.3, 2.17, 2.24, 5.1, 5.5, 5.6,
    6.3, 6.5, 8.1, 9.1, 9.5, 10.23, 12.2,
    12.23, 13.7**
Automation:
  hard, **1.3**
  soft, **1.3**

Back-drivability, **11.1, 11.2, 11.10, 11.11,
    11.18, 15.4, 15.5**
Backlash, **3.27, 11.1, 11.2, 11.21, 11.22,
    11.24, 12.8**
Ball joint, **3.2, 3.5, 14.53**
Bandwidth, **3.16, 3.24, 4.12, 6.9, 12.3,
    12.4**
Beam, **4.7, 12.25, 15.5, 15.9**
Bearings, **2.9, 3.25, 3.26, 5.1** to **5.4, 6.5,
    7.22, 10.24, 12.14**
Binary code, **6.10**
Binary image, **2.14**
Binary light level, **12.16**
Brakes, **2.6, 2.19, 3.14, 5.4, 10.3, 10.17** to
    10.20, 11.2
Bus, **5.4, 8.9**

C language, **9.1**

Cables, **3.25, 3.26, 5.2, 5.4** to **5.6, 6.23,
    8.9, 11.1** to **11.4, 13.4**
Castigliano's theorem, **4.5**
Collision avoidance, **2.29**
Compliance, **2.11, 3.19, 3.20, 3.27, 15.9** to
    15.10
  active, **2.6, 2.7**
  passive, **2.6, 2.7, 15.4**
Constrained motion, **2.16, 2.24, 15.10**
Coordinate system:
  cartesian, **2.1, 2.20** to **2.23, 3.9, 3.12,
    6.6, 9.5, 14.5**
  cylindrical, **2.22, 3.10, 3.11**
  joint, **2.22, 14.28**
  spherical, **3.10**
  tool, **2.22, 14.28**
Coordination, **2.26, 2.28**
Counterbalancing, **3.15**
Coupling, **6.6, 6.7, 6.11, 6.12**

Damping, **6.12, 6.22**
Database, **2.2, 9.3, 9.7**
Dc drive, **3.14, 8.3, 10.1** to **10.28, 13.4**
Deadstick power switch, **2.19**
Debugging, **2.19, 3.27, 3.28**
Degrees of freedom, **3.2** to **3.6, 7.17, 7.19**
Denavit-Hartenberg matrices, **14.1** to
    14.64
Direct-drive devices, **1.2, 2.9, 3.14, 11.2,
    15.4, 15.5**
Drift, **12.3, 12.20**
Dual, **3.10**
Dual elbow kinematics, **3.11**
Duty cycles, **10.3**
Dynamic compliance properties, **2.7**
Dynamic coupling between joints, **6.4,
    6.6, 6.7**
Dynamic friction, **2.9**
Dynamic instability, **10.4**

Dynamic loads, **3.15**
Dynamic positioning, **6.2**
Dynamic response, **6.14, 6.15**
Dynamic simulations, **6.23, 7.2, 7.4 to 7.22**
Dynamics, **1.5, 3.19**

Elbow manipulator, **3.10, 3.11**
Electromagnetic interference (EMI), **3.25 to 3.27**
Encoder, **12.1 to 12.28**

Feedback, **2.4, 6.1 to 6.23, 15.6**
Finite-element method of structural analysis, **4.1 to 4.27, 7.12, 7.14, 7.15**
Flexibility:
    compliance and, **1.3, 2.3, 2.11, 3.23, 4.1, 5.6, 12.16, 7.3, 7.10, 7.19 to 7.22, 15.1, 15.4, 15.7**
    multiple purposes and, **2.31**
Flexible manufacturing system (FMS), **2.31, 8.7, 9.3**
Force control, **2.10, 2.12, 6.22, 12.1, 14.37**
Force feedback, **1.5**
Foundation, **1.4, 15.2**
Frequency, **3.16, 3.17, 3.24, 3.26, 4.2, 4.17 to 4.22, 6.5, 6.6, 6.9, 7.20, 12.2, 12.13, 15.3, 15.7**
Friction:
    Coulomb, **2.4, 2.9, 3.27, 6.4, 6.5, 6.11, 6.13, 6.17, 6.23, 10.1, 10.18 to 10.20, 11.1, 11.10, 11.28, 12.14**
    viscous, **6.4, 6.5**

Gantry, **3.12**
Graceful failure, **2.33, 2.34, 8.10**
Gravity, **2.19, 2.27, 6.4, 6.12, 6.14, 7.17, 10.21, 10.22, 11.2**
Gray-scale image, **2.14**
Grinding, **2.3, 12.23**
Gripper, **5.2, 5.3, 5.6, 5.7, 11.1, 11.5, 12.2, 12.18, 12.23, 13.1 to 13.11, 14.2**

Harmonic drive, **11.24, 11.26 to 11.28**
Homogeneous transformation, **7.5, 14.1 to 14.64**
Hooke's law, **4.5, 4.9**
Hoses, **5.2, 5.4, 5.6, 10.7**
Hydraulic actuators, **3.14, 10.3, 10.5 to 10.7, 15.5**

Hydraulic cylinder, **11.12**
Hydraulic friction brakes, **10.18, 10.19**
Hydraulic hoses, **5.4 to 5.7**
Hydraulic lines, **3.25**
Hydraulic pressure, **10.17, 13.2**
Hydraulic transmissions, **11.2**
Hysteresis, **12.3**

Inertia, **4.9, 5.3, 6.6, 6.7, 6.23, 8.6, 10.3, 10.8, 10.9, 10.21, 10.23 to 10.28, 11.1, 13.4**
Inertia tensor, **3.15**
Inspection, **2.13, 9.3**
Interface, **5.6, 5.7, 8.7, 13.1**
Interrupt, **2.25, 2.26, 9.8**

Jacobian, **14.1 to 14.64**
Joints:
    hinged, **3.5, 3.6, 11.12, 14.23**
    linear, **2.5, 3.5, 5.4, 10.7, 11.12**
    prismatic, **3.5, 3.6, 3.9, 6.4, 10.7, 14.19, 14.21, 14.22**
    revolute, **3.5**
Joystick, **2.18, 2.19, 9.3, 9.5**

Kinematic design, **2.10, 3.2, 3.13, 3.24**
Kinematic simulations, **6.23, 7.2 to 7.6**
Kinematics, **3.19, 3.26, 7.16**
Kinesthetic communication, **2.18**

Lagrangian, **6.8, 6.9, 7.6, 7.14**

Manufacturing automation protocol (MAP), **2.32**
Mean time between failure (MTBF), **2.2, 8.10**
Mobile robots, **1.4, 10.6, 15.1**
Modal analysis, **3.17**
Modal superposition, **4.17**
Mouse, **2.18**

Newton-Euler formulation, **6.8**
Numerical control, **1.4, 1.5, 15.3**

Obstacle avoidance, **3.18**
Open-chain kinematics, **1.4, 3.9**
Orientation, **2.13, 2.22**

Overshoot, 3.2, 6.2, 6.4

Painting, 2.3, 2.21, 6.2
Parallel processing, 12.16
Parallel programming, 2.27
Parallelism, 2.24, 2.25, 9.6 to 9.8
Pascal language, 9.1
Path control, 2.21, 14.64
Path following, 1.2, 2.1, 2.21
Payload, 2.1, 3.21, 3.24, 5.1, 5.3, 6.7, 6.8,
    7.1, 7.7, 7.11, 7.19, 8.4, 10.4, 15.2,
    15.3, 15.5
PID controllers, 3.21, 3.22, 3.24, 3.25,
    6.5, 6.17, 6.19, 6.20, 8.3
Piezoelectric effect, 3.11, 10.14 to 10.15,
    10.17, 12.12, 12.13
Pitch, 3.2, 14.6, 14.10 to 14.12
Pneumatic actuators, 3.14, 10.6, 10.7
Pneumatic brakes, 10.18, 10.19
Pneumatic hoses, 5.4 to 5.7
Pneumatic lines, 3.25
Pneumatic motive power, 13.2
Pneumatic pressure, 10.17
Point-to-point devices, 2.21, 3.21
Point-to-point motion, 15.3
Potentiometer, 6.10
Precision, 2.2, 2.3, 11.2, 12.9
Programmed logic arrays (PALs), 8.3
Proprioception, 12.1, 12.4, 12.9, 12.10
Proximity sensor, 2.16, 2.17
Pulse-width modulation (PWM), 7.1, 7.7,
    8.4

Quick-change interfaces, 5.6, 5.7, 13.1

Relative motion, 2.23
Remote center compliance (RCC), 2.8,
    6.2, 15.4
Repeatability, 2.2, 2.4 to 2.6, 12.3, 12.4
Resolution, 2.2, 2.4 to 2.6, 2.10, 2.12,
    2.17, 6.23, 12.2 to 12.5, 12.7, 12.13,
    12.16
Resonance, 6.4, 7.1, 10.2, 10.10
Rigid arm, 2.6, 2.7
Rigid design, 1.5, 4.1
Rigid limbs, 15.2 to 15.4
Rigid links, 4.12, 6.9, 6.23, 7.6, 7.11,
    7.21, 14.17

Rigidity in power interruption, 10.18
Rigidly connected objects, 2.27
Roll, 3.2, 14.6, 14.10 to 14.12
Run-out, 5.3

Safety, 2.2, 2.3, 2.20, 2.30 5.2, 8.1, 8.9 to
    8.10, 10.6, 10.7, 15.2
Selective compliance assembly robot arm
    (SCARA), 3.12, 3.13, 10.11, 11.2,
    15.2, 15.4
Sensing, 1.5, 2.13, 2.18, 6.3, 8.1, 8.9, 9.1
Servomechanism technology, 2.10 to
    2.12, 3.23 to 3.27, 6.1 to 6.23, 10.5,
    10.6, 10.12, 12.4, 12.11, 12.12, 15.5
Simulation, 7.1 to 7.22, 9.6, 9.8
Soft automation, 1.3
Springs, 3.15, 4.11
Stability, 3.27, 6.4, 6.22, 7.10, 10.4, 10.5,
    12.3, 12.11, 15.7
Stepping motors, 3.14, 3.21, 6.10, 10.2,
    10.9, 10.10, 13.4
Stiffness matrix, 4.6 to 4.25, 7.15
Symbolic communication, 2.18
Symbolic programming, 9.5
Synchros, 12.7 to 12.8

Tachometer, 6.9, 6.11, 6.23, 12.11
Tactile sensor arrays, 2.15, 6.3, 12.18 to
    12.19, 12.21
Teleoperator, 1.2, 2.20, 3.1, 9.3, 10.4
Tendon, 5.2, 11.2, 11.3, 13.4
Touch, 1.5, 2.12, 2.13, 2.15, 12.18 to
    12.19
Track, 2.13, 2.22, 6.2, 14.32
Training, 2.19, 2.20, 9.5, 9.6
Truss, 3.2, 3.6, 4.3, 4.12
Twist, 2.22

VAL control system, 2.2, 2.12, 2.22, 2.23,
    2.27, 8.9, 9.6, 14.6, 14.14, 14.51
Vision, 1.5, 2.13, 2.19, 2.22, 2.23, 12.14
    to 12.16

Welding, 2.3, 2.21, 2.22, 2.24, 5.2
Wrist, 2.11, 3.2, 3.9, 5.2, 6.23, 10.24,
    12.13, 12.23, 14.47, 14.53 to 14.54

Yaw, 3.2, 14.6, 14.10 to 14.12